Biomarkers of Toxic Metals

Vanda Maria Falcão Espada Lopes de Andrade
Life Quality Research Centre (CIEQV), IPSantarém/IPLeiria
Rio Maior, Portugal and Research Institute for Medicines (iMed.ULisboa)
Faculty of Pharmacy, Lisbon University
Lisboa, Portugal

CRC Press
Taylor & Francis Group
Boca Raton London New York

CRC Press is an imprint of the
Taylor & Francis Group, an **informa** business

A SCIENCE PUBLISHERS BOOK

First edition published 2023
by CRC Press
6000 Broken Sound Parkway NW, Suite 300, Boca Raton, FL 33487-2742

and by CRC Press
4 Park Square, Milton Park, Abingdon, Oxon, OX14 4RN

Library of Congress Cataloging-in-Publication Data (applied for)

ISBN: 978-1-032-03938-1 (hbk)
ISBN: 978-1-032-03940-4 (pbk)
ISBN: 978-1-003-18981-7 (ebk)

DOI: 10.1201/9781003189817

Typeset in Times New Roman
by Radiant Productions

Preface

These days there is an increased exposure of the human population to metals, given our "chemical-dependent" way of life. The importance of biomarkers to prevent and control these exposures, and the ensuing effects, is of undoubtable relevance. Moreover, the areas of toxicology are broad, as well are the spectrum of implications for health induced by the exposure to these toxics. This book can be of interest for researchers, toxicologists, physicians, pharmacologists and persons working in governmental regulatory agencies and other public health fields.

The first part of this book cover general aspects of biomarkers of toxic metals. The first chapter focuses on anthropogenic emissions of metals to the environment, human exposure through air, water and food chain, and consequent public health impacts. An overview of biomarkers, their types, applications and relevance for human health will be represented in Chapter two. The following chapter will be devoted to metal biomarkers in different application fields, including environmental, occupational and clinical toxicology, highlighting their contribution to public health. Parts 2 and 3 cover biomarkers of xenobiotic metals, aluminum, arsenic, cadmium, chromium, lead, mercury and biomarkers of essential metals with potential for toxicity, copper, iron, magnesium, manganese and zinc, respectively. In these chapters, sources of exposure to metals, regulated limit values in several environmental sections, essentiality, deficiency/toxic effects as well as biomarkers of status/exposure, susceptibility and effect will be shown. In part 4, novel approaches to metal biomarkers are focused. Specifically, the recognition of our "real life" chronic exposure to chemical mixtures (metal mixtures), rather than to a single metal, and the ensuing advances achieved so far in the field of biomarkers, is explored. The final chapter describes multi-biomarkers approaches in the scope of metal toxicity, with relevance to "omics" technologies, describing new insights in the increase of accuracy and predictive power of biomarkers, as well as their applications to risk assessments on metals and metal mixtures.

Contents

Part II: Biomarkers of Xenobiotic Metals

Part 4: Novel Trends in Metals Biomarkers

Part I

General Considerations

CHAPTER 1

Metals

Pollution and Health Impacts

Metals as Natural Resources

Chemical nature of metals

Of the 103 chemical elements in the periodic table, approximately 80 are metals (Delfino and Saccone 2009). Metals are "elements which conduct electricity, have a metallic lustre, are malleable and ductile, form cations, and have basic oxides" (Duffus 2002). They are usually crystalline solids, and in most cases, have a simple crystal structure distinguished by a close packing of atoms and a high degree of symmetry. Except for mercury (Hg), all metals may occur in a solid state. Typically, atoms of metals contain less than half the full complement of electrons in their outermost shell (Britannica 2020).

Metals differ from other elements, and more generally from non-metallic solids, due to their characteristic metallic bonds which is the basis of their solid structure (Delfino and Saccone 2009). In these bonds, the metallic material releases electrons from their outer shells and the electrons get dispersed between metal cations, as a sea of delocalized electrons; the electrostatic interactions between electrons and cations are specifically called "metallic bonds." It is due to the metallic bonding that metals have an ordered structure, while their ability to conduct electricity is because the electrons can move. The high melting and boiling points of these materials are attributed to the strength of the metallic bonding (Dunee 2012).

A group of metals commonly called "heavy metals" is generally mentioned in literature (Dunee 2012). According to a report of the International Union of Pure and Applied Chemistry (IUPAC) there is an incoherent use of this term, and this reflects in a considerable amount of the inconsistency in scientific literature (Duffus 2002). To illustrate, chemists define heavy metals as natural elements characterized by their high atomic mass and density. Usually a density of at least 5 g/cm^3 is a criteria to define a heavy metal and differentiate it from other "light metals" (Järup 2003). However, when providing the density criteria, in the periodic table of elements "heavy metals" would occupy columns 3–16, of the periods 4 to 6; those positions encompass transition metals, post-transition metals and lanthanides. Meanwhile, according to other definitions, "heavy metals" have an atomic mass higher than 23 or an atomic number exceeding 20, which is also highly error prone, as both alternative definitions cause the inclusion of nonmetals (Koller and Saleh 2018). Currently, there is no clear chemical definition of what a heavy metal is, in most cases density is taken to be the defining factor (Järup 2003).

Toxicologists use the term "heavy metals" to group metals and metalloids associated with contamination and potential toxicity, regardless of their atomic weights (Duffus 2002). The most frequently encountered and pronounced ones include arsenic (As), cadmium (Cd), chromium (Cr), lead (Pb), Hg, copper (Cu), iron (Fe), manganese (Mn) and zinc (Zn) (Dunee 2012, Altunatmaz et al. 2019); along with aluminium (Al) and magnesium (Mg), these elements will be shown in this book. Under the same toxicological perspective, several authors also mention other elements such as cobalt (Co), nickel (Ni), tin (Sn), silver (Ag), and selenium (Se) as "heavy metals" (Dunee 2012, Altunatmaz et al. 2019). However chemically Se shows a borderline metalloid or nonmetal behaviour and is classified as a nonmetal despite sometimes being classified as a metalloid (Gupta 2018). Since there is no basis for deciding which metals should be included in the category "heavy metals," it is not surprising that in one of the most widely used textbook in toxicology, "Casarett and Doull's Toxicology", this term is never used. Some authors defend that the classification of metals be removed, since using terms such as "heavy metals," have no sound terminological or scientific basis. It would be necessary to review the use that has developed for this term, paying particular attention to its relationship to fundamental chemistry; such a classification would permit a better interpretation of the biochemical basis for toxicity. Even so, legal regulations still specify a list of "heavy metals" to which they apply (Duffus 2002).

Metals in Nature and their uses by man

Metals are present naturally on the Earth's crust since its formation (Briffa et al. 2020). The vast majority occur in ores (mineral-bearing substances), though a few, such as Cu, gold (Au), platinum (Pt), and Ag, frequently occur in the free state, as they do not readily react with other elements (Britannica 2020). Natural causes contribute to metal entrance in several environmental parts , and include volcanic activity, metal corrosion, metal evaporation from soil and water, sediment resuspension, soil erosion and geological weathering (Koller and Saleh 2018, Briffa et al. 2020).

Humans have been using metals in many different areas for thousands of years. Pb has been employed since at least 5,000 yr , with their early applications including building materials, pigments for glazing ceramics and pipes for transporting water. Pb acetate was used to sweeten old wine in ancient Rome and is believed that some people might have consumed as much as 1 g of Pb per day. Romans also allegedly used Hg to alleviate teething pain in infants and this metal was later used as a remedy for syphilis from the 1300s to the late 1800s. Claude Monet used Cd pigments extensively in the mid-1800s, although the scarcity of the metal limited their use in artists' materials until the early 1900s (Järup 2003). Since the middle of the 19th century profound changes occurred regarding the use of metals, which increased largely for more than 100 yr, resulting in considerable emissions to the environment (Järup 2003, Qin et al. 2021). Nowadays human activities are exerting immense impacts on the environment on all scales and in ways that overcome natural processes. While such activities are very recent in the geological timescale, their influence on the global environment have been dramatic (Karaouzas et al. 2021). Men use more than three-quarters of metals to some extent at the current time and there is an exponential increase of usage in several applications (Delfino and Saccone 2009, Tchounwou et al. 2012). Such events raised relevant environmental and public health concerns, which at the end of the 20th century ended in several political resolutions to diminish metal emissions in developed countries; as an example, emissions of metals fell by over 50% between 1990 and 2001 in the U.K. However, metal emissions continue to occur and are even increasing in undeveloped areas of the world (Järup 2003, Tchounwou et al. 2012).

Environmental Pollution by Metals

According to the Environmental European Agency (2012), pollution "is the introduction of substances or energy into the environment, resulting in deleterious effects of such a nature as to endanger human health, harm living resources and ecosystems, and impair or interfere with amenities and other legitimate uses of the environment".

Metals are typical pollutants due to their environmental persistence since they cannot be broken down (they can only be transferred from one chemical state to another), bioaccumulation and negative effects on human health through the food chain, and high toxicity (Yu et al. 2020, Qin et al. 2021). In the environment, their adverse effects are not only related to their total concentration but also depend strongly on chemical partitioning (Yu et al. 2020).

Anthropogenic activity is the main cause of the impacts caused by metal pollution, for which the primary sources are mining, smelting, foundries and other industries that are metal based. Agriculture is the secondary source of metal pollution, attributed to the use of pesticides, and fertilizers (Tchounwou et al. 2012, Jaishankar et al. 2014, Briffa et al. 2020). Other sources include the combustion of fossil fuels, pharmaceutical industry, domestic effluents, waste dumps, livestock and chicken manure, automobiles and roadworks; unintended release including shipwrecks, oil spills, mining and fires (Tchounwou et al. 2012, Ali et al. 2019, Briffa et al. 2020).

Metals may enter the hydrosphere, lithosphere, atmosphere and biosphere in several ways (Briffa et al. 2020). Due to environmental compartments turn over , metals released into the atmosphere will return to the land and cause contamination of waters and soils; due to their persistence, they either accumulate in biota or leach down into ground waters (Ali et al. 2019). Additionally, metals in wastewater pollute rivers, groundwater and soil, and can enter water bodies through soil erosion (Qin et al. 2021, Yang et al. 2021). Different physicochemical and climatic factors affect their overall dynamics and biogeochemical cycling (Ali et al. 2019). The metal's movement depends on temperature, flow and direction of surface waters, circulation of air masses and wind speed (Briffa et al. 2020).

Air

The atmosphere performs several functions that have allowed humans to survive and develop anywhere on the Earth's surface and unfortunately, atmospheric pollution with metals is one of the most serious problems facing humanity and other life forms on our planet today (Hassanien 2011, Amin et al. 2018). Metals in air can function as catalysts in atmospheric transformations to form secondary pollutants, contributing to economic loss through corrosion of materials, and in association with total suspended Particulate Matter (PM) induce meteorological and geophysical effects such as scattering solar radiation back into space and reducing visibility. Fallout or washout of airborne trace metals can initiate a host of ecological effects in water and soil (Lee and von Lehmden 1973). Unfortunately, some atmospheric metals may be associated with excess mortality, even in areas with low levels of air pollution (Lequy et al. 2019). The most significant anthropogenic sources of metals to the air are the metals industry, mining process, construction, fossil fuels' combustion, electricity and heat production, road transportation, petroleum refining, phosphate fertilizers in agricultural areas, incineration activities and windblown soil dust (Hassanien 2011, Amin et al. 2018). The industrial sector has been the target of many studies as the activities of industries continuously introduce contamination risks for workers and people living in surrounding zones (Amin et al. 2018). In urban areas metals are commonly reported pollutants; road dust building and road materials, vehicle brake, tyre wear and tailpipe emissions are the main non-point sources of metal pollution in cities (Naderizadeh et al. 2016, Messager et al. 2021). Vehicle emissions are one of the main sources of metal contamination in urban environments, reaching waterways directly through atmospheric deposition and indirectly through rain runoff, and by leaching or picked up as dust after depositing and building on impervious surfaces (Messager et al. 2021).

Metals occur in air in distinct phases, as solids, gases or adsorbed to PM with aerodynamic sizes ranging from below 0.01 μm to 100 μm, and larger. Larger particles tend to settle to the ground by gravity in a matter of hours, whereas the smallest particles can stay in the atmosphere for weeks and are mostly removed by precipitation (Hassanien 2011). PM constitutes a most important air pollutant which may create a risk for human health, particularly PM with metals, which can cause serious reactions in the respiratory tract and accumulate in human tissues (Kunt et al. 2021). There are two major categories of PM according to their size: PM10 and PM2.5; the fine PM2.5 carries a higher burden of toxic metals than the coarser PM10 (Srithawirat 2016, Popoola et al. 2018). PM2.5 (emitted from fossil fuel combustion, motor vehicle exhausts and wood burning) are demonstrated to be associated with several toxic metals, such as As, Cd, Pb, and Zn (Hassanien 2011).

The concentration of metals in air shows spatial variation dependent on distances to the sources and the relative importance of local sources; speed and direction of the wind as well as seasonal variations are additional factors (Amin et al. 2018). Due to long-range transboundary air pollution, environmental persistence and potential for global atmospheric transfer, atmospheric emissions of metals affect even the most remote regions (WHO 2007, Jiries et al. 2018).

Metal pollution in air can be assessed using different methods such as the analysis of suspended dust, dry deposition, wet precipitation, dew composition, or using plants, lichens and tree bark as bio-indicators (Jiries et al. 2018). Quite interestingly, the concentrations of airborne PM and their metals content may directly reflect the environmental pollution level in the atmosphere of the studied area (Arshad et al. 2015). Both PM metal content and the determination of an Enrichment Factor (EF) can be used to access metal air pollution. EF estimations can determine whether an element has an additional contamination from the anthropogenic activities and are widely used to identify the anthropogenic source of metallic elements (Fang et al. 2006, Arshad et al. 2015). The EF for each metal (i) can be expressed as $EF_i = (i/j)$ air / (i/j) crust, where EF_i is the enrichment factor of species i; j is a reference element for the crustal material, (i/j) air is the ratio of the amount of species i to that of species j in a sample, and (i/j) crust is the ratio of species i to species j in the crust (Fang et al. 2006).

Soil

Metals naturally enter into soils from the parent material; factors affecting their presence and distribution are the composition of the parent rock, degree of weathering, physical, chemical and biological characteristics of soil and climatic conditions (Ali et al. 2019). Owing to agricultural practices, metallurgical processes and urban discharges, the pollution of soils with metals has become globally ubiquitous (Yang et al. 2021). In many regions, an additional factor for soil pollution is the use of containing metals sewage sludge to fertilize and irrigate croplands. Sewage sludge is produced in vast amounts in Japan (70 million ton), China (30 million ton), and the USA (six million ton) (Gall et al. 2015). Soils may become contaminated with metals in urban areas from heavy vehicular traffic on roads; soil samples in urban areas have elevated concentrations of Pb, out of which 45–85% is bioaccessible . The bioavailability of metals in soils is particularly important for their fate in the environment and for their uptake in plants. Each metal has a different bioavailability, which is dependent on the metal speciation and on the different physicochemical properties of soils (Ali et al. 2019).

It is exceedingly difficult to recover a soil polluted by metals and the impacts are quite severe, as metals tend to accumulate significantly shifting the soil ecosystem by lowering the soil quality (Qin et al. 2021, Yang et al. 2021). Besides, farming in metal contaminated soils poses a troubling threat to agricultural productivity, food safety and human health (Qin et al. 2021). Metal contents in soils have been frequently found to exceed their standard values worldwide, even though not all metals come from anthropogenic sources and their background concentrations vary broadly. It is also worth noting that the standard values vary from country to country, common regulatory guidance values differing by five to seven orders of magnitude (Qin et al. 2021). Despite such lack of harmonization, it is vital to monitor and characterize the impact of metals on soil. Different authors have proposed various assessment methods, of each single factor indices that are most generally used; these are based on the ratio of metal concentrations in soil to the corresponding background values or quality guidelines; the single factor index (Pi) and the geo-accumulation index (Igeo) are the most popular ones (Yang et al. 2021).

Sediments

Metals adsorb to suspended particles in water that then settle as sediment (Algül and Beyhan 2020). In recent years, the increasing levels of metals detected in sediment beds has become a major concern, with several studies revealing that, namely, marine sediments are highly polluted (Perumal et al. 2021). When the chemistry of the aquatic system changes (for example the pH or the redox conditions), metals can release from sediments to the overlying water causing secondary pollution and cause significant damages. In addition, continuing deposition of metals in sediments can lead to contamination of groundwater (Algül and Beyhan 2020). Thus, sediments quality can indicate the status of water pollution (Ali et al. 2019). Factors such as temperature, hydrodynamic conditions, redox state, content of organic matter and microbes or salinity, the chemical composition of the sediments, grain size and content of total organic matter, can affect metals' adsorption, desorption and subsequent concentrations in sediments. A particularly important determinant is pH, since a lowering in pH increases the competition between metal ions and H^+ for binding sites in sediments, may result in the dissolution of metal complexes, thereby releasing free metal ions into the water column (Ali et al. 2019).

Given the importance of assessing sediments' pollution, several indices were proposed to quantify the pollution and eco risk of metals in sediments. Such indices can be summarized as (1) total concentration-based indices, such as EF, Igeo, contamination factor, potential ecological risk index or sediment quality guidelines; and (2) chemical partitioning-based indices, including concentration enrichment ratio and risk assessment code (Yu et al. 2020).

Water

Water is the "life-blood of the biosphere" and since it is a universal solvent, dissolves different organic and inorganic chemicals; however, it also dissolves environmental pollutants such as metals (Briffa et al. 2020). Metals can enter aquatic systems through natural, but also anthropogenic pathways. These includes, domestic sewage, agricultural run-off, industry, mining and smelting, transportation and energy production related activities from villages towns and cities (Ali et al. 2019, Briffa et al. 2020, Yu et al. 2020). The release of industrial effluents without treatment into the aquatic bodies is a major source of pollution of surface and groundwater water (Ali et al. 2019). Both industrial and domestic wastes usually reach the sewage system and even in treated water, metals occur in high concentrations in raw sewage and are not degraded in sewage treatment. Stringent regulations have been set up due to problems caused by sewage dumping into rivers and seas without previous treatment

(Briffa et al. 2020, Dendievel et al. 2020). Once in the water, the distance travelled by metals in rivers depends on the currents, stability and physical state of the metal (Briffa et al. 2020). Rivers and lakes can act both as sinks and secondary sources of metals for the adjacent marine environment, with river inputs constituting the dominant transport pathway of metals from the land to the sea; in this view, metal pollution within an estuary is also of significant environmental concern (Cruz et al. 2020, Karaouzas et al. 2021). When metals enter the sea and oceans, wind and currents transport the pollutant further (Briffa et al. 2020).

All aquatic ecosystems are very vulnerable to pollution. The pollution of water bodies with metals is a worldwide problem and it is acknowledged that the existing freshwater resources are gradually becoming polluted and unavailable (Ali et al. 2019, Hosseini et al. 2020). Such a matter is of critical concern since once metal discharges end up in aquatic ecosystems, they quickly change their physical, chemical and biological properties, posing great risks to the residents' health, economic losses, social stability and ecological safety (Liu et al. 2021).

Among several pollution indices applied to estimate the pollution of water, Heavy metal Pollution Index (HPI), Heavy metal Evaluation Index (HEI) and Contamination Index (CI) are the most effective ways of rating water quality; these indices are considered to provide an overall quality of the water regarding metals content (Hosseini et al. 2020).

Food chains

Metals are transferred from the abiotic environment to living organisms and accumulate in biota at different trophic levels (Ali and Khan 2019). Adverse effects of toxic metals can occur at all levels of biological organization and affect ecological interactions such as predation, parasitism, competition as well as the structure of communities and ecosystems. All organisms have the same challenges when it comes to dealing with metals, needing some metallic elements as nutrients, but on the other hand being faced with the need of avoiding toxicity from both xenobiotic metals and essential metals when present in excessive levels. It is for this reason that mechanisms for ensuring metal homeostasis, detoxification and excretion are largely conservative, especially at the cellular level (Dar et al. 2019). However, as living organisms possess various mechanisms for the accumulation of sufficient amounts of essential metals from their environment, these same mechanisms can facilitate the uptake of non-essential metals.

Bioaccumulation and biomagnification

The uptake and retention of a metal (or any other chemical) by an organism is termed as bioaccumulation, which depends on the metal bioavailability and on the organism's capacity for uptake and excretion of this metal; when uptake exceeds excretion, bioaccumulation occurs (Landrum and Fisher 1999, Mann et al. 2011). When compared with lipophilic forms, water-soluble metal species have more mobility, bioavailability and are the most toxic metal fraction (Ali and Khan 2019). On the other hand, they usually do not bioaccumulate as much as lipophilic forms as can be dissolved in bodily fluids and excreted. Lipophilic metals passively diffuse through the lipid bilayer of biological membranes and moreover, can suffer absorption and storage in the body rather than being eliminated along with other waste products (Mann et al. 2011, Zhao et al. 2016).

The term trophic transfer (also called biotransference) refers to the passage of a contaminant in food chains, from one trophic level to the next (Ali and Khan 2019). In the aspect of food web pollution, another term—biomagnification—is also of particular importance. Biomagnification is the increase in concentration of a pollutant in an organism compared to that in its food. Some pollutants, biomagnify along food chains because successive trophic levels consume massive quantities of biomass (food) to obtain the resources required for metabolic functioning. Hence, if that biomass is contaminated, the contaminant will be taken up in enormous quantities by the consumer. In a food chain when a consumer is eaten by another consumer organism, the fat tissue is digested, and the contaminant is stored in the tissues of the latter one. It is in this way that a contaminant builds up in the fatty tissues of subsequent consumers and their concentration becomes higher with each trophic level. There is ongoing discussion as to whether biomagnification of metals occurs within aquatic and terrestrial systems (Mann et al. 2011). According to some authors all non-essential metals do not biomagnify, whereas other authors are of the opinion that for metals such as Cd or Pb, biomagnification occur in specific food chains. Despite such debates it is well-known that trophic transfer, bioaccumulation and biomagnification of hazardous metals have important implications on wildlife and human health. Primary producers are especially important in trophic transfer of metals because they bridge metal fluxes between abiotic and biotic components of ecosystems (Ali and Khan 2019). Metals entering the food chain through fish can then affect predators such as bigger fish, birds and mammals, some of which migrate and transport the pollutant to different ecosystems. Remarkably, the processes of trophic transfer of metals in food chains may contribute to their long-distance transport throughout the environment; the significance of long-range transport of pollutants, such as Hg, by migratory animals demonstrated earlier (Wania 1998, Ali and Khan 2019, Briffa et al. 2020).

Aquatic and terrestrial food chains

In aquatic ecosystems, metals have a long persistence, bioaccumulation and biomagnification, thus causing toxic effects at points far from the source of pollution. Metals adsorbed into sediments, may bioaccumulate to benthic (bottom-dwelling) organisms and subsequently to the food chain (Ali et al. 2019, Karaouzas et al. 2021). The bioaccumulation of metals in marine and freshwater fish is well known and its assessment is important to assess potential risks of contaminated fish consumption to human health. Carnivorous fish tend to have higher concentrations of metals in their bodies, when compared to herbivorous organisms, because they are at the top of food chains (Mann et al. 2011, Ali and Khan 2019). Some researchers consider that aquatic food chains are less at risk than terrestrial food chain when it comes to biomagnification of metals and metalloids (Mann et al. 2011).

In the case of terrestrial environments, many studies have focused on metal accumulation on individual taxa, with fewer studies examining the effects of metals at the communities' level or their ability to transfer through a food chain. But it is known that the risk of metals entering a food chain depends on the mobility of the metal and its availability in the soil. When metals detach from soil particles and enter the soil solution, they become bioavailable and with potential to accumulate in plants and other soil-dwelling organisms (Gall et al. 2015). Metals may also enter the higher terrestrial food chain after beginning their trajectories through aquatic food chains (Mann et al. 2011).

Metals can adversely affect soil organisms through microbial processes and soil-microbe interactions (Gall et al. 2015). In plants, excessive levels of metals can disrupt critical physiological processes and result in toxicity, and moreover not all plants exclude excess metals by roots. Other plants can tolerate and even hyperaccumulate various metals (Mann et al. 2011, Gall et al. 2015). An interesting hypothesis is that metal accumulation by plants is a defence strategy to discourage consumption by herbivores, for which similarly, some species have evolved to use metals bioaccumulated in the ingested plant biomass as a defence against subsequent predation. The transfer of metal from prey to predator is dependent on two highly variable factors. First, the bioavailability of the sequestered metal. Second, the capacity of predators to minimize net assimilation of metals contained within their prey; the digestive physiology of vertebrates and some invertebrate predators provides an effective barrier against metal assimilation and accumulation (Mann et al. 2011). Birds ingest metals through food, water and through geophagy. Metal levels in birds of different feeding habits are described as: carnivorous > omnivorous and insectivorous > grainivorous. Studies reporting bioaccumulation of metals in mammals are less common, compared to other animal groups, such as invertebrates or fishes (Ali and Khan 2019).

Human Exposure to Metals

Sources

Metals present in air, water and soils untimely end up in drinking water and foods, which along with air, represent major sources of exposure for humans (Gall et al. 2015).

Several decades ago, convincing scientific data emerged relating specific air pollutants to health effects, and the results of such studies provided arguments for setting limit values for specific pollutants in ambient air (WHO 2007). Some airborne PM containing metals are considered as a public health concern, since they can enter human lungs through the respiratory system (Arshad et al. 2015). The toxicity of the particulates depends on their size, the larger size particles (PM10) being easier to eliminate through coughing, sneezing, and swallowing (Hassanien 2011, Arshad et al. 2015). PM 2.5 are the ones considered especially dangerous since they can be transported over long distances, can easily penetrate the lungs into the alveoli, have the potential to penetrate tissues and cannot be moved out easily by exhalation (Amin et al. 2018).

Contaminated drinking water is a leading cause for chronic metal poisoning of millions of people, with 1.6 million children dying each year from various diseases (Fernández-Luqueño et al. 2013). Drinking water contaminated with As, Cd, Ni, Hg, Cr, Zn and Pb is becoming a major health concern for public and health care professionals (Rehman et al. 2018).

In recent decades, the adverse effects of metal contaminants on crop grown on metal-contaminated agricultural soils have threatened food security and human health (Gall et al. 2015, Qin et al. 2021). As most metals in soil can accumulate in crops and then transfer to humans through the food chain, many soil to plant transfer indices are used to evaluate the environmental safety risks of metals in soil. Among them, the bio-concentration factor (expressed as concentration in crops/concentration in soil) is an essential parameter. The highest bio-concentration factor values are found in leafy vegetables, followed by tuberous, whereas the lowest values are found in horticulture crops and fruits. The potential ecological risk index is another parameter which reflects the sum of the risk factors for all hazardous metals in a soil sample and the biotic responses; ecological risk indexes can be classified as low (< 50), moderate (50 to 100), considerable (100 to 200), or intense/high (> 200) (Qin et al. 2021). In some livestock diets metals can be added as supplementary trace elements to promote health and growth. Nevertheless, most metals consumed by livestock are excreted in urine or faeces, which will then be present in manure that is applied to land (Gall et al. 2015). Unintentional contamination of feedstuffs and ingredients with metals also poses a major problem as metal can accumulate into meat, egg and milk products (Elliot et al. 2017).

Routes

Metals can enter a human body through ingestion, inhalation and skin contact; their toxicity will depend upon the amount absorbed by the body, the duration and the route of exposure (Amin et al. 2018, Scharninghausen 2019). Therefore, metal exposure assessment includes the characterization of the exposure routes (EPA 2007). People can be exposed to metals by ingesting contaminated food or water and during outdoor activities, where they can easily come into contact with airborne metals through inhalation of fine particulates and dermal contact with particulate fallout; dust particles containing metals can also be found indoors (Amin et al. 2018). These particles may also dissolve in sweat and eventually enter the bloodstream through the skin pores; contact of metals with the skin is an important route of exposure, which is not well characterized (Hostýnek et al. 2008, Ugwu and Ofomatah 2020). Humans could also ingest dusty materials form air and soil into their mouth or drink water with dust contamination (Ugwu and Ofomatah 2020). Infants and children can have enhanced exposures to metals through surface dust as they crawl and play near the dust and often mouth their hands (e.g., finger sucking) and objects in their environment (EPA 2007).

Toxicity of Metals

Currently we are witnessing increasing global concerns regarding metals' negative enduring impacts on health (Koller and Saleh 2018). The main threats to human health are associated with exposures to the xenobiotic metals Cd, Hg, Pb and to the metalloid As (Järup 2003). Continuous exposure to metals may lead to internal imbalance in the body, as when they start accumulating in the organism, are used as substitutes of essential elements; calcium (Ca) is replaced by Pb, Zn by Cd and the majority of trace elements are replaced by Al. Stored metals destroy major metabolic processes, and the activity of various hormones and essential enzyme's function are also influenced. Metals creates antioxidant imbalance and whatever the molecular pathway that is influenced by them, Reactive Oxygen Species (ROS) are generated. ROS accumulation produces Oxidative Stress (OS) that may lead to neurological disorders, damage of the kidney function, endocrine abnormalities and different kinds of cancer , among other effects (Rehman et al. 2018). Some metals/metalloids which are described as serious toxicants due to their carcinogenicity include Cd, As, Cr^{6+} and Ni, classified as category 1 (carcinogenic to humans) according to the International Agency for Research on Cancer (IARC) (Kim et al. 2015). Other metals are reported as suspected to be carcinogenic (Co and Hg), mutagenic [As, Vanadium (V)], teratogenic (As), allergenic (Ni) or endocrine disruptors (Ag, Cu, Zu). Some are well known to cause neurological and behavioural alterations especially in children (Pb), induce central nervous system damage (Hg, Mn and Pb), bone marrow damage and osteoporosis (Cd and Pb), are hepatotoxic and/or nephrotoxic (Cd, Hg and Mn) or negatively affect the immune system (Pb) (Mohammad et al. 2020, Puzas et al. 2004, Koller and Saleh 2018).

The relation between exposure to metals from the environment and human health effects is illustrated in Fig. 1.

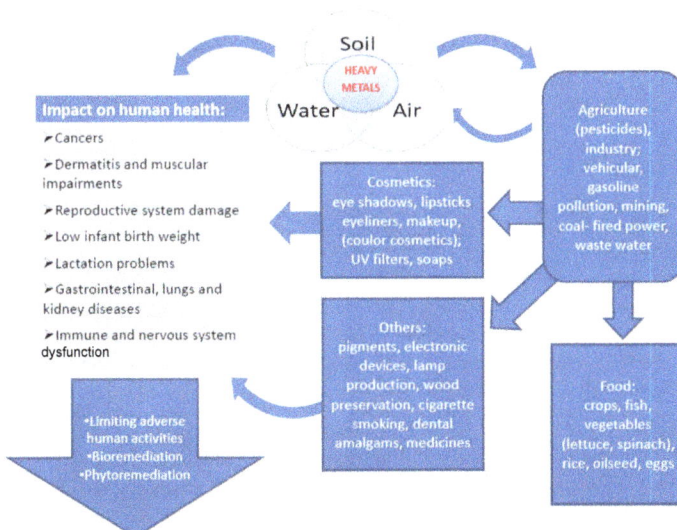

Figure 1. Routes of exposure, the impact of toxic metals on human health (Witkowska et al. 2021).

Final Remarks

Several aspects of pollution by metals were discussed in this chapter informing how this is generating great apprehension today. Metals are typical pollutants due to their environmental persistence, bioaccumulation and toxicity but even so, their anthropogenic emissions are still increasing in some areas of the world. Emissions by metal-based industries and agriculture were examined as well as drinking water and food pollution as major sources for human exposures. Adverse consequences for the environment and for human health have resulted from metals' accumulations in several environmental parts (including air, water, soils and the food chain). Exposure to metals contributes to the onset and/or progression of many diseases, which include cancers and neurological disorders. in the light of these facts and despite the current challenges and difficulties, policies and plans to control metal pollution should certainly continue and even be reinforced.

References

Algül, F. and M. Beyhan. 2020. Concentrations and sources of heavy metals in shallow sediments in Lake Bafa, Turkey. Sci. Rep. 10(1): 11782.

Ali, H. and E. Khan. 2019. Trophic transfer, bioaccumulation, and biomagnification of non-essential hazardous heavy metals and metalloids in food chains/webs—Concepts and implications for wildlife and human health. Hum. Ecol. Risk Assess. 25(6): 1353–1376.

Ali, H., E. Khan and I. Ilahi. 2019. Environmental chemistry and ecotoxicology of hazardous heavy metals: Environmental persistence, toxicity, and bioaccumulation. J. Chem. 2019: 6730305.

Altunatmazi, S.S., D. Tarhan, F. Aksu, N.P. Ozsobaci, M.E. Or and U.B. Barutuçu. 2019. Levels of Chromium, Copper, Iron, Magnesium, Manganese, Selenium, Zinc, Cadmium, Lead and Aluminium of honey varieties produced in Turkey. Food Sci. Technol. 39(2): 392–397.

Amin, N., A. Azid, M. Sani, K. Yusof, M. Samsudin, N. Rani et al. 2018. Heavy metals in the air: Analysis using Instrument, air pollution and human health-a review. pollution and human health-a review. J. Appl. Fund. Sci. 14(4): 490–494.

Arshad, N., Z. Hamzah, Ab.K. Wood, A. Saat, and M. Alias. 2015. Determination of heavy metals concentrations in airborne particulates matter (APM) from Manjung district, Perak using energy dispersive X-ray fluorescence (EDXRF) spectrometer. AIP Conference Proceedings. 1659: 050008.

Briffa, J., E. Sinagra and R. Blundell. 2020. Heavy metal pollution in the environment and their toxicological effects on humans. Heliyon 6 (9): e04691.

Britannica, Encyclopædia Britannica, Inc. 2020. "Metal". Encyclopaedia Britannica. Britannica, Chicago, USA.

Cruz, T.C. G.N. Nayaka, A.K. Tiwari and M.R. Nasnodkara. 2020. Assessment of metal pollution and bioaccumulation of metals by edible bivalve *Polymesoda erosa* in the Zuari Estuary, west coast of India. Mar. Poll. Bull. 158: 111415.

Dar, M.I., I.D. Green and F.A. Khan. 2019. Trace metal contamination: Transfer and fate in food chains of terrestrial invertebrates. Food Webs. 20: e00116.

Delfino, S. and A. Saccone. 2009. Chemistry of Metals. pp. 88–127. *In*: I. Bertini [ed.]. Chemistry of Metals, Inorganic and Bio-organic Chemistry (Vol. 1). United Nations Education Scientific and Cultural Organization, Paris, France.

Dendievel, A.M., B. Mourier, A. Dabrin, H. Delile, A. Coynel, A. Gosseta et al. 2020. Metal pollution trajectories and mixture risk assessed by combining dated cores and subsurface sediments along a major European river (Rhône River, France). Environ. Int. 144: 106032.

Duffus, J. 2002. "Heavy metals" a meaningless term? (IUPAC Technical Report). Pure Appl. Chem. 74(5): 793–807.

Dunee. 2012. Difference Between Metal and Heavy Metal. Science & Nature/Science/Chemistry/Inorganic Chemistry. https://www.differencebetween.com/difference-between-metal-and-vs-heavy-metal/. Accessed at 20 th June 2021.

Elliott, S., A. Frio and T. Jarman. 2017. Heavy metal contamination of animal feedstuffs—a new survey. J. Appl. Anim. Nutr. 5: E8.

[EEA] European Environmental Agency, Helpcenter definition—Pollution, 2012. EEA, Copenhagen, Denmark.

[EPA] Environmental Protection Agency. 2007. Framework for Metals Risk Assessment. Environmental Protection Agency, Washington, USA.

Fang, G.C., Y.S. Wu, S.Y. Chang, S.H. Huang, and J.Y. Rau. 2006. Size distributions of ambient air particles and enrichment factor analyses of metallic elements at Taichung Harbor near the Taiwan Strait. Atm. Res. 81(4): 320–333.

Fernández-Lueño, F., F. López-Valdez, P. Gamero-Melo, S. Luna-Suárez, E.N. Aguilera-González, A.I. Martínez et al. 2013. Heavy metal pollution in drinking water—a global risk for human health: A review. Afr. J. Environ. Sci. Technol. 7(7): 567–584.

Gall, J.E., R.S. Boyd and N. Rajakaruna. 2015. Transfer of Heavy Metals Through Terrestrial Food Webs: A Review. Environ. Monit. Assess. 187(4): 201.

Gupta, P.K. 2018. Illustrated Toxicology. Academic Press, Elsevier, London, UK.

Hassanien, M.A. 2011. Atmospheric Heavy Metals Pollution: Exposure and Prevention Policies in Mediterranean Basin. pp. 287–307. *In*: L. Simeonov, M. Kochubovski and B. Simeonova [eds.]. Environmental Heavy Metal Pollution and Effects on Child Mental Development. NATO Science for Peace and Security. Series C: Environmental Security, Vol. 1. Springer, Dordrecht, Netherlands.

Hosseini, H., A. Shakeri, M. Rezaei, M.D. Barmaki and M.R. Mehr. 2020. Water chemistry and water quality pollution indices of heavy metals: a case study of Chahnimeh Water Reservoirs, Southeast of Iran. Int. J. Energ. Water Res. 4: 63–79.

Hostýnek, J.J., R.S. Hinz, C.R. Lorence, M. Price and R. H. Guy. 2008. Metals and the Skin. Crit. Rev. Toxicol. 23(2): 171–235.

Jaishankar M., T. Tseten, N. Anbalagan, B.B. Mathew and K.N. Beeregowda. 2014. Toxicity, mechanism, and health effects of some heavy metals. Interdiscip. Toxicol. 7(2): 60–72.

Järup, L. 2003. Hazards of heavy metal contamination. Br. Med. Bull. 68: 167–182.

Jiries, A., A.H. Ziadat and R. Al-Atwi. 2018. Atmospheric pollution with heavy metals at Tabouk City-KSA. Adv. Sci. Eng. Technol Int. Conferences (ASET). Dubai 2018: 1–5.

Karaouzas, I., N. Kapetanaki, A. Mentzafou, T.D. Kanellopoulos and N. Skoulikidis. 2021. Heavy metal contamination status in Greek surface waters: A review with application and evaluation of pollution indices. Chemosphere. 263: 128192.

Kim, H.S., Y.J. Kim and Y.R. Seo. 2015. An Overview of Carcinogenic Heavy Metal: Molecular Toxicity Mechanism and Prevention. J. Cancer Prev. 20(4): 232–240.

Koller, M. and H.M. Saleh. 2018. Introducing Heavy Metals. pp. 3–11. *In*: H.M. Saleh [ed.]. Heavy Metals. IntechOpen, London, UK.

Kunt, F., Z.C. Ayturan, F. Yümün, I. Karagönen, M. Semerci and M. Akgün. 2021. Measurement and evaluation of particulate matter and atmospheric heavy metal pollution in Konya Province, Turkey. Environ. Monit. Assess. 193(10): 637.

Landrum, P.F. and S.W. Fisher. 1999. Influence of Lipids on the Bioaccumulation and Trophic Transfer of Organic Contaminants in Aquatic Organisms. pp. 203–234. *In*: M.T. Arts and B.C. Wainman [eds.]. Lipids in Freshwater Ecosystems. Springer, New York, NY, USA.

Lequy, E., J. Siemiatycki, S. Leblond, C. Meyer, S. Zhivin, D. Vienneau et al. 2019. Long-term exposure to atmospheric metals assessed by mosses and mortality in France. Environ. Int. 129: 145–153.

Lee Jr., R.E. and D.J. von Lehmden. 1973. Trace Metal Pollution in the Environment. J. Air Poll. Contr. Assoc. 23(10): 853–857.

Liu, J., R. Liu, Z. Yang and S. Kuikka. 2021. Quantifying and predicting ecological and human health risks for binary heavy metal pollution accidents at the watershed scale using Bayesian Networks. Environ. Poll. 269: 116125.

Mann, R.M., M.G. Vijver and W.J.G.M. Peijnenburg. 2011. Metals and Metalloids in Terrestrial Systems: Bioaccumulation, Biomagnification and Subsequent Adverse Effects. pp. 49–74. *In*: F. Sánchez-Bayo, P.J. van den Brink and R.M. Mann [eds.]. Ecological Impacts of Toxic Chemicals. Bentham Science Publishers Ltd., Sharjah, United Arab Emirates.

Messager, M.L., I.P. Davies and P.S. Levin. 2021. Low-cost biomonitoring and high-resolution, scalable models of urban metal pollution. Sci. Tot. Environ. 767: 144280.

Mohammad, N.A., O.M. Mehdi, F. Omid, M. Leila, and H. Rezal. 2020. Manganese-Induced Nephrotoxicity Is Mediated through Oxidative Stress and Mitochondrial Impairment. J. Ren. Hepat. Disord. 4(2): 1–10.

Naderizadeh, Z., H. Khademi and S. Ayoub. 2016. Biomonitoring of atmospheric heavy metals pollution using dust deposited on date palm leaves in southwestern Iran. Atmósfera. 29(2): 141–155.

Perumal, K., J. Anthony and S. Muthuramalingam. 2021. Heavy metal pollutants and their spatial distribution in surface sediments from Thondi coast, Palk Bay, South India. Environ. Sci. Eur. 33:63.

Popoola, L.T., S.A. Adebanjo and B.K. Adeoye. 2018. Assessment of atmospheric particulate matter and heavy metals: a critical review. Int. J. Environ. Sci. Technol. 15: 935–948.

Puzas, J.E., J. Campbell, R.J. O'Keefe and R.N. Rosier. 2004. Lead Toxicity in the Skeleton and Its Role in Osteoporosis. pp. 363–376. *In*: M.F. Holick and B. Dawson-Hughes [eds.]. Nutrition and Bone Health. Nutrition and Health. Humana Press, Totowa, USA.

Qin, G., Z. Niu, J. Yu, Z. Li, J. Ma and P. Xiang. 2021. Soil heavy metal pollution and food safety in China: Effects, sources and removing technology. Chemosphere. 267: 129205.

Rehman, K., F. Fatima, I. Waheed and M.S.H. Akash. 2018. Prevalence of exposure of heavy metals and their impact on health consequences. J. Cell. Biochem. 119(1): 157–184.

Scharninghausen, J. 2019. Protecting Against Heavy Metals. Directorate of Assessments and Prevention, Workplace Safety Division, U.S. Army Combat Readiness Center, Fort Rucker, Alabama, USA.

Srithawirat, T. 2016. Indoor PM10 and its heavy metal composition at a roadside residential environment, Phitsanulok, Thailand. Atmósfera. 29(4): 311–322.

Tchounwou, P.B., C.G. Yedjou, A.K. Patlolla and D.J. Sutton. 2012. Heavy Metals Toxicity and the Environment. EXS. 101: 133–164.

Ugwu, K.E. and A. C. Ofomatah. 2020. Concentration and risk assessment of toxic metals in indoor dust in selected schools in Southeast, Nigeria. SN Apll. Sci. 3: 43.

Wania, F. 1998. The Significance of Long-Range Transport of Persistent Organic Pollutants by Migratory Animals. WECC Report 3/98. Wania Environmental Chemists Corp. Toronto, Canada.

Witkowska, D., J. Słowik and K. Chilicka. 2021. Heavy Metals and Human Health: Possible Exposure Pathways and the Competition for Protein Binding Sites. Molecules. 26(19): 6060.

[WHO] Joint WHO/, Convention Task Force on the Health Aspects of Air Pollution, Health risks of heavy metals from long-range transboundary air pollution. 2007. WHO, Geneve, Switzerland.

Yang, H., F. Wang, J. Yu, K. Huang, H. Zhang and Z. Fu. 2021. An improved weighted index for the assessment of heavy metal pollution in soils in Zhejiang, China. Environ. Res. 192: 110–246.

Yu, Z., E. Liu, Q. Lin, E. Zhang, F. Yang, C. Wei and J. Shen. 2020. Comprehensive assessment of heavy metal pollution and ecological risk in lake sediment by combining total concentration and chemical partitioning. Environ. Poll. 269: 116212.

Zhao, C.M., P.G.C. Campbell and K.J. Wilkinson. 2016. When are metal complexes bioavailable? Environ. Chem. 13: 425–433.

CHAPTER 2
Biomarkers, an Overview

The Concept of Biomarker

The use of biomarkers (BMs) goes back to the very beginning of medical treatment. The practice of uroscopy (examining a patient's urine for signs of disease) dates to the 14th century when doctors would regularly examine the colour and sediment of their patient's urine. Only much later was the word "biomarker" first used by Karpetsky, Humphrey and Levy in a 1977 edition of the Journal of the National Cancer Institute. This word is a little over 40 yr old, being a portmanteau of "biological marker" (Strimbu and Tavel 2010, Fathi et al. 2014).

Defining biomarkers

Currently the World Health Organization (WHO) defines BM as "any substance, structure, or process that can be measured in the body or its products and influence or predict the incidence of outcome or disease" (Strimbu and Tavel 2010, Fathi et al. 2014). There are several other definitions of BM in literature, such as the one given by the United States National Academy of Sciences Committee on Biological Markers; they define a BM as "an alteration in cellular or biochemical components, processes, structure or functions that is measurable in a biological system or sample", despite not being a measure of the disease, a disorder or condition itself (Ladeira and Viegas 2016). These definitions in a practical way, means that as body temperature is a well-known BM for fever; blood pressure is used to determine the risk of a stroke; and cholesterol values are risk indicators for coronary and vascular disease (Califf 2018). A more comprehensive definition is "a characteristic that is objectively measured and evaluated as an indicator of normal biological processes, pathogenic processes or pharmacologic responses to a therapeutic intervention" (Fathi et al. 2014). Here, BMs do not only indicate the incidence or the outcome of a disease, but also the effects of treatments or interventions. As an example, if a subject is taking an anti-cholesterol medication, cholesterol levels can be measured in a follow-up appointment to determine whether the medicine is working; in other words, whether the medication has lowered cholesterol and reduced the risk for having a heart attack (Califf 2018).

Contexts of biomarkers

In clinical, laboratory or toxicological contexts, BMs can be regarded with different implications. Clinically BMs refers to a broad subcategory of medical signs—that is, objective indications of the medical state observed from outside the patient—which can be measured accurately and reproducibly. It should be noted that medical signs stand in contrast to medical symptoms, which are limited to those indications of health or illness perceived by patients themselves. In fact, while BMs are by definition objective and quantifiable characteristics of biological processes, they do not necessarily correlate with a patient's experience and sense of wellbeing (Strimbu and Tavel 2010). Some authors even mention that BMs should be distinct from direct measures of how a person feels, functions or survives, which should be a different category of measure known as Clinical Outcome Assessment (COA) (Califf 2018). COA can be defined as a well-defined and reliable assessment of patients' symptoms, overall mental state or how they function. This is a complex process, which includes: the patient-reported outcome measures; observer-reported outcome measures given by someone other than the patient or a health professional such as a parent or a caregiver; clinician-reported outcome measures; and performance outcome measures which are standardized task(s) actively undertaken by a patient according to a set of instructions (FDA 2020). Therefore, distinguishing BMs from COAs is very important, since COAs measure outcomes that are directly important to the patients, BMs serve a number of purposes, one being to link a measurement to a prediction of COAs (Califf 2018).

In the laboratory BMs may include basic chemistries to more complex tests on biological fluids, cells and other tissues, as well as histologic, radiographic or physiologic characteristics (Fathi et al. 2014). An emerging field of interest is represented by those studies that integrate imaging parameters with molecular BMs for improving patients' diagnoses and prognoses (Canese et al. 2020).

The term BM was introduced in the field of toxicology, environment and occupational health, since the past 20 yr (Aitio et al. 2007). BMs have been used to identify biological changes due to toxic chemicals and in the assessment of environmental health, as part of an integrated approach (Ladeira and Viegas 2016). In a report regarding the validity of BMs in environment risk assessment, WHO stated that a true definition of BMs "includes almost any measurement reflecting an interaction between a biological system and a potential hazard, which may be chemical, physical, or biological. The measured response may be functional and physiological, biochemical at the cellular level, or a molecular interaction" (Strimbu and Tavel 2010). Currently most research on BMs in the field of toxicology is concerned with markers which can increase the ability to identify long-term risks due to toxicant exposure and identify early markers of toxicity (Ladeira and Viegas 2016).

Monitorization of Environment and Human Health

Before describing different BMs categories, it is vital to allude two related concepts: "Environmental Monitoring" and "Biological Monitoring". Both can help in the assessment of exposure to specific chemicals, characterization of exposure pathways and potential risks and their mitigation; thus, can serve as elements of health surveillance programs (Matatiele et al. 2019).

Environmental monitoring

Environmental monitoring is a systematic and repeated observation of harmful factors that affect the environment which living beings inhabit, according to predetermined standards or rules. Therefore, constituting a tool to assess environmental conditions and trends, supporting policy development and its implementation, and developing information for reporting to national policymakers, international forums and the public. To provide an example, the environmental monitoring of biodiversity demonstrates the extent to which areas important for conserving biodiversity, cultural heritage, scientific research, recreation, natural resource maintenance and other environmental values are protected from incompatible uses (UNECE 2021).

Under a toxicological perspective environmental monitoring can be defined as measurements of concentrations of chemical substances in various exposure media such as air, water and food, aiming at the protection of the environment and human health (Hambach et al. 2013, Artiola and Brusseau 2019). The process includes the collection of environmental media samples (air, water, soil) for chemical analysis, and may include real-time monitoring using devices that detect exposures to hazardous agents (Covaci 2014). For example, in occupational settings environmental monitoring can be performed as stationary sampling of industrial air, and such monitoring may be preferred when the source identification is important (Aitio et al. 2007).

A limitation of environmental monitoring is that while it involves the measurement and control of airborne chemicals to know an external dose, other aspects of the problem, such as skin absorption, ingestion and non-work-related exposure, remain undetected and therefore, uncontrolled (Foà and Alessio 2011). To illustrate, estimations of inhalation exposure can be made regarding the 'intake', or 'external load' based on known concentrations in ambient air; but in fact, the actual 'uptake' or 'internal load' can only be approximated by means of environmental monitoring methods (Evelo and Henderson 1992). Biological monitoring may help to fill these gaps.

Biological monitoring

Biological monitoring (or biomonitoring) can be explained as: "a repeated controlled measurement of chemicals or BMs in subjects' fluids or tissues (i.e., urine or blood), or other accessible samples to look for evidence of exposure to chemical, physical or biological risk factors in the workplace and/or the general environment, or to assess the interaction (absorption, early health effects, susceptibility) between physical, chemical or biological agents and the human organism" (Manno and Sanolo 2004, Holstege and Hardison 2014, Ladeira and Viegas 2016, Brucker et al. 2020). Biological monitoring relies on these facts: due to absorption, distribution, metabolism and excretion processes, a certain internal dose of a toxic agent (the net amount of a pollutant absorbed in or passed through the organism over a specific time interval) is effectively delivered to the body, becoming detectable in body fluids; biochemical and cellular events occur as result of the chemical interaction with a receptor in the critical organ (the organ which, under specific conditions of exposure, exhibits the first or the most important adverse effect). In this manner biological monitoring allows both the internal dose and the elicited biochemical and cellular effects that can be measured (Evelo and Henderson 1992, Foà and Alessio 2011).

Biological monitoring of chemicals and their effects is more common than physical or biological agents. It can be conducted evaluating the unchanged chemical in the body, a metabolite of the original chemical, an enzymatic alteration, a physiological effect or a secondary clinical finding (Holstege and Hardison 2014). Hence biological monitoring can be divided into monitoring of exposure and monitoring of effect, for which indicators of internal dose and of effect are used, respectively (Mayeux 2004).

Biological monitoring of exposure is used to assess the health risk through the evaluation of the internal dose, reaching an estimate of the biologically active body burden of the chemical in question (Foà and Alessio 2011). However, for chemicals that are excreted rapidly, cross-sectional biomonitoring data reflect only recent exposure, for which repetitive sampling to characterize long-term exposure patterns are necessary (Mayeux 2004). The rationale of biological monitoring of exposure is to ensure that a person's exposure does not reach levels capable of eliciting adverse effects (Foà and Alessio 2011).

An effect is termed "adverse" when there is an impairment of functional capacity, a decreased ability to compensate for additional stress, a decreased ability to maintain homeostasis or an enhanced susceptibility to other environmental influences (Foà and Alessio 2011). Biological monitoring of effects counts on the fact that changes in physiological functions of the organism can be indicative of changes in the state of health as consequence of a certain chemical exposure. Such information on the state of health or early impairment of health, is very useful in the prevention of overt intoxications due to the continuation of overexposures (Evelo and Henderson 1992). While biological monitoring using exposure BMs considers interindividual differences in absorption, biological monitoring using BMs of effect examines additionally interindividual differences in susceptibility; unfortunately, only few effect BMs have been validated so far (Santonen et al. 2015).

Besides the earlier mentioned advantages of biological monitoring over environmental monitoring, several others exist. Different from environmental samples, biological samples reveal integrated effects of repeated exposure, directly reflect the total body burden, precisely reflect biological effects resulting from all routes of exposure (inhalation, absorption through the skin and ingestion), as well as interindividual variability in exposure levels, metabolism and excretion rates. Such data reflect additionally modifying influences in physiology, bioavailability, bioaccumulation and persistency. Moreover, biomonitoring can demonstrate the association between the body burden of pollutants and/or contaminants and the respective health effects, in epidemiological studies. The power of biological monitoring to identify spatial and temporal trends in human exposures has successfully provided relevant exposure and risk information, as well as allowed focusing on the protection of susceptible populations, such as children and pregnant mothers (Ladeira and Viegas 2016).

Biological monitoring also has limitations, one being the fact that biological indicators of internal dose allow assessing the degree of exposure, but do not furnish data to measure the actual amount present in a critical organ. Another frequent limitation is the absence of knowledge regarding possible interferences in the metabolism of the substances being monitored, by others to which the organism is simultaneously exposed (Mayeux 2004).

Categories of Biomarkers

BMs are used in biological monitoring to assess specific exposures and predict the risk of adverse health effects, in both an individual and on a populational basis (Jeddi et al. 2021). A few categories of BM have been defined according to their putative applications; however, a single BM may meet multiple criteria for different uses. BMs may be used to assess the exposure (absorbed amount or internal dose), effect(s) of chemicals and susceptibility of individuals; they may be applied whether exposure has been from dietary, environmental or occupational sources. BMs may also be used to elucidate cause-effect and dose-effect relationships in health risk assessment, in clinical diagnosis and for monitoring purposes (WHO 1993, Brucker et al. 2020). The traditional and most generally accepted classification of BMs (and the one used in modern biological monitoring programs) divides them into three main categories: BMs of exposure, BMs of susceptibility and BMs of effect (Hambach et al. 2013).

Biomarkers of exposure

BMs of exposure are the first ones to be used in human biomonitoring studies and continue to be closely related to biological monitoring (Alimonti and Mattei 2008, Steckling et al. 2018). They can be defined as "an exogenous substance or its metabolite or the product of an interaction between a xenobiotic agent and some target molecule or cell, that is measured in a compartment within an organism" (Mussali-Galante et al. 2013, Steckling et al. 2018). These BMs can be identified in a number of human specimens such as blood, urine, deciduous teeth or hair, and be used to confirm and assess the exposure of individuals or populations to a particular substance, providing a link between the external exposure and the internal dose (WHO 1993, Steckling et al. 2018). The Environmental Protection Agency (EPA 2021) considers a 3rd group of BMs of exposure denominated endogenous surrogates. Since a chemical or class of chemicals may result in an endogenous response which is highly characteristic of that chemical or class, measures of that response can be used as a surrogate in lieu of a direct measurement of the chemical or metabolite concentration, when sufficient additional information is available. However, when considering that many factors can influence endogenous responses, this type of exposure BM is accompanied by many uncertainties, which must be identified and discussed (EPA 2021). BMs of exposure can also be divided into markers of internal dose and markers of effective dose (Ladeira and Viegas 2016). The simplest indicator of internal dose is for example, the blood concentration of a chemical agent measured following exposure. Diversely, the productive dose is an indication of the biologically active fraction of xenobiotics, which can interact with cellular macromolecules at the target site (Miraglia et al. 2004). Effective doses of BMs are considered mechanistically related to disease outcomes and thus in theory, are best able to link exposure (external and internal) to disease outcomes (Links et al. 1995, IMUSC 2001).

Many benefits are attributed to the use BMs of exposure, since they can reflect bioavailability, provide an integrated measure of chemical uptake, the latest important being in the case of agents that exhibit large route-dependent differences in absorption (Ladeira and Viegas 2016). Another advantage is that over a history of exposure, they estimate the actual "internal" dose of the exposure (Links et al. 1995, Mayeux 2004).

Biomarkers of susceptibility

BMs of susceptibility reflect intrinsic characteristics of an organism that make it more or less susceptible to the adverse effects of an exposure to a specific a chemical, physical or biological agent (Mayeux 2004, Ladeira and Viegas 2016). Evidence indicates that both inherited and acquired genetic susceptibility, epigenetic modifications, as well as alterations in physiological structures and functions induced by age, pathological conditions and lifestyle factors, may lead to different phenotypic expressions from xenobiotic exposures (Iavicoli et al. 2016). Considering this, susceptibility BMs are normally used to identify 'at-risk' individuals in the following situations: where risk is inherited as indicated by genetic markers; or the more usual, when it is acquired (e.g., they indicate a disorder that renders people more susceptible to an environmental exposure) (Kelly and Vineis 2014).

Inherited susceptibility

Differences among subjects in their qualitative and quantitative responses to chemical exposures may occur due to their genetic make-up, which will vary depending on the presence of DNA sequence variations (polymorphisms) within critical genes. A general view is emerging that there are dozens of genes that modulate the response to environmental exposures and that polymorphisms in these genes results in differences among subjects regarding the uptake, absorption and metabolism of pollutants; this infers enzymes of activation and detoxification, repair enzymes, as well as changes in target molecules for toxic chemicals (Miller et al. 2001, Mayeux 2004). Two processes involving susceptibility to chemical agents are considered: toxicokinetic and toxicodynamic; BMs of susceptibility may be of either type, with the determination of relative enzyme activities or the presence or absence of other gene products often used as susceptibility BMs. Namely, enzymes involved in xenobiotic metabolism can be particularly important in the overall mechanism of action of xenobiotics, genetic polymorphisms in metabolic enzymatic activity being a common basis for interindividual differences in toxicity (Ladeira and Viegas 2016).

Namely, it is the relationship between enzymes that activate (e.g., cytochrome P450, N-acetyltransferases) or detoxicate (e.g., glutathione S-transferases) carcinogens found in cigarette smoke and the risk of lung and bladder cancer. The analysis of polymorphisms in these genes can clarify the distribution of exposure risk in human populations and estimate risks in susceptible subpopulations (Miller et al. 2001).

It should be highlighted that BMs of susceptibility do not represent stages along the dose-response mechanistic sequence; instead, they represent conditions that alter the rate of transition between the stages or molecular events (Ladeira and Viegas 2016). Furthermore, genetic variants may be related to susceptibility but are not deterministic, an example being the knowledge that most adult-onset degenerative diseases of the nervous system are likely to be a composite of related characteristics, heritable and environmental (Mayeux 2004, Ginsberg et al. 2014).

Acquired susceptibility

Previous exposure to an environmental agent, the physiological state of the individual (i.e., starvation, disease state) and developmental or age-related processes controlling pathways that impact genetic expression, constitute environmental factors that may influence susceptibility and the ensuing response to a chemical exposure (Hagger et al. 2006). Interesting examples concerning acquired susceptibility resulting from the subject physiological state, are the results of studies which revealed that stroke victims are more susceptible to the effects of air pollution with respect to cognitive functions, or the belief that hypertension (often regarded as a marker of vascular dysfunction) may mediate an association between air pollution and depressive symptoms (Wang and Yang 2018).

In general terms the knowledge on susceptibility BMs provides insights into disease biology, allowing for better disease prediction and prognosis models to be built, and enabling the identification of vulnerable people (Brucker et al. 2020).

Biomarkers of effect

Effect BMs are defined as measurable biochemical, structural, functional, behavioural or any other kind of alterations in an organism that, according to its magnitude, can be associated with established or potential health impairment or disease (Ladeira and Viegas 2016). Such alterations can be elicited as a result of interactions of the organism with a host of different environmental factors (including chemical, physical and biological agents), at the level of the whole organism, at the level of organ function, at the level of tissue and individual cells and at the subcellular level (Ladeira and Viegas 2016, Brucker et al. 2020).

In toxicology, BMs of effect can be seen as indicators of a change in biologic function in response to an exposure; thus, they more directly relate to insight into the potential for adverse health effects than BMs of exposure (EPA 2021). The effect of BMs can be divided in those of altered structure (e.g., genotoxic damage in peripheral blood samples) and those of altered function (e.g., changes in lung function measured by spirometry) (Links et al. 1995). BMs of effects may also suggest results from exposures to chemical mixtures or aggregate exposure (i.e., exposure to the same chemical from multiple exposure routes). They can also help in identifying early effects in humans at low toxic doses, establish dose–response relationships, explore mechanisms and increase the biological plausibility of epidemiological associations; in this manner, the effect of BMs are important tools to improve risk assessment (Jeddi et al. 2021).

An interesting classification provided by EPA (2021) examines three sub-groups of effect BMs: bioindicators, undetermined consequences and exogenous surrogates.

Bioindicators

An ideal BM of effect has an explicitly known mechanism that links the marker and an adverse outcome, and when these mechanisms are known enough, bioindicators can be developed providing a high degree of confidence in predicting the potential for adverse effects in an individual or population. Specifically, when cellular or molecular initiating events can be identified as critical steps in an adverse outcome pathway, mainly the ones preceding progressive structural or functional damage on the molecular and cellular levels, they can be markers for precursor events; and can lead to clinically detectable adverse outcomes, to support early detection and prevention and reversible clinical response (Ladeira and Viegas 2016, Brucker et al. 2020, EPA 2021).

Undetermined consequences

Undetermined consequences BMs provide more limited and uncertain indications of the potential for adverse effects because the events or deterministic linkages in an adverse outcome pathway are less well known (EPA 2021). An example would be markers of oxidative stress, for which evaluations have been associated with a number of adverse outcomes, but the explicit relationships have yet to be defined. These BMs can be used in conjunction with others to improve the specificity and sensitivity of an overall set of markers.

Exogenous surrogates

The concept of exogenous surrogates is based on the rational that some chemicals have well known adverse effects, which are accompanied by other effects that can be used as surrogate indicators of the main adverse effect of interest. An example is the measurement of urinary para-nitrophenol as an indicator of the potential for toxicity due to methyl parathion-induced acetylcholinesterase inhibition (EPA 2021). Methyl parathion is detoxified by phase II reactions to para-nitrophenol and dimethyl phosphate, that occur in 86% of para-nitrophenol urinary excretion by 8 hr postexposure (Hryhorczuk et al. 2002).

Other biomarkers

Along with the accepted classification of exposure, susceptibility and effect BMs, other categories exist. These are mostly used in clinical practice and include: diagnostic BM, which are used to detect or confirm the presence of a disease or condition of interest or identifies an individual with a subtype of the disease; monitoring BMs, which can be measured to assess the status of a disease or medical condition for evidence of exposure to a medical product or environmental agent or to detect an effect of a medical product or biological agent (Fathi et al. 2014).

Predictive, prognostic and safety biomarkers

Additional classifications consider predictive BMs, prognostic BMs and safety BMs. A predictive BM is defined by finding that the presence or change in the BM, predicts an individual or group of individuals to be more likely to experience a favourable or unfavourable effect, from the exposure to a medical product or environmental agent; preventive measures should be taken by people with elevated levels of these BMs. A prognostic BMs categorizes individuals by the degree of risk for disease occurrence or progression, informing the natural history of the disorder in that particular patient in the absence of a therapeutic intervention (Fathi et al. 2014, Califf 2018). These are different from susceptibility BMs because they deal with the transition from the healthy state to disease; they are also different from predictive BMs, as predictive BMs discriminate those who will respond or not respond to therapy or to an environmental agent, whereas prognostic BMs are associated with differential disease outcomes (Califf 2018). Safety BMs are measured before or after an exposure to a medical intervention or environmental agent, to indicate the likelihood, presence or extent of a toxicity as an adverse event. They can be used to monitor a population for exposure to an environmental risk or to monitor a population after an exposure (Califf 2018).

Surrogate markers

A surrogate marker is an additional concept which has been used in therapeutic trials. While typically an effect of BMs is a laboratory measurement that reflects the activity of a disease process, a "surrogate marker" can be defined as "a laboratory measurement or physical sign that is used in therapeutic trials as a substitute for a clinically meaningful endpoint that is a direct measure of how a patient feels, functions or survives, and is expected to predict the effect of the therapy." The primary difference between a BM and a surrogate marker is that a BM is a "candidate" surrogate marker (Katz 2004). To give a better explanation, for a BM be qualify as a surrogate not only must it be correlated with the outcome, but also the change in the BM must "explain" the change in the clinical outcome. The overwhelming majority of BMs are not valid surrogates (Califf 2018).

Imaging and molecular biomarkers

Other classifications are based on the type of tests or examinations performed and include imaging BMs and molecular BMs. Imaging BMs such as Computerized Axial Tomography, Positron Emission Tomography or Magnetic Resonance Imaging, produce intuitive, multidimensional results yielding both qualitative and quantitative data (Fathi et al. 2014). Molecular BMs are no imaging BMs with biophysical properties, which allow measurements in biological samples (e.g., plasma, serum, cerebrospinal fluid, bronchoalveolar lavage, biopsy) (Laterza et al. 2007).

It should be highlighted that although the different types of BMs are considered as separate and alternative for classification purposes, it is not always possible to attribute them to a single category. In fact, the allocation of a BMs to one type or to another, sometimes depends on its toxicological significance and on the specific context in which the test is being used (Mayeux 2004, Ladeira and Viegas 2016). The relationship among classical types of BMs, exposure and consequent outcomes is illustrated in Fig. 2.

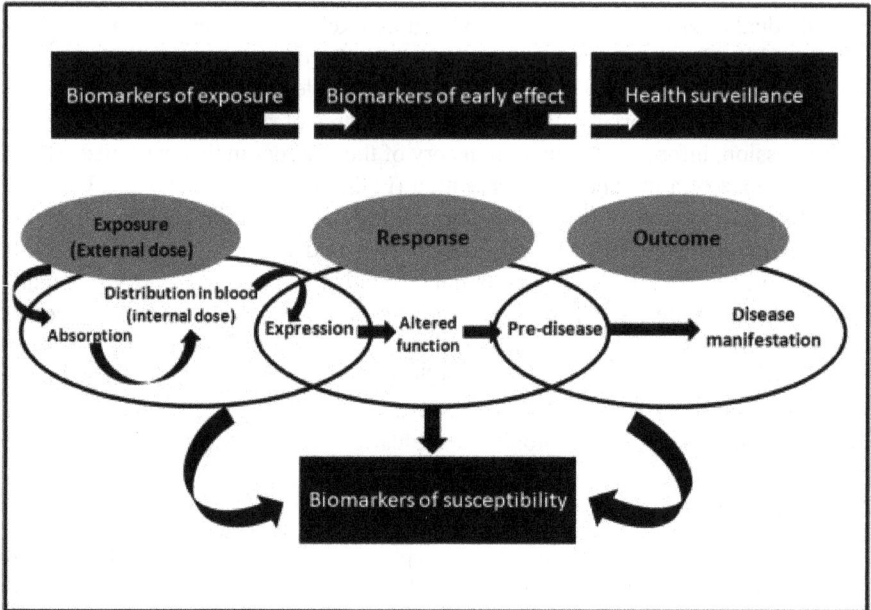

Figure 2. Relationships between external and disease effects and the role of biomarkers of exposure, susceptibility and effects in the monitoring of each stage (Flora 2014).

Biomarkers' Strengths and Weakness

Certain universal characteristics are important for any BM, while considering that a valid BM is a BM that is measured in an analytical test system with well-established performance characteristics, and for which there is an accepted scientific framework or body of evidence that elucidates the physiologic, toxicologic, pharmacologic or clinical significance of the test results (Hunter et al. 2010). If there are certain desirable qualities in a BM, on the other hand, sources of error and limitations may also exist.

Qualities

Relevance is an important quality for a BM, and this term refers to a BM's ability to appropriately provide clinically suitable information on questions of interest to the public, healthcare providers or health policy officials (Strimbu and Tavel 2010). A BM must have several other qualities, such as being safe and easy to perform. This means that it must be as non-invasive as possible, using preferentially external body fluids or blood. The tests should be performed at the bedside or with a (relatively) simple laboratory test using a rapid and reliable standardized platform (Ray 2014). Additionally, a BMs should preferably be able to identify subtypes and causes of disease, be sensitive enough for early detection and have the ability to account for all or most of the variations in a physiological state (Fathi et al. 2014, Ladeira and Viegas 2016). Laboratory reliability or repeatability is also crucial, as laboratory errors can lead to misclassification of exposures or disease (Mayeux 2004).

Validity is another quality although evaluating the accuracy of a BM is a complex process and not typically black or white, but instead a spectrum. Some authors even reject the term validation, considering it unsuitable since it suggests that there can be a complete biological understanding of the relationship between a given BMs and a clinical endpoint, which is not possible (Strimbu and Tavel 2010). For those who defend validation, three aspects of measurement validity are considered: 1) content validity, which shows the degree to which a BM reflects the studied biological phenomenon; 2) construct validity, which pertains to other relevant characteristics of the disease or trait, for example other BMs or disease manifestations, and 3) criterion validity, which shows the extent to which the BM correlates with the specific disease. Criterion validity is usually measured by sensitivity, specificity and predictive power (Mayeux 2004). A high sensitivity is the capability to correctly identify a high proportion of true cases, whereas high specificity is the capability to correctly identify a high proportion of true non-cases (Grund and Sabin 2010). The use of the statistical tool Receiver-Operator Characteristic (ROC) curves provides the tools necessary to determine the best BM choice in terms of sensitivity and specificity and is available on many software statistical packages. Currently ROC is the most widespread methodology used in BMs validation when there are two conditions or disease states, e.g., non-diseased and diseased subjects (Mayeux 2004, Franco-Pereira et al. 2020). However, sensitivity and specificity tell one the accuracy of the test but not the probability of disease, whereas in a different way odds ratios obtained from statistical logistic regression provide the advantage of describing associations of BMs with clinical events (Mayeux 2004, Grund and Sabin 2010).

Sources of errors and limitations

There are factors which might affect the measurement of BMs other than those errors that occur in the laboratory; these include problems with the collection equipment or in the transportation of specimens to the laboratory or improper storage of the samples. Most laboratories and large-scale studies establish a quality assurance and quality-control program to reduce these errors (Mayeux 2004). Confounders are also relevant sources of error, when there is failure to identify factors that may alter the measurement of the BM. They can include age, gender, weight, diet and other metabolic factors of a subject or other diseases; for example, smoking is a confounder in lung cancer studies (Mayeux 2004, Forshed 2017, Andrea et al. 2019).

When designing a study, data on potential confounders should be collected and included in the analysis of the relation between the BM and the outcome of interest (Mayeux 2004).

Variability can constitute a relevant limitation of a BM because the repeatability of a BM measurement determines its association with disease outcomes in epidemiological studies (Thyagarajan et al. 2016). Variability occurs regardless of whether the BM represents an exposure or effect modifier, a surrogate of the disease or an indication of susceptibility. It includes methodological variability, within-individual variability and between-individual variability, all of them being important determinants for the ability to detect BM-disease associations (Thyagarajan et al. 2016).

The methodological variability encompasses process (pre-analytical) variability such as variability in blood drawing, biological stability of the BM, field-centre processing (including centrifuging and freezing), shipping, laboratory assay (analytical) variability such as the batch of laboratory kits used and post-analytical variability (e.g., errors in data transmission) (Mayeux 2004, Thyagarajan et al. 2016). Methodological variability can be reduced by better analytical techniques and standardization of biospecimen collection and processing procedures (Thyagarajan et al. 2016).

Concerning intraindividual variability, the fact is that an individual at any given time is a complex configuration of characteristics; some of which change from moment to moment, day to day, week to week, whereas others are relatively stable (Nesselroade and Ram 2004). Individual variability is determined by the characteristics of the population being studied and likely will differ across various epidemiological studies (Thyagarajan et al. 2016). Additional factors may explain group variability for example, the fact that some workers always wear protective equipment whereas others may not (Mayeux 2004).

Cost is a common limitation particularly in epidemiologic studies including thousands of subjects unless the laboratory procedure is automated and relatively simple (Mayeux 2004).

Final Remarks

This chapter described how the integration of environmental monitoring and biological monitoring could improve health surveillance programs, helping in the assessment of exposure to chemicals, characterization of exposure pathways, potential risks and their mitigation; biomarkers used in biological monitoring can additionally help to fill several gaps encountered in environmental monitoring processes. Among current definitions and classifications for BMs, the most generally accepted divide them into three main categories: exposure, susceptibility and effect. Several desirable qualities are expected in a BM, on the other hand, sources of error and limitations could also exist. Given the importance of BMs in providing insights regarding relationships among environmental exposures, human biology and disease, as well as their proven contribution to improve the health of people, continuous research to improve and develop news ones is certainly necessary .

References

Aitio, A., A. Bernard, B.A. Fowler and G.E. Nordberg. 2007. Biological Monitoring and Biomarkers. *In*: G. Nordberg, B. Fowler and M. Nordberg [eds.]. Handbook on the Toxicology of Metals. Academic Press, Cambridge, USA.

Alimonti, A. and D. Mattei. 2008. Biomarkers for human biomonitoring. *In*: M.E. Conti [ed.]. Biomarkers for Human Biomonitoring, Theory and Applications. WIT Press, Southampton, UK.

Andrea, S.E., L. Brockbals and T. Kraemer. 2019. Metabolomic Strategies in Biomarker Research–New Approach for Indirect Identification of Drug Consumption and Sample Manipulation in Clinical and Forensic Toxicology? Front. Chem. 7: 319.

Artiola, J.F. and M.L. Brusseau. 2019. The role of environmental monitoring in pollution science. pp. 149–162. *In*: M.L. Brusseau, I.L. Pepper and C.P. Gerba [eds.]. Environmental and Pollution Science. Academic Press, Cambridge, USA.

Brucker, N., S. Nunes do Nascimento, L. Bernardini, M. Feiffer Charão and S.C. Garcia 2020. Biomarkers of exposure, effect, and susceptibility in occupational exposure to traffic-related air pollution: A review. J. Appl. Toxicol. 40(6): 722–736.

Califf, R.M. 2018. Biomarker definitions and their applications. Exp. Biol. Med. 243(3): 213–221.

Canese, R., A. Bazzocchi, G. Blandino G, Carpinelli, C. De Nuccio, M. Gion et al. 2020. The role of molecular and imaging biomarkers in the evaluation of inflammation in oncology. Int. J. Biol. Markers 35(1): 5–7.

Covaci, A. 2014. Environmental fate and behavior. pp. 372–374. *In*: P. Wexler [ed.]. Encyclopedia of Toxicology. Academic Press, Cambridge, USA.

Evelo, C.T.A. and P.T. Henderson. 1992. Biological effect monitoring. pp. 268–277. *In*: H.M. Bolt, F.A. de Wolff and P. T. Henderson [eds.]. Archives of Toxicology, Vol 15. Springer, Berlin, Germany.

[EPA] Environmental Protection Agency, Defining Pesticide Biomarkers, 2021. EPA, Washington, USA.

Fathi, E, S.A. Mesbah-Namin and R. Farahzadi. 2014. Biomarkers in Medicine: An Overview. Br. J. Med. Res. 4(8): 1701–1718.

[FDA] United States Food and Drug Administration, Clinical Outcome Assessment (COA): Frequently Asked Questions, 2020. FDA, New Hampshire, USA.

Flora, J.S. 2014. Metals. pp. 485–519. *In*: R.C. Gupta [ed.]. Biomarkers in Toxicology. Academic Press, Cambridge, USA.

Foà, V. and L. Alessio. 2011. *In*: R. Lauwerys [ed.]. Enciclopaedia of Occupational Health and Safety. Biological Monitoring—General Principles. International Labour Organization, Genebra, Switzerland.

Forshed, J. 2017. Experimental design in clinical 'omics biomarker discovery. J. Proteome Res. 16: 3954–3960.

Franco-Pereira, A.M., C.T. Nakas and M.C. Pardo. 2020. Biomarker assessment in ROC curve analysis using the length of the curve as an index of diagnostic accuracy: the binormal model framework. AStA. Adv. Stat. Anal. 104: 625–647.

Ginsberg, G.L., R.R. Dietert and B.R. Sonawane. 2014. Susceptibility based upon chemical interaction with disease processes: Potential implications for risk assessment. Curr. Envir. Health Rpt. 1: 314–324.

Grund, B. and C. Sabin. 2010. Analysis of biomarker data: logs, odds ratios, and receiver operating characteristic curves. Curr. Op. HIV AIDS 5(6): 473–479.

Hagger, J.A., B.M. Jones, D.R. Paul Leonard, `R. Owen and T.S. Galloway. 2006. Biomarkers and integrated environmental risk assessment: are there more questions than answers? Integr. Environ. Assess. Manag. 2(4): 312–329.

Hambach, R., D. Lison, P.C. D'Haese, J. Weyler, E. De Graef, A. De Schryver et al. 2013. Co-exposure to lead increases the renal response to low levels of cadmium in metallurgy workers. Toxicol. Lett. 222(2): 233–238.

Holstege, C.P. and L.S. Hardison. 2014. Medical surveillance. pp. 180–181 *In*: P. Wexler [ed.]. Encyclopedia of Toxicology. Academic Press, Cambridge, USA.

Hryhorczuk, D.O., M. Moomey, A. Burton, K. Runkle, E. Chen, T. Saxer et al. 2002. Urinary p-nitrophenol as a biomarker of household exposure to methyl parathion. Environ. Health Perspect. 110(6): 1041–1046.

Hunter, D.J., E. Losina, A. Guermazi, D. Burstein, M.N. Lassere and V. Kraus. 2010. A pathway and approach to biomarker validation and qualification for osteoarthritis clinical trials. Curr. Drug Targets 11(5): 536–345.

Iavicoli, I., V. Leso and P.A. Schulte. 2016. Biomarkers of susceptibility: State of the art and implications for occupational exposure to engineered nanomaterials. Toxicol. Appl. Pharmacol. 299: 112–124.

[IMUSC] Institute of Medicine (United States) Committee to Assess the Science Base for Tobacco Harm Reduction. 2001.Exposure and Biomarker Assessment in Humans. *In*: K. Stratton, P. Shetty, R. Wallace et al. [eds.]. Clearing the Smoke: Assessing the Science Base for Tobacco Harm Reduction. National Academies Press, Washington, USA.

Jeddi, M.Z., N.B. Hopf, S. Viegas, A.B. Price, A. Paini, C. van Thriel et al. 2021. Towards a systematic use of effect biomarkers in population and occupational biomonitoring. Environ. Int. 146: 106257.

Katz, R. 2004. Biomarkers and surrogate markers: an FDA perspective. NeuroRx. 1(2): 189–195.

Kelly, R.S. and P. Vineis. 2014. Biomarkers of susceptibility to chemical carcinogens: The example of non-Hodgkin lymphomas. Br. Med. Bull. 111(1): 89–100.

Ladeira, C. and S. Viegas. 2016. Human Biomonitoring—An overview on biomarkers and their application in occupational and environmental health. Biomonitoring 3: 15–24.

Laterza, O.F., R.C. Hendrickson and J.A. Wagner. 2007. Molecular Biomarkers. Drug Inf. J. 41(5): 573–585.

Links, J.M., T.W. Kensler and J.D. Groopman. 1995. Biomarkers and mechanistic approaches in environmental epidemiology. Annu. Rev. Public Health. 16: 83–103.

Manno, M. and N. Sannolo. 2004. Toxicological significance of biological markers. G. Ital. Med. Lav. Ergon. 26(4): 270–277.

Matatiele, P., L. Mochaki, B. Southon, B. Dabula, P. Poongavanum and B. Kgarebe. 2019. Environmental and biological monitoring in the workplace: A 10-year South African retrospective analysis. AAS Open Res. 1: 20.

Mayeux, R. 2004. Biomarkers: potential uses and limitations. NeuroRx. 1(2): 182–188.

Miller, M.C., H.W. Mohrenweiser and D.A. Bell. 2001. Genetic variability in susceptibility and response to toxicants. Toxicol. Letters 120(1–3): 269–280.

Miraglia N., G. Assennato, E. Clonfero, S. Fustinoni and N. Sannolo. 2004. Biologically effective dose biomarkers. G. Ital. Med. Lav. Ergon. 26(4): 298–301.

Mussali-Galante, P., E. Tovar-Sánchez, M. Valverde and E. Rojas del Castillo. 2013. Biomarkers of exposure for assessing environmental metal pollution: from molecules to ecosystems. Rev. Int. Contam. Ambie. 29(1): 117–140.

Nesselroade, J.R. and N. Ram. 2004. Studying intraindividual variability: what we have learned that will help us understand lives in context. Res. Hum. Develop. 1(1-2): 9–29.

Ray, C.A. 2014. Biomarker accuracy: Exploring the truth. Bioanalysis 6(3): 269–271.

Santonen, T., A. Aitio, B.A. Fowler and M. Nordberg. 2015. Biological monitoring and biomarkers. pp. 155–171. *In*: F.G. Nordberg, B.A. Fowler and M. Nordberg [eds.]. Handbook on the Toxicology of Metals. Academic Press, Cambridge, USA.

Steckling, N., A. Gottic, S. Bose-O'Reilly, D. Chapizanis, D. Costopoulou and F. De Vocht et al. 2018. Biomarkers of exposure in environment-wide association studies—Opportunities to decode the exposome using human biomonitoring data. Environ. Res. 164: 597–624.

Strimbu, K. and J.A. Tavel. 2010. What are biomarkers? Curr. Opin. HIV AIDS 5(6): 463–466.

Thyagarajan, B., A.G. Howard, R. Durazo-Arvizu, J.H. Eckfeldt, M.D. Gellman, R.S. Kim et al. 2016. Analytical and biological variability in biomarker measurement in the Hispanic Community Health Study/Study of Latinos. Int. J. Clin. Chem. 463: 129–137.

[UNECE] United Nations Economic Commission for Europe, Environmental Monitoring and Reporting, Introduction, 2021. UNECE, Geneva, Switzerland.

Wang, Q. and Z. Yang. 2018. Does chronic disease influence susceptibility to the effects of air pollution on depressive symptoms in China? Int. J. Ment. Health. Syst. 12: 33.

[WHO] World Health Organization & International Programme on Chemical Safety, Biomarkers and risk assessment: concepts and principles. 1993. WHO, Geneva, Switerzland.

CHAPTER 3

Application Fields of Metal Biomarkers

Risk Assessment of Metal Contamination

The environment and it's areas have been severely polluted by metals, thus threatening it's ability to foster the health of humans, animals and plants (Masindi and Muedi 2018). It is essential to control metals in the air, water, soil and in human populations and with this in mind , governmental agencies protecting public health are required to review, quantify and regulate chemicals such as metals, in a way that will protect and enhance public health and the environment (NRC 2007). Risk assessment is a crucial tool for such control and can be defined as a process or method which consists in the identification of hazards and risk factors that have the potential to cause harm, and in finding appropriate ways to eliminate the hazard or control the risk when the hazard cannot be eliminated (CCOHS 2022). The final objective of risk assessment is to facilitate scientific and data-informed decision making; risk assessment is increasingly expected to provide more precise predictions of actual risk (NRC 2007). During a risk assessment process biomarkers (BM) are used in hazard identification, which is crucial to gauge whether a toxic is hazardous (WHO 2001). In addition, the use of BMs as functional measures of exposure to stressors, which can be expressed at the sub-organismal, physiological or behavioural level, can surrogate measures of biological impact (Hagger et al. 2006).

Chemical risk assessment is conducted in several scientific fields which include many application areas of toxicology. Such areas comprise of environmental toxicology, and most directly focus on human health, occupational toxicology, food toxicology or and medical/forensic toxicology (NCR 2007, Casarett and Doull 2013). Ecological risk assessment is more concerned about populations of organisms (e.g., individual species of fish in a river) or ecological integrity (e.g., if the species living in the river will change over time), whereas human health risk assessment usually considers protecting life of individual human beings (CIDA 2021). Both approaches share concerns regarding metal contamination and toxicity, and both apply BMs (WHO 2001, McCarty et al. 2002, Masindi and Muedi 2018).

Notably, a collaborative partnership is developed with the World Health Organization's/WHO) International Programme on Chemical Safety (IPCS), the Organization for Economic Cooperation and Development (OECD) and the U.S. Environmental Protection Agency (EPA). This partnership aims to integrate assessment approaches for ecological risks and for human health, since it was considered that it was time to move towards a more integrated, "holistic" approach. Ecological and human health risks are interdependent, since humans depend on nature for food and water, resources have been declining, due to effects of toxic chemicals, such as metals (Chambers et al. 2002, Järup 2003). In addition, ecological damage may result in increased human exposures to contaminants; assessments that do not integrate health and ecological risks are likely to overlook important approaches of action involving interactions between effects on the environment and effects on humans (Chambers et al. 2002, Suter et al. 2003, Sall et al. 2020). So far, ecological and human health risk assessment have developed independently for a long time. For practical reasons the next sections of this chapter will represent the application of BMs in ecological and human health risk assessments separately.

Substitute by biomarkers in ecological risk assessment

Large amounts of metals end up in the environment as a result of ever-increasing anthropogenic activities and economic development, making monitoring and analysis of metal concentrations in the environment necessary for pollution assessment and control of the exposures (Ali et al. 2019, Jiang et al. 2019). In view of this, environmental analysis provides useful information about the distribution, principal sources and fate of these toxics in the environment and their bioaccumulation in food chains and is used to assess the risk posed by these elements to wildlife and to human health (Somerset 2011, Ali et al. 2019). During such processes, in particular non-destructive sampling techniques and the use of environmental BMs should be chosen to avoid loss of biota due to analysis (Ali et al. 2019). In the past 20 yr considerable efforts have been made in developing and applying BMs to be used in ecotoxicology and ecological risk assessment. These efforts resulted partly from a need for early warning indicators that would respond before measurable effects on individuals and populations occur, and requiring a tool to identify the causes of observed population- and community-level effects. In such a perspective, the objectives of developing BM are the possibility of using them to indicate that organisms have been or are being exposed to certain chemicals or that organisms are suffering or likely will suffer future impairments of ecological relevance (Forbes et al. 2006).

Several authors consider an overall classification for ecotoxicological standard BMs: gold standard, silver standard and bronze standard. The first category includes BMs which can be used to diagnose a problem without the need for chemical analysis (e.g., eggshell thinning); silver standards have wider applications than gold standard measures, but presumably, would need to be used in combination with chemical analyses (e.g., induction of mixed-function oxidases, DNA adducts and immune responses); bronze standards are those that have been applied successfully, to a greater or lesser extent, but are the ones requiring more research before they could be used in formal environmental assessments (thyroid function, retinol, porphyrin and stress proteins are included in this category) (Forbes et al. 2006).

Under the ecotoxicological risk assessment scope, sentinel organisms and molecular biomarkers are additional concepts, which are worth describing.

Sentinel organisms

Meaningful assessments of metal pollution can be obtained by measuring metal concentrations in selected species of the resident biota. Some species, sentinel species, are models which can serve as good indicators to assert an ecosystem's health or stress's evidence during monitoring programs; nonhuman organisms are often more largely exposed to environmental contaminants and may be more sensitive, suggesting potential sources of human hazards (Abdel-Halim 2018, Suter et al. 2003). Therefore, different plants, animals, as well as other living beings, have been used as biological indicators (or bioindicators) to assess and monitor metal contamination and pollution in the environment through mortality, developmental defects, reproductive effects, carcinogenicity, neurotoxicity, immunotoxicity, behavioural changes and other endpoints (Abdel-Halim 2018, Ali et al. 2019). In this way, those indicator organisms may quantitatively and/or qualitatively record the damages (Maresca et al. 2018).

Mosses and lichens have been used in biomonitoring studies on air pollution caused by toxic metals. Depending on their chemical form and bioavailability, toxic metals affect plants by impairing their anatomy, ultrastructure and molecules or by adversely affecting their physiology and biochemistry. Ultrastructure damage in plants is also a marker closely related to air metal pollution; specifically, vascular plants such as *Robinsonecio gerberifolius* constitute an example of a sentinel organism used to evaluate the effect of metals (Maresca et al. 2018, Martínez-Pérez et al. 2021). Earthworm species bioaccumulate metals, are widely recognized as suitable organisms to monitor the effects of xenobiotics in contaminated soils (Abdel-Halim 2018). Microalgae have ecological significance attributing their position at the base of the aquatic food webs. These algae are used as sentinel organisms in environmental studies to evaluate the toxicity of different chemicals or pollution discharges, and particularly the inputs of metals (Salamat et al. 2013, Abdel-Halim 2018). In the marine environment BMs measured in fish are worldwide-recognized tools for the assessment of pollution impacts, some of them being incorporated in environmental monitoring programs. Fishes can direct metals from the surrounding medium and accumulate pollutants in the body, with the effects becoming apparent when concentrations in tissues attain a threshold level. Many experimental and field studies showed that the fish liver is the target organ for the accumulation of many metals, and that this organ is highly active in the uptake and storage of these toxics, due to its role in storage, redistribution, detoxification and transformation of contaminants (Omar et al. 2014). Additionally, metals cause histopathology and therefore the appraisal of these changes in target organs can give a clear picture of metal toxicity (Khan et al. 2020). Fishes living in metal contaminated freshwater exhibit histopathological changes in the liver, evinced by cytoplasmic vacuolation, necrosis or sinusoid dilation. Experimental exposure of fishes to metals also resulted in observing necrosis in renal tubular cells, tubular haemorrhage, oedema, damaged glomerulus, collecting duct damage and blood congestion (Mahboob et al. 2020).

Histopathological examinations of the gills, which is the site of direct metal uptake from water, already revealed deterioration in their histoarchitecture associated with higher concentrations of metals in these organs (Omar et al. 2014).

Molecular biomarkers

The most promising BMs investigated so far rely on the induction or increase of certain molecular endpoints in response to xenobiotic or stressor exposures (Chambers et al. 2002).

Oxidative stress

In the case of metals, their toxic effects are related with the production of Reactive Oxygen Species (ROS) in biological systems, which can trigger redox-sensitive pathways that lead to different alterations affecting various cellular processes; these are mainly the functioning of membrane systems through lipid peroxidation (LPO), protein carbonylation, DNA damage, activation of kinase cascades and transcription factors, which ultimately affect cellular viability (Radwan et al. 2010, Maresca et al. 2018). The potential of ROS to damage tissues and cellular components is called Oxidative Stress (OS), which is a general response to toxicity induced by many contaminants. The use of OS BMs is of interest for assessing the impact of pollutants, such as metals, including LPO evaluation for which the concentrations of the subproduct malondialdehyde is recognized as a marker of LPO.

Immuno-competence is indispensable for the maintenance of the complete health of living beings, suggesting that the damage of innate immunity may be more exceptional in fish compared to mammals (Khan et al. 2020). Studies on immunomodulation by metals, individually or in mixtures, in exposed rainbow trout have been reported by Sanchez-Dardon et al. (1999). The respiratory burst assay is an indicator of the innate immune system in fish and during the process, cytokines are released and an inflammatory response occurs. Concomitantly a generation of different reactive oxygen species such as the singlet oxygen, superoxide anion radical, hydroxyl radical and hydrogen peroxide, occurs. Nitric oxide synthase, total leukocyte count and differential leukocyte count and globulin concentrations are additional immunological indexes in these studies (Robinson 2008, Yang et al. 2016, Khan et al. 2020).

Antioxidant defences

Changes in the activities of antioxidant defence systems have also been found in many species of animals and plants in response to different factors, which include metals. There are efficient systems of enzymatic and non-enzymatic antioxidants working in synergy for scavenging ROS in different areas inside cells. Among them superoxide dismutase (SOD) is the first line of defence against ROS, it dismutates the highly reactive superoxide anion radical into hydrogen peroxide. Further, another enzyme named catalase (CAT) breaks hydrogen peroxide to water. Glutathione S-Transferase (GST) is one more enzyme belonging to a family of detoxifying enzymes, able to catalyze reactions of binding xenobiotics with the antioxidant glutathione (GSH);

GST plays an important role for the neutralization of lipid hydroperoxides generated by metal exposure. The levels of these molecules, using plants as sentinel organisms, have been used in ecotoxicological studies on metal pollution (Maresca et al. 2018). Other works using *Theba pisana* land snails as sentinel species, included the determination of reduced GSH levels, CAT and GST. According to the authors antioxidant defence components in the digestive gland of this sentinel organism, were sensitive parameters which could be useful as bio monitors for urban metal pollution (Radwan et al. 2010).

Metallothioneins

Metallothioneins (MTs) are additional BMs which have been used in ecotoxicology. They are small cysteine-rich proteins that play important roles in metal homeostasis and in protecting against metal toxicity (Si and Lang 2018). It is known that increased MT synthesis is associated with the escalating capacity for binding metals and, in this way, protection against metal toxicity (Roesijadi 1994). Overexpression of metalloproteins mRNA has also been detected in fish and shellfish exposed to high concentrations of zinc, aluminium, copper, lead and cadmium, and considered to be a compensatory mechanism (Abdel-Halim 2018).

Enzymes

Enzymes are very sensitive molecules which are easily affected even by small changes either in the internal or external medium; metals in the medium may bind with the enzymes of organisms and modify their activity at different levels. For example, an enhancement of lactate dehydrogenase activity (LDH) in fish tissues suggests a rapid conversion of pyruvate into lactate and this leads to glycogenolysis, which increases under pollutant stress (Perumalsamy and Arumugam 2013). Positive results were attained after using zinc chloride as a test substance, to evaluate LDH activity as an effect criterion in toxicity tests with the small planktonic crustacean *Dhania magna* (Diamantino et al. 2001). Succinate dehydrogenase (SDH) is an oxidative enzyme which is highly affected by the action of metals. Their measurements have been used as a marker for detecting the presence of the tricarboxylic acid (TCA) cycle in tissues (Moorthikumar and Muthulingam 2011). SDH is a key enzyme for this cycle participating in tail end oxidations and serving as a linking component of electron phosphorylation. The shift of an aerobic to an anaerobic condition causes a reduction in TCA cycle enzymes, especially SDH; it is acknowledged that the deposition of aerobic sediments into an anaerobic environment may release metals (Butler 2011, Perumalsamy and Arumugam 2013). The exposure to nickel chloride induces decreased SDH activity in the gill, liver, kidneys, brain and muscles of the fish *Labeo rohita*. Similar reports exist for other fish species exposed to Pb or Cu, attributed to metal-triggered metabolic shift from aerobiosis to an anaerobiosis (Moorthikumar and Muthulingam 2011).

The fish *Oreochromis niloticus* on exposure to metals, also exhibit changes in the levels of some other liver and kidney enzymes generally released into the circulation as result of some disease; these include aspartate aminotransferase, alkaline phosphatase and alanine aminotransferase (Khan et al. 2020).

Genotoxicity and epigenetic changes

Metals are also known to cause genotoxicity. Micronuclei and single cell gel electrophoresis are genotoxicity assays considered as the most rapid and reliable for ecotoxicological evaluations (Khan et al. 2020). Metal-OS induced DNA damage in plants measured by the Comet assay can detect DNA single strand breaks and alkali-labile damage in individual cells and is also a good tool for the assessment of genotoxicity in polluted environments (Lapuente et al. 2015, Maresca et al. 2018). These tests have been applied to a wide range of experimental models: bacteria, fungi or cells cultures. For practical/technical reasons, blood is the most common choice, although tissues/cells like gills, sperm cells, early larval stages, coelomocytes, liver or the kidneys have been also used (Lapuente et al. 2015).

The field of epigenetics has gained increasing importance in ecotoxicological studies, since different classes of environmental contaminants (including metals) were previously linked to epigenetic effects (Brander et al. 2017, Abdel-Halim 2018). Three major epigenetic mechanisms, DNA methylation, histone modification and noncoding RNAs (ncRNAs), are shared across most taxa. So far, two of the best studied invertebrates regarding epigenetics are *Daphnia* species and the Pacific oyster (*Crassostrea gigas*), and in spite of invertebrate methylation levels being lower than in vertebrates, there are many similarities between aquatic invertebrates and fish (Brander et al. 2017). Epigenetic marks on a particular gene may serve as BMs for exposure to specific stressors or toxic outcomes (Abdel-Halim 2018).

Challenges in ecological risk assessment

BMs are demonstrated as novel strategies in ecotoxicological trails for risk assessment and policy, but the extent to which they can provide unambiguous and ecologically relevant indicators of exposure to or effects of toxicants, remains controversial (Forbes et al. 2006, Abdel-Halim 2018). Most BMs used in ecotoxicology have their origins in human toxicology, in which they have proved to be very useful as measures of human exposure to specific chemicals or as early warning indicators of specific diseases or syndromes. The fact that the biology of humans is relatively well understood, it improves the likelihood that BMs responses will be interpreted correctly, but it is not the case for ecological risk assessment (Forbes et al. 2006). A fundamental problem in ecological risk assessment is of how to extrapolate data produced for a small number of "standard" test species, to predict impacts for many species from different phyla present in an ecosystem, which may differ in their physiology, diet and life-history traits (Spurgeon et al. 2020). As the targets of protection in ecological risk assessments most often are populations, communities and ecosystems (only rarely individuals), BM response must be soundly and consistently linked to responses at these higher levels. Moreover, confounding nonchemical influences on the BM response, such as temperature, nutritional state or reproductive condition, must be well understood so that the response can be calibrated appropriately. Differences in BM response among populations of a species, as a result of geographical influences, other habitat parameters, genetic differences and/or exposure history, must be understood entirely (Forbes et al. 2006).

Biomarkers in Human Health Risk Assessment

Human health risk assessment incorporates several major components, which are described below . Hazard identification deals with the confirmation that a chemical subjected to appropriate circumstances can cause an adverse effect; dose-response assessment establishes the quantitative relationship between dose and effect in humans; exposure assessment identifies and defines the exposures that occur or are anticipated to occur (WHO 2001). Risk characterization is the final step of the risk assessment process and integrates the results of hazard identification, hazard characterization and exposure assessment, to estimate potential carcinogenic risks and noncarcinogenic health effects associated with exposure to chemicals (Sullivan et al. 2005, Kamrin 2014, Schwela 2014). The output is a quantitative or semiquantitative estimate (which include resulting uncertainties) of the probability of occurrence and severity of adverse effect(s)/event(s) in a given population under defined conditions (Hull et al, 2020). The risk-assessment process is applied to several areas of toxicology, these include occupational, clinical, forensic and in food and consumers products.

Occupational risk assessment

For centuries the work environment has contributed as a significant risk of adverse health effects due to exposure to chemical hazards (Casarett and Doull 2013). Most occupational exposures to metals are associated with working with metal and machining, including but not limited to fabrication, smelting, welding, forges, foundries, grinding, refinishing and repair work (APHC 2019). Inhalation is usually by far the most significant exposure risk. Skin contact and absorption is probably of concern only with respect to those metals which are implicated in causing dermatitis. Poor hygiene practices, contamination of food and smoking in the workplace all contribute to the risk of ingestion of metals or their compounds (Graham 1985). Biomonitoring to assess exposure to chemicals in the workplace has been used and recognized for a long time as a valuable tool to indicate the need for preventive actions to protect workers' health, by reducing toxic effects and preventing the development of diseases related to these exposures (Brucker et al. 2020). Nevertheless, it is the opinion of some authors that BM's complete potential for occupational safety and health has yet to be realized. The last 30 yr have focused primarily on the development of technologies, validation and elucidation of processes and procedures for molecular epidemiology, analytical chemistry and bioinformatics (WHO 2001, NRC 2007, Schulte and Hauser 2012). Despite the importance of the achievements attained so far, the true promise of BMs will be realized when they are able to be used not only to assess known hazards and apprise control decisions, but also when they identify new hazards and groups of workers at increased risk of occupational disease (Schulte and Hauser 2012).

Biomarkers of exposure

Basically accessible body fluids and tissues have been sampled at one time or another to assess exposure to industrial chemicals using BMs; the most used are blood and urine (Graham 1985). The determination of metals in blood and urine is an important occupational environmental toxicology screening procedure, which reflects recent body burden, have been used to establish limit values for human biomonitoring parameters (Riaz et al. 2017, Brucker et al. 2020). BMs of exposure establish safe biomonitoring levels for metals and other chemicals; they are namely the American Conference of Governmental Industrial Hygienists (ACGIH) Biological Exposure Indices (BEIs) or the German biological tolerance values (BAT), among several others, promulgated by various organizations and government standards (Morgan 1997, Drexler et al. 2008, MIAC 2008).

Biomarkers of susceptibility

While everyone should be protected from exposure to toxic substances at work; extensive variability in the human response to workplace exposures has been observed knowing that genes can have multiple variations (polymorphisms), which may contribute to some of this variability (NIOSH 2009). Hence, BMs of genetic susceptibility in the context of workplace exposures could have a vital role in protecting more vulnerable workers due to their genetic background. However, it should be noted that when targeting risk groups within a worker population with genetic BMs, various ethical issues arise, and for now, the practice of genetic screening of worker populations has generally not been recommended or is even sanctioned by governments and authoritative organizations (Bertazzi and Mutti 2008).

However results do exist from various research groups, namely taking into consideration the relevance of metal-OS induction and that an important oxidative response mechanism is via the GSH metabolism pathway. Polymorphisms in the genes of the glutathione S-transferases (GSTs) superfamily have often been assessed in air pollution exposure studies. Among various GST polymorphisms, three isoenzymes, Glutathione S-Transferase Theta 1 (GSTT1), Glutathione S-Transferase Mu 1 (GSTM1), and Glutathione S Transferase Pi 1 (GSTP1), have led to special interest due to their role in the detoxification of air pollution agents (Brucker et al. 2020).

Biomarkers of effect

A great deal of importance is given to early detection of occupational health damage, with studies focused on air pollution in the workplace evaluating BMs of effect related to toxic mechanisms involved in the inhalation of gases and particulates (Brucker et al. 2020). Welding fumes are an example, since they contain a complex mixture of metals. Fumes are formed when a metal is heated above its boiling point and its vapours condense into very fine particles (solid particulates) (Persoons et al. 2014, Fortoul et al. 2015, Brucker et al. 2020, CCOHS 2022). Manganese, zinc, lead, copper, nickel and barium are metals which can adsorb to the surface of these particulates, potentially contributing to effects associated with Particulate Matter (PM) exposure (Potter et al. 2021). It is demonstrated that ultrafine (PM 0.1) and fine-size ambient particles (PM 2.5) can enter the circulatory system and cross the blood–brain barrier or enter through the optic nerve, and then upregulate inflammatory markers and increase ROS in the brain (Wang et al. 2019). In addition, PMs and the gases of welding fumes, are most likely to easily enter the respiratory system, leading again to ROS production with consequent macromolecule damage, as well as activation of inflammatory mediators capable of exacerbating lung pulmonary inflammation, the induction of increased blood coagulability and endothelial dysfunction (Brucker et al. 2020).

Numerous occupational reports exist on metal-induced OS, inflammatory response or lung disease (Cohen 2004, Palmer et al. 2006, Liu et al. 2007, Søyseth et al. 2013, Ulrik et al. 2013, Wyman and Hines 2018). In agreement with evidence on metal-induced OS, studies on occupationally exposed taxi drivers, bus drivers and motorcyclists, describe changes in related effects of BMs, such as increased plasmatic levels of MDA and protein carbonyls (PCO) resulting from metal induced OS stress (Pirinccioglu et al. 2010, Brucker et al. 2020). It is known that ROS formation depletes antioxidant defences, the analysis of erythrocytes from subjects occupationally exposed to air pollutants showed decreased SOD and CAT activity and diminished GSH levels, when compared to non-occupationally exposed subjects (Brucker et al. 2020).

Some studies also confirm the effect of metals on the immune system, namely lead-induced alteration in the inflammatory marker C-reactive protein (hs-CRP) and changes in immunoglobulins levels (Marth et al 2001, Sirivarasai et al. 2013). Traffic police officers exposed to PM2.5 similarly exhibited augmented hs-CRP levels as well as immunoglobulins modifications, with increased IgM, IgG and IgE and decreased IgA concentrations. Considering the potential role of these BMs in assessing atherosclerotic disease, the results provided an association between PM2.5 exposure and the risk of development and/or progression of cardiovascular disease in these officers (Brucker et al. 2020).

Clinical and forensic assessments

In clinical/forensic practice, acute/sub-acute metal poisoning diagnostics may require the use of BMs. Metal poisoning can occur as result of industrial exposure, air or water pollution, foods, medicines or improperly coated food containers, among other sources (NORD 2008). The diagnosis of any metal poisoning requires previous appropriate exposure history and clinical findings consistent with poisoning by that metal, and a patient should only undergo specific metal testing if there is concern for a specific poisoning (ACMT/AACT 2013). Symptoms and physical findings associated with metal poisoning may vary according to the metal accumulated, whereas those from acute poisoning may differ from those associated with chronic toxicity (Verma 2018, Griffin 2020, NORD 2008).

Metal levels

Arsenic, cadmium, mercury and lead are among metals involved in the most common intoxications. In acute exposure to arsenic, urinary levels of arsenic are BMs of exposure more relevant than hematic measurements, as arsenic is rapidly cleared from the blood; excretion of more than 200 µg in a 24 hr urine collection is suggestive of arsenic overload. Acute exposure to cadmium is usually determined by measuring the 24-hr urinary cadmium excretion (Lentini et al. 2017). The metal accumulates in the kidneys following exposure and is slowly released into the urine, usually proportionally to the levels found in the kidneys. Due to this, urinary cadmium levels are thought to reflect long-term exposure, while blood Cd concentrations reflects a combination of both long-term and more recent exposures (Vacchi-Suzzi et al. 2016). A case report on acute self-poisoning by ingestion of cadmium described blood cadmium concentrations after 2 d of 15.1ng/mL, and urine concentrations of 8.4 ng/mL (Hung and Chung 2004). Recent exposure to mercury is reflected by blood and urine mercury concentrations. A mercury concentration >45 mg/dL in blood indicate acute poisoning; however, since blood mercury levels rapidly decrease measurement needs to take place right after the exposure (Ye et al. 2016, Lentini et al. 2017). In turn, urinary mercury is very stable representing a quick way of identifying those exposed to mercury. Organic mercury, such as methylmercury, is usually excreted by faeces and thus, urine concentrations cannot reflect organic mercury levels of the body (Ye et al. 2016). Blood lead concentrations are currently regarded as the most reliable index of exposure to this metal; over 95% of blood lead is bound to the erythrocytes and is in dynamic equilibrium with plasma lead (Sakai 2000). The concentrations of lead in the complete blood are indicators of recent exposure, with blood lead levels of 50 µg/dL or more found in cases of severe toxicity (Lentini et al. 2017).

Blood analysis

Other tests for metal intoxication diagnosis or to gauge its severity are prevalent (Lentini et al. 2017). Complete blood count is useful, since the haemolytic action of metals is known for a long time; such action is associated with the development of peroxidative processes in erythrocyte membranes (Ribarov and Benov 1981). In arsenic intoxication breakdown of the haemoglobin of red blood cells usually occurs (NORD 2021). Basophilic stippling of red blood cells are inclusions of aggregated ribosomes, which has been noted in lead intoxication since 1899 and considered a classic laboratory sign of lead poisoning (Cheson et al. 1984). Nowadays it is known that this BM is an inconstant finding in lead intoxication, since is not specific for lead toxicity and may be observed also in arsenic toxicity, sideroblastic anaemia and thalassemia, among other conditions (Adal 2020).

Kidney function tests

Kidney function tests are used as BMs of metal acute intoxication because reabsorption of metals occurs in the first zone of the proximal tubule, in the loop of Henle and terminal segments. Moreover, acute kidney injury can be induced by metals cytotoxicity, with cellular membrane disruption and uncoupling of the mitochondrial respiration pathway concomitant with the release of numerous apoptosis signals, such as cytokines and reactive oxygen species. Such events may result in acute tubular necrosis, and haemoglobinuria and proteinuria (Lentini et al. 2017). This is particularly true for nephrotoxic metals, cadmium intoxications are a well-known example (NORD 2021).

Imaging

Abdominal radiographs are also indicated in acute ingestions. Radio-opacities demonstrated in the gastrointestinal tract should be cleared by whole-bowel irrigation prior to instituting chelation therapy. Large, retained gastric foreign bodies (e.g., bullets, shotgun cartridges, fishing sinkers, curtain weights) may cause metal toxicity and should be removed endoscopically if they do not pass, if serum concentrations are concerning or increasing, or if the patient becomes symptomatic (Adal 2020). An x-ray of the abdomen may show ingested arsenic, which is not penetrable by x-rays (radiopaque) (Griffin 2020). Cases of patients who have injected elemental mercury subcutaneously and developed mercury toxicity have been documented; radiographs of the suspect areas showing large subcutaneous deposits of radio-opaque material were helpful in confirming the diagnosis and need for surgical intervention to limit the exposure (Adal 2020).

Metal poisoning diagnosis usually involves the integration of several tests' results. To illustrate several general laboratory tests to evaluate mercury intoxication include a complete blood cell count, electrolyte assays and renal and hepatic function tests. Electrocardiography (ECG), Pulmonary Function Test (PFT), cardiovascular monitoring, electroneuromyography and neuropsychological tests are also used is this evaluation (Ye et al. 2016).

Risk Assessment of Food and Consumer Products

Food

Metals can occur as residues in food, because of human activities such as farming, industry, car exhausts or from contamination during food processing and storage. Thus, people can be exposed by ingesting contaminated food (and water) (EFSA 2021). The concentrations of metals found in food depend on many factors, which include the levels of the elements in air, water and soil used to grow crops; in crops, the levels of metals vary and are dependent on factors such as, natural geographical differences, past or current contamination, the type of the food crop, how much "uptake" there is of specific elements from the environment, as well as the industrial, manufacturing and agricultural applied processes. In addition, the properties of specific metals, the amount of each food intake or a person's age and developmental stage, are all key factors that modulate how metals in food affects individual health (FDA 2021b).

The difficulty to eliminate metals from diets is irrefutable because they are naturally absorbed by florae used in animal food production (FDA 2021a). Aiming to control the exposure to metals, several governmental agencies around the world routinely test food and animal food, including pet food (EFSA 2021, FDA 2021a). The European Food Safety Authority (EFSA) has received requests from the European Commission States to provide risk assessments on several metals as contaminants, which comprise arsenic, cadmium, chromium, lead, mercury, nickel and uranium. Contaminant levels must be kept as low as can reasonably be achieved following recommended good working practices and maximum levels must be set for for a few to protect public health. Food containing a contaminant to an amount unacceptable from the public health viewpoint, is not to be placed on the market (EFSA 2021).

It should also be noted that understanding the risk that harmful metals pose in our food supply is complicated, due to the fact that exposure to metals comes from many different foods. Combining all the foods we eat, even low levels of harmful metals from individual food sources, can sometimes add up to a level of concern (FDA 2021b). While exposure assessment is a fundamental part of the risk assessment paradigm, it can often present a number of challenges and uncertainties. New approaches are required to accurately assess consumers' exposure and better apprise the risk assessment. Such novel approaches may include the use of BMs, instead of not just focusing on the occurrence in food and consumption patterns (i.e., on external exposure metrics) (Rietjens et al. 2018). Food metabolome is defined as the part of the human metabolome directly derived from the digestion and biotransformation of foods and their constituents (Scalbert et al. 2014). With the advent of metabolomics, picturing the totality of metabolite profiles opens the prospect of collecting comprehensive exposure information on specific dietary chemicals and their interaction in the body. Monitoring selected BMs in body fluids or tissues as quantitative exposure indicators, allows the determination of external exposure levels based on these BMs of internal exposure; such approaches are defended as more reliable methods than food frequency questionnaires (Rietjens et al. 2018).

Consumer products

Metals are present in virtually every area of modern consumerism, from construction materials to cosmetics/personal care products, medicine products, processed foods, fuels, etc. As regards to cosmetic products, zinc oxide is widely used in sunscreens, diaperointments, moisturizers, shampoos and concealers; iron oxides are often used colourants in eye shadows, blushes and concealers; arsenic, cadmium and lead are additional metals which can be present as ingredients in cosmetics (Odukudu et al. 2015). These products may also contain various undisclosed chemical constituents unknown to the consumers and with any or insufficient health safety information (Sardar et al. 2019). While the skin provides a protective barrier, toxics may penetrate it and become systemically available. The possibility of inhalation exposure from cosmetic products may also occur when they are applied via a spray. Topically applied products to mucous membranes dispensing the possibility of enhanced availability and in the case of lip products, they could provide an opportunity for oral ingestion.

Risk assessment of metals in consumer products is important, since prevention measures can be accepted into public laws, reducing the negative impact that they can have on human health (Odukudu et al. 2013). Accurate exposure estimation is a critical and particularly, a challenging step in these risk assessments. Once the relationship between blood, urine or tissue BMs and internal concentrations of the compound of interest has been understood, extrapolation to external exposure is possible. In this manner BM-based assessment of exposure can constitute important tools to evaluate exposure of consumers (Bolt 2018).

Final Remarks

In this chapter risk assessment was presented as a crucial tool for governmental agencies charged to regulate chemicals, such as metals, to protect the environment and public health. Several aspects of both environmental risk assessments and human health risk assessments were explored under the optics of BMs utilization. However, despite ecological injuries could result in increased human exposures to contaminants, ecological and human health risk assessment have developed independently for a long time. As reiterated by regulatory organisms, it is now time to consider a move to a more integrated, "holistic" approach.

References

[ACMT/AACT] The American College of Medical Toxicology and The American Academy of Clinical Toxicology, Ten Things Physicians and Patients Should Question, 2015. American Board of Internal Medicine (ABIM) Foundation, Philadelphia, USA.

Abdel-Halim, Y.K. 2018. Biomarkers in ecotoxicological research trails. J. Forens. Sci. Toxicol. Remedy Publications LLC 1(1): 1005.

Adal, A. .2020. Which standard lab tests are performed in the workup of heavy metal toxicity? Medscape. https://www.medscape.com/answers/814960-121145/what-is-the-role-of-imaging-studies-in-the-workup-of-heavy-metal-toxicity.

Ali, H., E. Khan and I. Ilahi. 2019. Environmental chemistry and ecotoxicology of hazardous heavy metals: environmental persistence, toxicity, and bioaccumulation. J. Chem. 2019: 6730305.

[APHC] Army Public Health Center, Occupational Heavy Metal Exposures. 2019. APHC, Aberdeen Proving Ground, USA.

Bertazzi, A. and A. Mutti. 2008. Biomarkers, Disease mechanisms and their role in regulatory decisions. pp. 243–254. *In:* C. Wild, P. Vineis and S. Garte [eds.]. Molecular Epidemiology of Chronic Diseases. John Willey and Sons, N. Y., USA.

Bolt, H.M. 2018. Biomarker monitoring for food contaminants. Arch. Toxicol. 92: 1021–1022.

Brander, S.M., A.D. Biales and R.E. Connon. 2017. The role of epigenomics in aquatic toxicology. Environ. Toxicol. Chem. 36(10): 2565–2573.

Brucker, N., S. Nunes do Nascimento, L. Bernardini, M. Feiffer Charao and S.C. Garcia. 2020. Biomarkers of exposure, effect, and susceptibility in occupational exposure to traffic related air pollution: A review. J. Appl. Toxicol. 40: 722–736.

Butler, B.A. 2011. Effect of imposed anaerobic conditions on metals release from acid-mine drainage contaminated streambed sediments. Water Res. 45(1): 328–336.

Casarett and J. Doull. 2013. Casarett and Doull's Toxicology: The Basic Science of Poisons. McGraw-Hill Education, New York, USA.

[CCOHS] Canadian Centre for Occupational Health and Safety. 2022. Welding—Fumes And Gases. Canadian Centre for Occupational Health and Safety, Hamilton, Canada.

Chambers, J.E., J.S. Boone, R.L. Carr, H.W. Chambers and D.L. Straus. 2002. Biomarkers as Predictors in health and ecological risk assessment. Hum. Ecol. Risk Assess. 8(1): 165–176.

Cheson, B.D., W.N. Rom and R.C. Webber. 1984. Basophilic stippling of red blood cells: A nonspecific finding of multiple etiology. Am. J. Ind. Med. 5(4): 327–334.

[CIDA] Canadian International Development Agency, Canadian Persistent Organic Pollutants (POPs) Trust Fund, A brief comparison of Human Health and Ecological Risk Assessments, 2021. CIDA, Gatineau, Canada.

Cohen, M.D. 2004. Pulmonary immunotoxicology of select metals: Aluminum, Arsenic, Cadmium, Chromium, Copper, Manganese, Nickel, Vanadium, and Zinc. J. Immunotoxicol. 1(1): 39–69.

de Lapuente, J., J. Lourenço, S.A. Mendo, M. Borràs, M.G. Martins, P.M. Costa et al. 2015. The Comet Assay and its applications in the field of ecotoxicology: a mature tool that continues to expand its perspectives. Front. Genet. 6: 180.

Diamantino, T.C., E. Almeida, M.V.M. Amadeu Soares and L. Guilhermino. 2001. Lactate dehydrogenase activity as an effect criterion in toxicity tests with *Daphnia magna* straus. Chemosphere. 45(4-5): 553–560.

Drexler, H., T. Göen and K.H. Schaller. 2008. Biological tolerance values: Change in a paradigm concept from assessment of a single value to use of an average. Int. Arch. Occup. Environ. Health. 82(1): 139–142.

[EFSA] European Food Safety Authority, Metals as contaminants in food, 2021.EFSA, Parma, Italy.

[FDA] Food and Drug Administration, Chemical Hazards, 2021a. FDA, Silver Spring, USA.

[FDA] Food and Drug Administration, Metals and Your Food, 2021b. FDA, Silver Spring, USA.

Forbes V.E., A. Palmqvist and L. Bach. 2006. The use and misuse of biomarkers in ecotoxicology. Environ. Toxicol. Chem. 25(1): 272–80.

Fortoul, T.I., V. Rodriguez-Lara, A. Gonzalez-Villalva, M. Rojas-Lemus, L. Colin-Barenque, P. Bizarro-Nevares et al. 2015. Health effects of metals in particulate matter. *In*: F. Nejadkoorki [ed.]. Current Air Quality Issues. IntechOpen, London, UK.

Graham, B.W.L. 1985. Exposure to heavy metals in the workplace. J. R. Soc. N. Z. 15(4): 399–402.

Griffin. 2020. Heavy Metal Poisoning. WebMD. https://www.webmd.com/a-to-z-guides/what-is-heavy-metal-poisoning.

Hagger, J.A., M.B. Jones, D.R.P. Leonard, R. Owen and T.S. Galloway. 2006. Biomarkers and integrated environmental risk assessment: Are there more questions than answers? Integr. Environ. Assess. Manag. 2(4): 312–329.

Hull, R., G. Head and G.T. Tzotzos. 2020. Principles of risk assessment. pp. 99–125. *In*: Genetically Modified Plants, Assessing Safety and Managing Risk. Academic Press, Cambridge, USA.

Hung, Y.-M. and H.-M. Chung. 2004. Acute self-poisoning by ingestion of cadmium and barium. Nephrol. Dial. Transplant. 19(5): 1308–1309.

Järup, L. 2003. Hazards of heavy metal contamination. Br. Med. Bull. 68: 167–82.

Jiang, K., B. Wu, C. Wang and Q. Ran. 2019. Ecotoxicological effects of metals with different concentrations and types on the morphological and physiological performance of wheat. Ecotoxicol. Environ. Saf. 15(167): 345–353.

Kamrin, M.A. 2014. Risk Characterization. pp. 818–820. *In*: P. Wexler [ed.]. Encyclopedia of Toxicology. Academic Press, Cambridge, USA.

Khan, M.S., M. Javed, M.T. Rehman, M. Urooj and Md. I. Ahmad. 2020. Heavy metal pollution and risk assessment by the battery of toxicity tests. Sci. Rep. 10: 16593.

Lentini, P., L. Zanoli, A. Granata, S. Santo, S.P. Castellino and R. Dell'Aquila. 2017. Kidney and heavy metals—The role of environmental exposure (Review). Mol. Med. Rep. 15(5): 3413–3419.

Liu, H.H., Y.C. Wu and H.L. Chen. 2007. Production of ozone and reactive oxygen species after welding. Arch. Environ. Contam. Toxicol. 53(4): 513–518.

Mahboob, S., A. Khalid, H.F. Al-Ghanim, F. Al-Balawi, Z. Al-Misned and Z. Ahmed. 2020. Toxicological effects of heavy metals on histological alterations in various organs in Nile tilapia (*Oreochromis niloticus*) from freshwater reservoir. J. King Saud Univ. Sci. 32(1): 970–73.

Maresca, V., L. Fusaro, S. Sorbo, A. Siciliano, S. Loppi, L. Paoli et al. 2018. Functional and structural biomarkers to monitor heavy metal pollution of one of the most contaminated freshwater sites in Southern Europe. Ecotoxicol. Environ. Saf. 163: 665–673.

Marth, E., S. Jelovcan, B. Kleinhappl, A. Gutschi and S. Barth. 2001. The effect of heavy metals on the immune system at low concentrations. Int. J. Occup. Med. Environ. Health. 14(4): 375–386.

Martínez-Pérez, M., F. Arenas-Huertero, J. Cortés-Eslava, O. Morton-Bermea and S. Gómez-Arroyo. 2021. Robinsonecio gerberifolius as a sentinel organism for atmospheric pollution by heavy metals in several sites of Mexicocity and its metropolitan area. Environ. Sci. Pollut. Res. Int. 28(24): 31032–31042.

Masindi, V. and K.L. Muedi. 2018. Environmental contamination by heavy metals. *In*: H.E.-D.M. Saleh and R.F. Aglan [eds.]. Heavy Metals. IntechOpen, London, UK.

McCarty, L.S., M. Power and K.R. Munkittrick. 2002. Bioindicators versus biomarkers in ecological risk assessment. Hum. Ecol. Risk Assess. 8(1): 159–164.

[MIAC] Mining Industry Advisory Committee, Risk-based health surveillance and biological monitoring, 2008. Department of Consumer and Employment Protection, Perth, Australia.

Moorthikumar, K. and M. Muthulingam. 2011. Impact of heavy metal nickel chloride on enzyme succinate dehydrogenase of Freshwater fish *Labeo rohita* (Hamilton). Int. J. Curr. Res. 3(7): 115–119.

More, A.S., C.S. Ranadheera, Z. Fang, R. Warner and S. Ajlouni. 2020. Biomarkers associated with quality and safety of fresh-cut produce. Food Biosci. 34: 100524.

Morgan, M.S. 1997. The biological exposure indices: A key component in protecting workers from toxic chemicals. Environ. Health Perspect. 105(1): 105–115.

[NIOSH] National Institute for Occupational Safety and Health. 2009. Genetics in the Workplace, Implications for Occupational Safety and Health. Department of Health and Human Services, Centers for Disease Control and Prevention, National Institute for Occupational Safety and Health, Washington, USA.

[NORD] National Organization for Rare Disorders, Heavy Metal Poisoning, 2008. NORD, Quincy, USA.

[NRC] National Research Council Committee on Applications of Toxicogenomic Technologies to Predictive Toxicology, Overview of Risk Assessment. 2007. *In*: National Research Council (US) Committee on Applications of Toxicogenomic Technologies to Predictive Toxicology [ed.]. Applications of Toxicogenomic Technologies to Predictive Toxicology and Risk Assessment. National Academies Press, Washington, USA.

Odukudu, F.B., J.G. Ayenimo, A.S. Adekunle, A.M. Yusuff and B.B. Mamba. 2013. Safety evaluation of heavy metals exposure from consumer products. Int. J. Consum. Stud. 38(1): 25–34.

Omar, W.A., Y.S. Saleh, S.M. Mohamed-Assem and S. Marie. 2014. Integrating multiple fish biomarkers and risk assessment as indicators of metal pollution along the Red Sea coast of Hodeida, Yemen Republic. Ecotoxicol. Environ. Saf. 110: 221–231.

Palmer, K.T., R. McNeill-Love, J.R. Poole, D. Coggon, A.J. Frew, C.H. Linaker et al. 2006. Inflammatory responses to the occupational inhalation of metal fume. Eur. Respir. J. 27: 366–373.

Persoons, R., D. Arnoux, T. Monssu, O. Culié, G. Roche, B. Duffaud et al. 2014. Determinants of occupational exposure to metals by gas metal arc welding and risk management measures: A biomonitoring study. Toxicol. Lett. 231(2): 135–141.

Perumalsamy, N. and K. Arumugam. 2013. Enzymes activity in fish exposed to heavy metals and the electro-plating effluent at sub-lethal concentrations. Water. Qual. Expo. Health. 5: 93–101.

Pirinccioglu, A.G., D. Gökalp, M. Pirinccioglu, G. Kizil and M. Kizil. 2010. Malondialdehyde (MDA) and protein carbonyl (PCO) levels as biomarkers of oxidative stress in subjects with familial hypercholesterolemia. Clin. Biochem. 43(15): 1220–1224.

Potter, N.A., Y.G. Meltzer, O.N. Avenbuan, A. Raja and J.T. Zelikoff. 2021. Particulate matter and associated metals: A link with neurotoxicity and mental health. Atmosphere 12: 425.

Radwan, M.A., K.S. El-Gendy and A.F. Gad. 2010. Biomarkers of oxidative stress in the land snail, *Theba pisana* for assessing ecotoxicological effects of urban metal pollution. Chemosphere. 79: 40–46.

Riaz, M.A., A.B.T. Akhtar, A. Riaz, G. Mujtaba, M. Ali and B. Ijaz. 2017. Heavy metals identification and exposure at workplace environment its extent of accumulation in blood of iron and steel recycling foundry workers of Lahore, Pakistan. Pak. J. Pharm. Sci. 30(4): 1233–1238.

Ribarov, S.R. and L.C. Benov. 1981. Relationship between the hemolytic action of heavy metals and lipid peroxidation. Biochim. Biophys. Acta Biomembr. 640(3): 721–726.

Rietjens, I.M.C.M., P. Dussort, H. Günther, P. Hanlon, H. Honda, A. Mally et al. 2018. Exposure assessment of process-related contaminants in food by biomarker monitoring. Arch. Toxicol. 92(1): 15–40.

Robinson, J.M. 2008. Reactive oxygen species in phagocytic leukocytes. Histochem. Cell Biology. 130(2): 281–297.

Roesijadi, G. 1994. Metallothionein induction as a measure of response to metal exposure in aquatic animals. Environ. Health Perspect. 102(12): 91–5.

Sakai, T. 2000. Biomarkers of lead exposure. Ind. Health. 38: 127–142.

Salamat, N.Z., A. Soleimani, A. Safahieh, A. Savari and M.T. Ronagh. 2013. Using histopathological changes as a biomarker to trace contamination loading of musa creeks (Persian Gulf). Toxicol. Pathol. 41: 913–920.

Sall, M.L., A.K.D. Diaw, D. Gningue-Sall, S.E. Aaron and J.-J. Aaron. 2020 Toxic heavy metals: Impact on the environment and human health, and treatment with conducting organic polymers, a review. Environ. Sci. Pollut. Res. 27: 29927–29942.

Sanchez-Dardon, J., I. Voccia, A. Hontela, S. Chilmonczyk, M. Dunier, H. Boemans et al. 1999. Immunomodulation by heavy metals tested individually or in mixtures in rainbow trout exposed *in vivo*. Environ. Toxicol. Chem. 18: 1492–1497.

Sardar, S.W., Y. Choi, N. Park and J. Jeon. 2019. Occurrence and concentration of chemical additives in consumer products in Korea. Int. J. Environ. Res. Public Health. 16(24): 5075.

Scalbert, A., L. Brennan, C. Manach, C. Andres-Lacueva, L.O. Dragsted, J. Draper et al. 2014. The food metabolome: A window over dietary exposure. Am. J. Clin. Nutr. 99(6): 1286–1308.

Schulte, P.A. and J.E. Hauser. 2012. The use of biomarkers in occupational health research, practice, and policy. Toxicol. Lett. 213: 91–99.

Schwela, D. 2014. Risk assessment, uncertainty. pp. 158–164. *In*: P. Whexler [ed.]. Encyclopedia of Toxicology. Academic Press, Cambridge, USA.

Si, M. and J. Lang. 2018. The roles of metallothioneins in carcinogenesis. J. Hematol. Oncol. 11: 107.

Sirivarasai, J., W. Wananukul, S. Kaojarern, S. Chanprasertyothin, N. Thongmung, N. Ratanachaiwong et al. 2013. Association between inflammatory marker, environmental lead exposure, and glutathione S-transferase gene. Biomed. Res. Int. 2013: 474963.

Somerset, V. 2011. Environmental Monitoring using Voltammetric Trace Metal Analysis. Royal Society of Chemistry Environmental Chemistry Group. London, UK.

Søyseth, V., H.L. Johnsen and J. Kongerud. 2013. Respiratory hazards of metal smelting. Curr. Opin. Pulm. Med. 19(2): 158–162.

Spurgeon, D., E. Lahive, A. Robinson, S. Short and P. Kille Peter. 2020. Species sensitivity to toxic substances: Evolution, ecology and applications. Front. Environ. Sci. 8: 237.

Sullivan, P.J., F.J. Agardy and J.J.J. Clark. 2005. Living with the risk of polluted water. pp. 143–196. *In*: The Environmental Science of Drinking Water. Butterworth-Heinemann, Oxford, UK.

Suter II, G.W., T. Vermeire, W.R. Munns Jr. and J. Sekizawa. 2003. Framework for the integration of health and ecological risk assessment. Hum. Ecol. Risk Assess. 9(1): 281–301.

Ulrik, C.S., E.T. Würtz, Ø. Omland, V. Schlünssen, O.F. Pedersen, T. Aasen et al. 2013. Occupational exposure for dust and gases is an important risk factor for developing COPD. Ugeskr Laeg. 175(18): 1253–1256.

Vacchi-Suzzi, C., D. Kruse, J. Harrington, K. Levine and J.R. Meliker. 2016. Is urinary cadmium a biomarker of long-term exposure in humans? A review. Curr. Environ. Health Rep. 3(4): 450–458.

Verma, A. 2018. Forensic aspect of metal poisoning: A review. Int. J. Res. Appl. Sci. Eng. Technol. 6(I): 1089–1092.

Wang, Y., M. Zhang, Z. Li, J. Yue, M. Xu, Y. Zhang et al. 2019. Fine particulate matter induces mitochondrial dysfunction and oxidative stress in human SH-SY5Y cells. Chemosphere. 218: 577–588.

[WHO] World Health Organization, Biomarkers In Risk Assessment: Validity And Validation, 2001. WHO, Geneva, Switzerland.

Wyman, A.E. and S.E. Hines. 2018. Update on metal-induced occupational lung disease. Curr. Opin. Allergy Clin. Immunol. 18(2): 73–79.

Yang, P., S. Huang and A. Xu. 2016. The oxidative burst system in amphioxus. pp. 153–165. *In*: A. Xu [ed.]. Amphioxus Immunity. Academic Press, Cambridge, USA.

Ye, B.J., B.G. Kim, M.J. Jeon, S.Y. Kim, H.C. Kim, T.W. Jang et al. 2016. Evaluation of mercury exposure level, clinical diagnosis and treatment for mercury intoxication. Ann. Occup. Environ. Med. 28: 5.

Part II
Biomarkers of Xenobiotic Metals

CHAPTER 4

Aluminium

Sources of Exposure

Aluminium (Al) is the third most abundant element on the Earth's crust. The ability to separate Al metal from its ores on an industrial scale, has led Al to be the most widely used metal of the 21st century (Exley 2013). As a result there has been an increased exposure of the general population to Al, which most likely occurs through inhalation of ambient air as inhaled fumes/particles, medicinal products containing Al, consumption of food (mainly processed foods) and drinking water (ATSDR 2008b, Thenmozhi et al. 2016, Pandey and Jain 2017).

Air

Atmospheric Al concentrations show widespread temporal and spatial variations, with airborne Al levels ranging from 0.0005 µg/m³ over the Antarctica to more than 1 µg/m³ in industrialized areas. However, the contribution of air to the total exposure to Al is generally negligible (WHO 2010). Although, the exposure of those working in Al industries can be significantly higher compared with individuals not exposed to Al at work, meaning that the reference values derived for the general population may exceed in these workers (Klotz et al. 2017).

Workplace

Occupational exposure occurs during refining of the primary metal and in secondary industries that fabricate Al products (such as aircraft, automotive and metal products), where Al welding is carried out and during electrolysis in Al production (ATSDR 2008b, Klotz et al. 2017). An appropriate occupational source of exposure is mine work with bauxite, where ores contain Al_2O_3 (approximately 35–60% by weight), with miners being exposed daily to dust through smelting, grinding, crushing and various other activities. Long-term occupational exposure to bauxite dust is recognized as leading to adverse health conditions (Pingle et al. 2015). Workplace exposure limits have been set by the Occupational Safety and Health Administration (OSHA), with a legal airborne permissible exposure limit (PEL) of 5 mg/m³ as respirable dust and

15 mg/m^3 as total dust, averaged over an 8-hr work shift. The American Conference of Governmental Industrial Hygienists (ACGIH) also established a Threshold Limit Value (TLV), which is 1 mg/m^3 as a respirable fraction for Al, averaged over an 8-hr work shift (ATSDR 2008a, NJDH 2017).

Soil

Al is present in soils in a number of forms and bound to soil constituents, such as clay particles and organic matter. When the soil pH drops, Al becomes soluble and their amount in the soil solution increases (DPIRD 2018). Being among one of the four most common elements occurring in soils, the typical range is from 10,000 to 300,000 mg/kg, with naturally occurring concentrations varying over several orders of magnitude (Algattawi et al. 2018, DPIRD 2018). Industrial wastes into the environment represent the main source of soil contamination, with very high values reported near a cement factory in Libya (9761.95 mg/kg) and considered as unsuitable to be used for agriculture to ensure human health (Algattawi et al. 2018). Even though, it should be noted that the oral toxicity of Al compounds in soil is dependent on the chemical form, being insoluble compounds (such as Al oxides) that are much less toxic than soluble forms (such as Al chloride, nitrate, acetate or sulphate) (EPA 2003).

Water

Dissolved Al concentrations in waters with near-neutral pH values usually range from 0.001 to 0.05 mg/L but can rise to 0.5–1 mg/L in more acidic waters or waters rich in organic matter. However, at extreme acidity of waters affected by acid mine drainage, Al levels can be up to 90 mg/L (WHO 2010). Since Al salts are usually added as coagulants during water treatment to remove turbidity, organic matter and microorganisms, could lead to increased concentrations of Al in finished water (WAQB HC 2019). When concentrations become high, Al may be deposited in the distribution system, and when disturbance of these deposits occurs by changes in the flow rate, tap water may have increased Al levels, characterized by an undesirable colour and turbidity (WHO 2010). The Environmental Protection Agency (EPA) recommend a Secondary Maximum Contaminant Level (SMCL) of 0.05–0.2 mg/L for Al in drinking water, which are not based on levels that will affect humans, but rather on taste, smell or colour. A value of 0.2 mg/L has been decided by the Food and Drug Administration (FDA) for bottled water (ATSDR 2008a).

Food

Food is unquestionably the main source of Al intake, being the primary substance , the natural content of food caused by the uptake from the geologic surrounding. A secondary area is considered as the primary content and any possible contamination from aluminium articles that come into contact with food and additives, as well as veterinary drugs or fertilizers (Stahl et al. 2011). In a study measuring Al contents in different foods revealed that the highest mean Al content was in vegetables (16.8 mg/ kg), followed by fish and seafood (11.9 mg/kg) and roots and tubers (9.60 mg/kg). The food groups with the most notable contribution to the Tolerable Weekly Intake (TWI) were fruits, 18.2% for adults and 29.4% for children, and vegetables (32.5% for adults and children). Tea consumption can be responsible for increased Al in the diet because tea leaves have been reported to have higher Al levels than many other plants (Shaw et al. 2014, Hardisson et al. 2017). Nonetheless, in general terms the average weekly Al exposure resulting from food intake amounts to approximately 50% of the TWI obtained by the European Food Safety Authority (EFSA) (1 mg/ kg b.w./week) (Tietz et al. 2019). The relative exposure in children can be higher at up to 2.3 mg/kg b.w./week, and in some adults an estimated alimentary exposure to Al was determined as 0.2–1.5 mg/kg bw/week, which does not mean that there is an acute health hazard (Klotz et al. 2017).

Al has been used in food processing for over a century as first a firming agent, raising agent, stabilizer, anticaking agent and colouring matter (FEHD 2009, Geyikoglu et al. 2013, Hardisson et al. 2017, Klotz et al. 2017). The Food and Drug Administration (FDA) considers that Al used as food additives are generally safe, whereas some researchers defend that Al dietary intake can pose a health risk resulting from Al accumulation, namely in the brain, caused by long-term intake (ATSDR 2008a, Hardisson et al. 2017). Current apprehensions regard processed dairy products, flour, and infant formula, which may be high in Al when containing Al-based food additives (WHO 2010, Bohrer et al. 2014). Data from the French "Infant Total Diet Study" and the "Second French Total Diet Study" revealed that the TWI can be exhausted or slightly exceeded, for infants who are not exclusively breastfed and young children relying on specially adapted diets, such as soy-based, lactose free or hypoallergenic (Tietz et al. 2019). To present some values, for exclusively breastfed infants estimates of Al intake vary depending on the maternal dietary Al exposure; in human breast milk Al concentrations can range from 0.0092 to 0.049 mg/L. Non-breastfed infants can be exposed to higher levels, depending on its concentration in formulas; Al in soy-based infant formula can range 0.46–0.93 mg/L and values in milk-based infant formula are determined as 0.058–0.15 mg/L (ATSDR 2008a). Dietary intakes from formulas and infant foods are estimated for children aged six to 11 mon as 0.7 mg/d and for 2-yr-old as 4.6 mg/d (Bohrer et al. 2014).

Food-related materials and articles such as food packaging, Al foils, cooking utensils and baking trays, constitute additional sources of exposure to Al, which can be released to the food (Klotz et al. 2017). The Council of Europe passed a resolution for metals and alloys that come into contact with foodstuffs, where Specific Release Limits (SRL) for metals and alloys are proposed. An SRL specifies the maximal amount of metal ions (in mg) that may be transferred from a defined surface of the contact material to the food (in kg) or food simulant, is the SRL for the release of Al to foodstuffs specified to be 5.00 mg/kg foodstuff (Stahl et al. 2017a). Studies on migration of Al to food from camping dishes and utensils made of Al, showed that SRL is not exceeded when simulants for oil or for tap water are used; however, when using an aqueous solution of 0.5% citric acid, the limit clearly exceeded at 638 mg/L.

This result means that the TWI could be exceeded by 298% for a child weighing 15 kg, and for an adult weighing 70 kg it was equivalent to 63.8% (assuming a daily uptake of 10 mL marinade containing lemon juice over a period of 1 wk) (Stahl et al. 2017b).

For the most part, the intake of Al from water and food is low for the general population, when compared with the Al consumed by people taking Al-containing medication; it is estimated that exposure of the general European population to Al is 28.6-214.0 µg/kg b.w./d (ATSDR 2008b, Hardisson et al. 2017). This does not refute the importance of biomonitoring people and in these contexts, values such as TWI are important , as they are designed to be precautionary for long-term exposures for the general population (Klotz et al. 2017). However, different values exist according to different agencies. As mentioned, the TWI established by EFSA is 1 mg/kg b.w./ week, the TWI and the Provisional Tolerable Weekly Intake (PTWI), decided by the Joint Food and Agriculture Organization (FAO)/World Health Organization (WHO) Expert Committee on Food Additives is 2 mg/kg bw/wk, which is twice the EFSA value (Hardisson et al. 2017, Tietz et al. 2019).

Consumer products

 Al is also present in cosmetic products, which include antiperspirants, lipsticks, sun creams and toothpaste (Shaw et al. 2014, Klotz et al. 2017). Studies on these products revealed that the acute oral toxicity was moderate to low, and that in most blood samples taken after dermal application, Al concentrations were below the lower limit of quantification of the analytical method (SCCS 2019).

Additional sources

Al is largely used in medical/pharmaceutical applications, namely as an adjuvant in vaccines (approximately 0.85 mg/dose), where it is supposed to shock the recipient's immune defences into action, enhancing the immunogenicity for the pathogen(s) in the vaccine(s) (ATSDR 2008a, Shaw et al. 2014). For obvious reasons, vaccines are mostly used in nursing children, which is a motive for concern (Bohrer et al. 2014, Klotz et al. 2017).

The metal is used in other medicines, namely as an agent against pathological hyperhidrosis, in antacid agents, allergen injections, buffered analgesics, phosphate binders, antidiarrhoeal agents, first-aid antibiotic and antiseptics or anti-ulcerative medication (ATSDR 2008b, Klotz et al. 2017, Pandey and Jain 2017). Buffered aspirin may contain 10–20 mg of Al per tablet, while antacids have 300–600 mg of Al hydroxide per tablet or capsule or 5 mL per liquid dose. The FDA considers that Al concentrations used in antacids are generally safe, although in a risk assessment it was concluded that antacids can be a major source of Al exposure, advising against the long-term use of this type of product (ATSDR 2008a, Affourtit et al. 2020).

Fortunately, severe Al toxicity (serum Al level > 200 ng/mL) in patients on chronic haemodialysis is now uncommon, due to the removal of Al from water used for dialysis by reverse osmosis and deionization (Krewski et al. 2007, Tsai et al. 2018).

Even so controlling serum Al levels of these patients remains important, when taking into consideration that Al removal is not totally efficient, and possible additional sources of Al accumulation exist through commonly administered medications to dialysis patients; some of these medications are oral Al-containing phosphate binders and antacids (although efforts have been made to replace high-dose antacid therapy by alternatives) and certain other injectables. The National Kidney Foundation, Kidney Disease Outcomes Quality Initiative guidelines recommend that the baseline serum Al level should be below 20 ng/mL, and that Al levels and risk for Al toxicity should be assessed at least once per year in those patients (Tsai et al. 2018).

Toxicity

Al is relatively safe for humans when compared with other metals such as arsenic, cadmium or mercury (Ogawa and Kayama 2015); the metal is known to possess a low potential for producing adverse effects and only at relevant excessive concentrations of Al toxic manifestations are seen. Regardless of the duration of exposure, the toxicity attributed to Al is dependent on physiochemical properties (solubility, pH, bioavailability, etc.), type of Al preparation, route of administration and physiological status (such as the presence of a renal dysfunction) (Krewski et al. 2007).

Acute toxicity

There is little indication that Al is acutely toxic by oral exposure, the oral median lethal dose being (LD50) of Al nitrate, chloride or sulphate in mice and rats from 200 to 1,000 mg/kg b.w. In 1988 a population of about 20,000 individuals in England, was exposed for at least 5 d to increased levels of Al, which were accidentally distributed from a water supply facility using Al sulphate for treatment. Nausea, vomiting, diarrhoea, mouth ulcers, skin ulcers, skin rashes and arthritic pain were noted, but the symptoms were mostly mild and short-lived. No lasting effects on health are attributed to exposures from Al in drinking-water, although there are reports of damage to bone and central nervous system organs, forming an inadvertent human poisoning with excessive amounts of Al (Krewski et al. 2007, WHO 2010).

Chronic toxicity

Neurodegenerative diseases

Al has been reported as a major risk factor for the cause and development of Alzheimer's Disease (AD), amyotrophic lateral sclerosis and Parkinson's Disease (PD) (Thenmozhi et al. 2016). Yet the association between Al exposures and AD remains controversial, as some studies show that people exposed to high levels of Al may develop AD, while others have found this not to be true (ATSDR 2008a). Despite elevated Al content found in the brains of persons with AD, it remains unclear whether this is a cause or an effect of the disorder (Ogawa and Kayama 2015). Focusing to assess this association, Al exposure from drinking water has been extensively investigated in relation to the development of neurological disorders, including AD; it is proposed that there is enhanced bioavailability of Al when present in water (Krewski et al. 2007). In fact, epidemiological studies have shown that people living in districts with higher Al burden in drinking water are more likely to be diagnosed with AD (Shaw et al. 2014). Additionally, since regular consumers of antacids represent a unique subpopulation with heavy exposure to Al, they were also investigated. A prospective cohort study using observational data led the authors to conclude that the avoidance of this medication could prevent the development of dementia; this was supported by the indication that the experimental administration of antiacids increased the levels of β-amyloid in the brains of mice (Gomm et al. 2016). Evidence connecting the relationship between Al in food and the risk of AD is very minimal, possibly due to the difficulty in obtaining accurate exposure information in dietary studies. But so far, data from many works with these associations are difficult to interpret due to the large variation in study designs and highly variable quality of these studies (Krewski et al. 2007). Mechanisms by which Al could play a role in AD are proposed comprising both direct and indirect modes of potential action. In the direct mode, Al could potentiate the aggregation of molecules known to form pathologic lesions in AD, such as β-amyloid peptides, as already assessed *in vitro*. Whether Al would bind β-amyloid peptide at an appreciable rate *in vivo* is still unclear. Al might also impact the function of the nervous system in several ways through indirect modes of action. The induction of changes in cholesterol levels, which has been suggested in numerous studies as a potential modulator of Alzheimer-type amyloid formation, is an example. Elevated levels of markers of oxidative stress in animals exposed to Al, constitutes a potential mechanism by which long-term exposure to Al could be deleterious, and synergistically worsen cognitive abilities in individuals with pathologic abnormalities associated with AD (Krewski et al. 2007). Al is additionally able to modify hippocampal calcium signal pathways crucial to neuronal plasticity and, in this way, to the memory. Moreover, cholinergic neurons (which are affected by AD) are particularly susceptible to Al neurotoxicity, which alters the synthesis of the neurotransmitter acetylcholine. All these neurobiological effects are relevant in the presumed association between exposures to Al and AD (the Alzheimer's hypothesis) (Ogawa and Kayama 2015). However so far it is not known for certain if Al causes AD, what has certainly been achieved is that Al adversely impacts the nervous system. Accordingly, changes in neuropsychological tests and dose-dependent central nervous dysfunction have been observed in Al-handling workers (ATSDR 2008a, Ogawa and Kayama 2015). Significantly correlations were found between exposure to Al and a number of neuropsychiatric symptoms, such as loss of coordination and memory, and problems with balance (Krewski et al. 2007). Additional studies describe higher Al exposure associated with modified visual reaction and motor function (Hasan et al. 2020). It has also been shown that the probability of neurotoxic events is higher in children, and in people with kidney problems because the kidney plays a key role in the excretion of Al (Hardisson et al. 2017).

Cancer

There is conflicting evidence on Al induced carcinogenicity, but currently Al is not classified as a human carcinogen (Ogawa and Kayama 2015). Nevertheless, cancer hazards associated with specific context of exposures in Al production were evaluated by the International Agency on Research for Cancer (IARC 2012). However there is sufficient evidence in humans for carcinogenicity of occupational exposures, which leads to cancer of the bladder and lungs; thus, occupational exposures during Al production are established as carcinogenic to humans (Group 1) (IARC 2012).

Regarding the association of antiperspirants use and carcinogenic risks (such as breast cancer), only a few epidemiological studies with no clear results have been undertaken (Krewski et al. 2007, Ogawa and Kayama 2015). The Scientific Committee on Consumer Safety (SCCS) (2019) concluded that carcinogenicity is not expected at exposure levels which are achieved via cosmetic use, and that antiperspirant use had a minor impact on the body burden of Al due to its very low dermal bioavailability, in contrast to uptake via nutrition or vaccination. Some studies show that the frequent use of underarm cosmetic products results in accumulation of Al in the breast tissue. Increased incidence of breast cancer is observed in women reporting to use antiperspirants more than once daily, starting at an age below 30 yr. However, this study was mainly based on correlation analyses not providing causal links; as stated by the authors, a reverse causation effect, meaning that the breast tumour may accumulate Al, could not be excluded (SCCS 2019, Affourtit et al. 2020).

Other effects

Other chronic consequences of Al exposures are adverse respiratory tract effects, such as the ones reported in Al industry employees, with asthma-type of symptoms known as potroom asthma, which is the most exceptionally investigated respiratory effect (despite generally being associated with toxic chemicals other than Al in the workplace) (Krewski et al. 2007). Exposure to Al powder is also thought to be directly correlated with the development of pulmonary fibrosis in Al industry workers (Krewski et al. 2007, Ogawa and Kayama 2015). Other effects reported in workers exposed to Al alloys and Al dust include contact dermatitis and irritant dermatitis, haematological, musculoskeletal and genetic disorders (Krewski et al. 2007, Pingle et al. 2015).

There is a large number of literature on the adverse effects of non-occupational Al exposure in individuals with impaired renal function, which as referred, are typically exposed to Al through haemodialysis or medicinal sources. Anaemia, bone disease and dialysis encephalopathy are the most reported complications of Al exposure in those subjects (Krewski et al. 2007, Ogawa and Kayama 2015). Al may also accumulate in bones with high turnover replacing calcium and disrupting mineralization and bone cell growth; effects such as osteomalacia, have been found to be dose-dependent and time-dependent of Al exposure (Ogawa and Kayama 2015).

Biomarkers

Most people have small amounts of Al in their bodies, being the total body burden in healthy individuals 30–50 mg, with approximately 50% in the skeleton and 25% is in the lungs. (ATSDR 2008b). As described above the risk of Al overload may occur in certain situations which can lead to more or less severe toxic outcomes. Therefore, biomarkers (BM) are necessary to assess the status of the metal in the body and/or their effects.

Biomarkers of exposure

There have been several attempts to define BM for systemic Al overload and these include measurements of total Al in blood, urine, bone, hair, nails and sweat (Exley 2013, Hasan et al. 2020).

Aluminium levels in blood and urine

Only blood and urine have been tested to any significant degree, with serum, plasma or urine Al levels the most used measures of internal load and known to reflect short-term exposure, such as within the past several days or weeks (Hasan et al. 2020). Both blood and urine diagnoses have their limitations, since external exposure levels cannot be related very accurately with them; this is primarily because Al is very poorly absorbed by any route and its oral absorption in particular, can be quite affected by other concurrent intakes (ATSDR 2008b).

Whole blood seems to be a better indicator than serum as it is less prone to temporal factors associated with the redistribution of Al between many potential body parts (Exley 2013). Urine is the major route for excretion of systemic Al with up to 100 µg being excreted daily (Minshall et al. 2014). There is also an indication that high exposure levels are reflected in urine levels, but this cannot be well quantified since most of the Al may be rapidly excreted (ATSDR 2008b). Still, 24-hr urine as a composite sample can constitute an accurate representation of the ultra-filterable fraction of Al in the blood over this same period (Exley 2013).

Reference values for Al are < 5 µg/L in serum, with reports of healthy individuals exhibiting serum values ranging from 1 to 3 µg/L and < 15 µg/L in urine (ATSDR 2008b, Klotz et al. 2017). Since Al concentrations are especially likely to exceed in persons with an occupational exposure, a biological tolerance value in these contexts is of 50 µg/g creatinine in the urine (Klotz et al. 2017). So far there is wide variation in reference values provided by laboratories which may not represent "normal" ranges of a healthy population, making it difficult to interpret serum or urine Al ranges clinically. For laboratories using atomic absorption spectrometry, Al reference ranges can vary from < 5.41 µg/L to < 20 µg/L in serum, < 7.00 µg/L or 0 to 10 µg/L in plasma and 5 to 30 µg/L in urine; for those laboratories using the inductively coupled plasma mass spectroscopy, ranges can be from 0 to 6 µg/L or < 42 µg/L in serum, 0 to 10 µg/L or 0 to 15 µg/L in plasma and 0 to 7 µg/L or 5 to 30 µg/L in urine. Data on Al levels in children are considered warranted, since it is being increasingly requested for children with developmental issues (Zeager et al. 2012).

When associating levels of BMs of exposure with adverse outcomes, no health effects of Al at urinary concentrations below 55 µg/g creatinine are observed, at least with regards to direct harm to the lungs, bones, kidneys or concerning inflammation and oxidative DNA damage (Ogawa and Kayama 2015). Al levels in urine or blood are considered a measure for assessing Al- related neurotoxicity, with serum concentrations of 0.25–0.35 µmol/L (6.8–9.5 µg/L) or urinary Al levels of 4–6 µmol/L (108–162 µg/L), appearing to represent a threshold for observed adverse neurological effects (Ogawa and Kayama 2015, Klotz et al. 2017). Workers in the Al industry already exhibited declining performance in neuropsychological tests (attention, learning, memory) with approximately 13 µg/L of Al in plasma and concentrations exceeding 100 µg/g creatinine in the urine (Klotz et al. 2017).

Aluminium levels in other matrices

Recent data have suggested that perspiration might be a significant route of excretion of systemic Al, but there are no reliable data which relate to Al in sweat. In a study, the obtained Al values in sweat ranged from 329 to 5329 µg/L, which are higher than those that describe the daily excretion of Al in urine (100 µg/24-hr); notably men seemed to excrete more Al through sweat than woman (Minshall et al. 2014). The levels of Al in this fluid should certainly be focused on in future work aiming to elucidate its potential as BM of exposure.

Subjects exposed to excessive Al already showed a very large increase in hair Al, which is considered a useful indicator of the metal body burden in Al-induced conditions, such as dialysis encephalopathy (Yokel 1982). More recent studies showed that medical products containing Al could be detected in adult hair, while another work presented reliable data on the Al content of babies' hair after exposure to Al-adjuvanted vaccines (Bohrer et al. 2014).

Al levels have also been measured in nails, toenails or fingernails as potential BMs; when considering the growth rate of nails (and a recent evaluation of other metals in toenails), it was suggested that nail Al levels could reflect exposures from a longer time-period than blood or urine measurements, such as several months. However, several months still do not accurately reflect lifetime work exposure, which can last for years (ATSDR 2008a, Hasan et al. 2020).

In occupational contexts long-term exposure to Al has traditionally been evaluated using semi-quantitative methods, such as the determination of Cumulative Exposure Indices (CEIs). CEIs incorporate a combination of air sampling and work history data to summarize the total inhaled concentration over time. Cumulative workplace Al exposure assessed by CEIs was also associated with headaches and insomnia (Hasan et al. 2020). Bone Al levels are also a potentially useful quantitative BM of cumulative exposure as the metal is stored in the bones for long periods; it is estimated that the half-life of cortical bone Al is up to 29 yr in the levels found in healthy individuals 5 to 10 mg/kg (ATSDR 2008a, Hasan et al. 2020). Unfortunately, this highly invasive procedure is not a viable option to assess Al exposure for routine screenings or research studies. In alternative *in vivo* neutron activation analysis technology is a newer non-invasive method which could address this gap. Data in human populations which could establish the time of exposure reflected by bone Al are still lacking, but recent studies suggest that bone Al concentrations assessed using this technique reflect cumulative Al exposure over the past 15 yr. It is worth noting that the possibility of measuring cumulative Al exposure could also help address questions regarding associations between Al exposure and AD (Hasan et al. 2020).

Biomarkers of susceptibility

Only a little information is available regarding susceptibilities to Al toxicity which could be used as BMs of susceptibility.

Al inhibits biological oxidative stress management systems by interfering with the Glutathione S-Transferase (GST) detoxification system. GSTs are multi-gene isoenzymes encoded by three separate families of genes, constituted by eight classes; Glutathione S-transferase pi (GSTP1), mu (GSTM1) and theta (GSTT1) play important roles in detoxification of xenobiotics and polymorphisms in these genes that could affect biologic responses to metals. Mean blood Al concentrations were found to be associated with GSTP1 rs1695 polymorphism in autistic children (Rahbar et al. 2016).

Another study looked at Single Nucleotide Polymorphisms (SNPs) in a southern Chinese population in search for potential susceptibilities to Al toxicity and to understand the relationship between Al levels and genetic variants (Chen et al. 2021). Three SNPs, rs10224371, rs2316242 and rs10268004, were found in the DPP6 gene, which were significantly associated with Al concentrations in the plasma. Notably, DPP6, has been proved as a candidate gene for Amyotrophic Lateral Sclerosis (ALS) and a potential factor for mental diseases. The protein encoded by DPP6 is associated with the regulation of the potassium ion transport channel, with aberration causing neurodevelopmental impairment and mental diseases (Chen et al. 2021).

Biomarkers of effects

No simple, non-invasive test which can be used as BM of effects caused by Al, nor biochemical or histological change specific for Al exposures, has been identified so far (ATSDR 2008b). However, changes in biochemical parameters noticed in experimental and epidemiological works will be described as further potential BMs.

Oxidative stress

Al in nanoparticles is reported as inducing elevated levels of Reactive Oxygen Species (ROS) *in vitro* (Ogawa and Kayama 2015). Accordingly, in the liver, kidneys and brain of rats treated with $AlCl_3$, thiobarbituric acid reactive substances (TBARS) (a common marker of lipid peroxidation), increased, whereas the activities of the antioxidant enzymes glutathione S-transferase, superoxide dismutase, catalase and glutathione peroxidase concomitantly decreased (Newairy et al. 2009, Kumar et al. 2018). Similar evidence of oxidative stress was obtained through the observation of increased 8-hydroxydeoxyguanosine (8-OHdG) and DNA damage in Al smelters (Samir and Rashed 2018). In turn, aromatic DNA adducts are a form of DNA damage caused by covalent attachment of a chemical moiety to DNA, which can cause mutations that may give rise to cancer; thus, their levels are frequently used as BMs for chemical hazard exposure (Rozelle et al. 2021). The concentration of aromatic DNA adducts was found to be augmented in the lymphocytes of Al plant workers at different locations, at different times of the year and in different job categories (IARC 2012).

Another possible BM of effect is neopterin, which was already proposed for some human populations occupationally exposed to Al in mines. Neopterin is most known to constitute an early and valuable BM of cellular immunity and is a pyrazinopyrimidine compound soluble in the plasma or serum, produced by activated monocytes, macrophages and dendritic cells, following stimulation with γ-interferon (recognized for a long time for its role in antibody-producing responses) (Leibson et al. 1984, Pingle et al. 2015). Neopterin is also known as a gate keeper molecule due to its superior production before the onset of symptoms in adverse conditions. Due to these characteristics, it is used as a prognostic indicator for cell-mediated immunity, chronic infection, immune stimulation, and also serves as an indirect indicator for oxidative stress; in fact, it seems that neopterin plays an important role in the modulation of oxygen radical-mediated processes. Significantly high concentration of serum neopterin was found in experimental Al exposures, and in the serum of bauxite exposed workers. These results led to proposing the measurement of serum neopterin levels for early detection of health risks in workers exposed to bauxite dust in mines (Pingle et al. 2015).

Other endpoints

Severe pathological damages observed in liver samples on exposure to low doses of Al led to consider that Al can produce serious dysfunctions in this organ (Geyikoglu et al. 2013). The experimental administration of $AlCl_3$ is known to modify the activity of several liver enzymes, when measured in the plasma. These enzymes include aspartate aminotransferase (AST) and alanine aminotransferase (ALT) (which are the most sensitive indicators of hepatocyte injury), alkaline phosphatase (AlP) and acid phosphatase (AcP) (Sallam et al. 2005, Jeschke 2007, Geyikoglu et al. 2013).

Al has been shown to affect δ-aminolaevulinic acid dehydratase levels, an enzyme used in heme synthesis, in the blood of mice and in the bone marrow of rats; the effects were distinct according to the Al concentrations found in blood (Ogawa and Kayama 2015). Counts of red blood cells and haemoglobin levels, have been found to decrease significantly in Al-treated rats (Geyikoglu et al. 2013). In humans the heme precursor δ-aminolaevulinic acid also showed a significant difference between Al-handling and non-Al-handling workers (Ogawa and Kayama 2015).

Disturbed epididymal sperm parameters and disrupted steroidogenesis were also found in Al-exposed rats. The histological analysis of their testis showed dose-dependent degenerative and atrophic changes in the seminiferous germinal epithelium and in Leydig cells. A significant dose-dependent decline was additionally observed in germ cell count, seminiferous tubular diameter and Leydig cell nuclear diameter. The levels of serum testosterone, follicle stimulating hormone and luteinizing hormone also were decreased in a dose-dependent manner. However after 60 d of treatment these conditions regressed, the authors concluded that AlCl₃ intoxication exerted a partially reversible adverse impact (Pandey and Jain, 2017).

Final Remarks

This chapter illustrated how the general population is increasingly exposed to Al through diverse sources, mostly through the consumption of Al-containing medicines. However there is little indication that Al is acutely toxic by oral exposure, several adverse effects have proved to appear from chronic exposures, such as in the respiratory tract in occupational contexts or in the nervous system. Blood or urine Al are the BMs most used to access short-term exposure, whereas Al levels for nails are proposed BMs which can reflect exposures of several months; however, since several months still do not accurately reflect lifetime work exposure, methods using more advanced technologies have been developing to assess cumulative exposure. Very little data are available regarding susceptibilities to Al toxicities which could be used as BMs as well as any biochemical or histological change specific for Al exposures has been identified so far for BMs of effect. Such information is necessary for further investigation on susceptibility and effect BMs for Al, which could serve for chronic exposure conditions assessments on vulnerable persons, such as the ones on treatment of haemodialysis and/or chronic Al-containing medicines consumption.

References

Affourtit, F., M.I. Bakker and M.E.J. Pronk. 2020. Human health risk assessment of aluminium, RIVM report 2020-0001. National Institute for Public Health and the Environment, Bilthoven, Netherlands.

Algattawi, A.A., M. Fayez-Hassan and E. Khalil. 2018. Soil contamination with toxic aluminum at mid and West Libya. Toxicol. Int. 25(4): 198–203.

[ATSDR] Agency for Toxic Substances and Disease Registry, Aluminum CAS # 7429-90. 2008a. Department of Health and Human Services, Public Health Services, Washington, USA.

[ATSDR] Agency for Toxic Substances and Disease Registry, Toxicological Profile for Aluminum 2008b. Department of Health and Human Services, Public Health Services, Washington, USA.

Bohrer, D., M. Schmidt, R.C. Marques and J.G. Dórea. 2014. Distribution of aluminum in hair of Brazilian infants and correlation to aluminum-adjuvanted vaccine exposure. Clin. Chim. Acta. 428: 9–13.

Chen, T.-H., C.-C. Yang, K.-H. Luo, C.-Y. Dai, Y.-C. Chuang and H.-Y. Chuang. 2021. The mediation effects of aluminum in plasma and dipeptidyl peptidase like protein 6 (DPP6) polymorphism on renal function via genome-wide typing association. Int. J. Environ. Res. Public Health. 18: 10484.

[DPIRD] Department of Primary Industries and Regional Development. 2018. Effects of soil acidity. Department of Primary Industries and Regional Development's Agriculture and Food division, South Perth, Australia.

[EPA] Environmental Protection Agency. Ecological Soil Screening Level for Aluminum Interim Final (OSWER Directive 9285.7-60). 2003. Environmental Protection Agency, Office of Solid Waste and Emergency Response, Washington, USA.

Exley, C. 2013. Human exposure to aluminium. (Perspective) Environ. Sci.: Processes Impacts. 15: 1807–1816.

[FEHD] Centre for Food Safety of the Food and Environmental Hygiene Department) of the Government of the Hong Kong Special Administrative Region, Aluminium in food. Risk Assessment Studies, Chemical Hazard Evaluation, Report No. 35, 2009. FEHD, Hong Kong, China.

Geyikoglu, F., H. Türkez, T.O. Bakir and M. Cicek. 2013. The genotoxic, hepatotoxic, nephrotoxic, haematotoxic and histopathological effects in rats after aluminium chronic intoxication. Toxicol. Ind. Health. 29(9): 780–791.

Gomm, W., K. von Holt, F. Thomé, K. Broich, W. Maier, A. Fink et al. 2016. Association of proton pump inhibitors with risk of dementia: a pharmacoepidemiological claims data analysis. JAMA Neurol. 73(4): 410–416.

Hardisson, A., C. Revert, D. González-Weller, A. Gutiérrez, S. Paz and C. Rubio. 2017. Aluminium exposure through the diet. J. Food Sci. Nutr. 3: 020.

Hasan, Z., D. Rolle-McFarland, Y. Liu, J. Zhou, F. Mostafaei, Y. Li et al. 2020. Characterization of bone aluminum, a potential biomarker of cumulative exposure, within an occupational population from Zunyi, China. J. Trace. Elem. Med. Biol. 59: 126469.

[IARC] International Agency on Research for Cancer, IARC monographs on the evaluation of carcinogenic risks to humans, Occupational exposures during aluminium production. 2012. World Health Organization (WHO), Geneva, Switzerland.

Jeschke, M.G. 2007. The hepatic response to a thermal injury. *In*: D.N. Herndon [ed.]. Total Burn Care. Elsevier, Amsterdam, Netherlands.

Klotz, K., W. Weistenhöfer, F. Neff, A. Hartwig, C. van Thriel and H. Drexler. 2017. The health effects of aluminum exposure. Dtsch. Arztebl. Int. 114(39): 653–659.

Krewski, D., R.A. Yokel, E. Nieboer, D. Borchelt, J. Cohen, J. Harry et al. 2007. Human health risk assessment for aluminium, aluminium oxide, and aluminium hydroxide. J. Toxicol. Environ. Health Part B, Critical Rev. 10(1): 1–269.

Kumar, S., R.K. Chaitanya and V.R. Preedy. 2018. Assessment of antioxidant potential of dietary components. pp. 239–253. *In*: V.R. Preedy and R.R. Watson [eds.]. HIV/AIDS. Academic Press, Cambridge, EUA.

Leibson, H., M. Gefter, A. Zlotnik, P. Marrack and J.W. Kappler. 1984. Role of γ-interferon in antibody-producing responses. Nature 309: 799–801.

Minshall, C., J. Nadal and C. Exley. 2014. Aluminium in human sweat. J. Trace Elem. Med. Biol. 28(1): 87–88.

Newairy, A.S., A.F. Salama, H.M. Hussien and M.I. Yousef. 2009. Propolis alleviates aluminium-induced lipid peroxidation and biochemical parameters in male rats. Food Chem. Toxicol. 47(6): 1093–1098.

[NJDH] New Jersey Department of Health, Hazardous Substance Fact Sheet, Aluminium Oxide, 2017. NJDH, Trenton, USA.

Ogawa, M. and F. Kayama. 2015. A study of the association between urinary aluminum concentration and pre-clinical findings among aluminum-handling and non-handling workers. J. Occup. Med. Toxicol. 10: 13.

Pandey, G. and G.C. Jain. 2017. Aluminium chloride-induced testicular effects in rats: A histomorphometrical study. AJAST. 1(9): 46–52.

Pingle, S.K., L.R. Thakkar, A.A. Jawade, R.G. Tumane, R.K. Jain and P.N. Soni. 2015. Neopterin: A candidate biomarker for the early assessment of toxicity of aluminum among bauxite dust exposed mine workers. Indian J. Occup. Environ. Med. 19(2): 102–109.

Rahbar, M.H., M. Samms-Vaughan, M.R. Pitcher, J. Bressler, M. Hessabi, K.A. Loveland et al. 2016. Role of metabolic genes in blood aluminum concentrations of jamaican children with and without autism spectrum disorder. Int. J. Environ. Res. Public Health. 13(11): 1095.

Rozelle, A.L., Y. Cheun and S. Lee. 2021. DNA interstrand cross-links induced by the major oxidative adenine lesion 7,8-dihydro-8-oxoadenine. Nat. Commun. 12: 1897.

Sallam, S.M.A., M.E.A. Nasser, M.S.H. Yousef, A.M. El-Morsy, S.A.S. Mahmoud and M.I. Yousef. 2005. Influence of aluminum chloride and ascorbic acid on performance, digestibility, caecal microbial activity and biochemical parameters of rabbits. Res. J. Agr. Biol. Sci. 1(1): 10–16.

Samir, A.M. and L.A. Rashed. 2018. Effects of occupational exposure to aluminium on some oxidative stress and DNA damage parameters. Hum. Exp. Toxicol. 37(9): 901–908.

[SCCS] European Commission, Scientific Committee on Consumer Safety, Opinion on the safety of aluminium in cosmetic products—submission II, 2019. European Commission (EC), Brussels, Belgium.

Shaw, C.A., S. Seneff, S.D. Kette, L. Tomljenovic, J.W. Oller and R.M. Davidson. 2014. Aluminum-induced entropy in biological systems: Implications for neurological disease. J. Toxicol. 2014: 491316.

Stahl, T., H. Taschan and H. Brunn. 2011. Aluminium content of selected foods and food products. Environ. Sci. Eur. 23: 37.

Stahl, T., S. Falk, A. Rohrbeck, S. Georgii, C. Herzog, A. Wiegand et al. 2017a. Migration of aluminum from food contact materials to food—a health risk for consumers? Part I of III: Exposure to aluminum, release of aluminum, tolerable weekly intake (TWI), toxicological effects of aluminum, study design, and methods. Environ. Sci. Eur. 29(1): 19.

Stahl, T., S. Falk, A. Rohrbeck, S. Georgii, C. Herzog, A. Wiegand et al. 2017b. Migration of aluminum from food contact materials to food—a health risk for consumers? Part III of III: Migration of aluminum to food from camping dishes and utensils made of aluminum. Environ. Sci. Eur. 29(1): 17.

Thenmozhi, A.J., T. Raja, W. Raja, T. Manivasagam, U. Janakiraman and M.M. Essa. 2016. Hesperidin ameliorates cognitive dysfunction, oxidative stress and apoptosis against aluminium chloride induced rat model of Alzheimer's disease. Nutr. Neurosci. 20(6): 360–368.

Tietz, T., A. Lenzner, A.E. Kolbaum, S. Zellmer, C. Riebeling, R. Gürtler et al. 2019. Aggregated aluminium exposure: Risk assessment for the general population. Arch. Toxicol. 93: 3503–3521.

Tsai, M.H., Y.W. Fang, H.H. Liou, J.-G. Leu and B.-S. Lin. 2018. Association of serum aluminum levels with mortality in patients on chronic hemodialysis. Sci. Rep. 8: 16729.

[WAQB HC] Water and Air Quality Bureau, Health Canada, Aluminum in drinking water: Guideline technical document for consultation. 2019. WAQB HC, Ottawa, Canada.

[WHO] World Health Organization, Aluminium in drinking-water, Background document for development of WHO Guidelines for Drinking-water Quality. 2010. WHO, Geneva, Switzerland.

Yokel, R.A. 1982. Hair as an indicator of excessive aluminum exposure. Clin. Chem. 28(4 Pt 1): 662–665.

Zeager, M., A.D. Woolf and R.H. Goldman. 2012. Wide variation in reference values for aluminum levels in children. Pediatrics. 129(1): e142–147.

CHAPTER 5

Arsenic

Sources of Exposure

Metalloid Arsenic (As) is one of the largest chemical elements on the Earth's crust and is found throughout our environment (Chung et al. 2014). Among the types of naturally occurring As, inorganic As (iAs) is the most prevalent and includes trivalent ($^{3+}$) arsenite and pentavalent ($^{5+}$) arsenate [or As (III) and As (V), respectively)]. Examples of organoarsenicals, include monomethylarsonic acid (MMA), dimethylarsonic acid (DMA), trimethylarsonic acid and arsenobetaine (Hong et al. 2014) (Fig. 3).

The primary natural source of As is volcanic activity, which emerges from vegetation and wind-blown dusts that also contribute to the environmental burden. Most anthropogenic emissions come from agriculture or industrial processes, since As is primarily used as an insecticide and herbicide or as a preservative for wood, which is still used in industrial applications due to its germicidal power and resistance to rotting and decay; therefore, incineration of preserved wood products is an additional source of environmental As contamination (Chung et al. 2014, PAHO/WHO 2021). Other industrial uses comprise pharmaceuticals, glass, manufacturing of alloy, sheep dips, leather preservatives, pigments, antifouling paints, poison baits, textiles, paper, metal adhesives, microelectronics and optics (Chung et al. 2014, WHO 2018). In the general population (individuals who are not exposed to As through their work environment) the main source of exposure is likely to be oral from water, soil and contaminated agricultural and fish products. While exposure to trivalent or pentavalent iAs occurs through contaminated water, food, air or soil, exposure to organoarsenicals such as arsenobetaine takes place mainly through the consumption of marine animals (Hong et al. 2014).

Air

In air As is predominantly attached to particulate matter usually as a mixture of arsenite and arsenate, with a negligible amount of organic As species; when attached to these very small particles, the metalloid can stay there for many days and travel long distances. Human exposure generally occurs at very low concentrations ranging

Arsenic species	Structure	pKa
As(V)	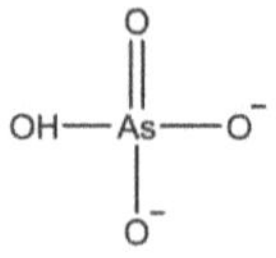	pKa₁ = 2.19 pKa₂ = 6.98 pKa₃ = 11.53
As(III)	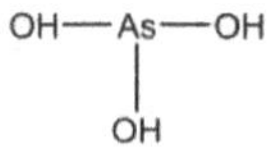	pKa₁ = 9.23 pKa₂ = 12.13 pKa₃ = 13.4
DMA	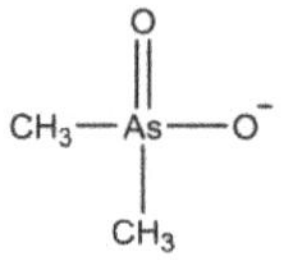	pKa = 6.2
MMA	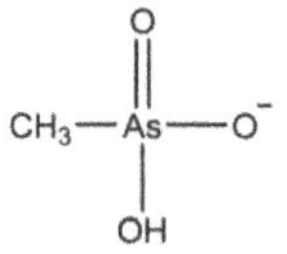	pKa₁ = 4.1 pKa₂ = 8.7
AsB	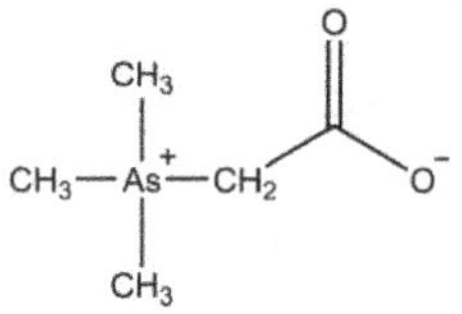	pKa = 2.18

Figure 3. Common arsenic species: inorganic arsenite [As (III)], inorganic arsenate [As (V)], monomethylarsonic acid (MMA), dimethylarsinic acid (DMA) and arsenobetaine (AsB) (Adapted from Reid et al. 2020).

from 0.4 to 30 ng/m³, it is estimated that approximately 40 to 90 ng of As per day are typically inhaled by humans; in unpolluted areas, approximately 50 ng or less As is inhaled per day (Chung et al. 2014).

Workplace

Unlike the general population, and despite improved awareness and stricter legal requirements, there is still potential for occupational exposure to iA in a range of industries (Farmer and Johnson 1990). These include manufacture and application of arsenical pesticides by farm workers, semiconductor manufacturing, employees involved in glass manufacturing, construction and mine workers exposed to As-containing soil, recyclers exposed to electronic or e-waste or workers who perform nonferrous smelting (Farmer and Johnson 1990, Baker et al. 2018, NIOSH 2019). Inhalation of As has been widely described as a major occupational exposure route, followed by ingestion, while dermal exposure is a minor one that would not pose a significant health risk in most settings (Hong et al. 2014, Baker et al. 2018).

A Permissible Exposure Level (PEL) for iAs is decided by agencies such as the Occupational Safety and Health Administration (OSHA), which is 10 μg/m³ of air averaged over an 8-hr period without the use of a respirator (Baker et al. 2018).

Water

Ground water, which is a major source of drinking water, can have elevated concentrations of As, as reported in Bangladesh, Vietnam, China, Taiwan, Argentina and Canada. These levels result from natural sources, erosion and leaching from geological formations and/or from anthropogenic sources due to the use of the metalloid for industrial purposes, mining activities, metal processing, pesticides and fertilizers (Chung et al. 2014, WHO 2018). At least 140 million people in 50 countries have been drinking water containing As at levels above the World Health Organization (WHO), Environmental Protection Agency (EPA) or Food and Drug Administration (FDA) provisional guideline value of 10 μg/L. Millions of people worldwide ingest drinking water contaminated with iAs at levels > 100 μg/L, and in a highly affected area of Bangladesh, 21.4% of all deaths were attributed to As levels above the limits settled by agencies (Hughes 2006). The United States National Research Council has stated that as many as 1 in 100 additional cancer deaths could be expected from a lifetime exposure to drinking-water containing 50 μg/L of As (Hughes 2006, Hong et al. 2014, Baker et al. 2018, WHO 2018).

Food

The irrigation of crops and the preparation of food with As contaminated water are relevant sources of exposure (Hong et al. 2014, WHO 2018). Studies using National Health and Nutrition Education Survey data have calculated that As exposure from food may exceed in some cases the quantity ingested from drinking water (Baker et al. 2018). It is estimated that the mean adult dietary intake for consumers from the U.K. or Australia is 63 μg/d , in the USA the total intake of As has been estimated at 57 μg/d and in Canada 49 μg/d (FAO/WHO 1999). An oral reference dose (RfD) for iAs is decided by the EPA as 0.3 μg/kg/d , considering estimates of As in food as 2 μg/d and a volume of consumed water of 4.5 L/d (Hughes 2006). Childhood is a period of heightened vulnerability to toxic-induced adverse effects and certain foods constitute a greater potential source of dietary iAs exposure; these include, infant formula, rice cereal and apple juice, which are mostly consumed by children (Baker et al. 2018). It is a matter of concern that infants and children's dietary patterns are often less varied than those of adults, that they consume more food relative to their body weight, and that those who have a gluten-free diet tend to eat more rice-based foods (Baker et al. 2018, FDA 2020).

For the general population, meat, poultry, dairy products, cereals, fish and shellfish can be dietary sources of As, although exposure from these foods is considered generally much lower, compared to exposure through contaminated groundwater

(WHO 2018). Foods of terrestrial origin generally contain concentrations of total As below 0.02 mg/kg w.w. and low concentrations of iAs. The exception are plants that accumulate As via root uptake from the soil (e.g., rice plant absorbs As from soil and water up to 10 times more than other food crops) or by absorption of airborne As deposited on the leaves (e.g., tea). Cereal and cereal products, and in particular rice and rice-based products, have the highest iA concentrations which are reported as 0.1–0.4 mg/kg d.w. Limit values settled for total As in cereals (including rice) are 1mg/kg, with the FDA issuing regulations to industry to not exceed iAs levels of 100 µg/L in infant rice cereal; the FDA places a high priority on monitoring levels of As in rice and rice products, since are highly consumed foods worldwide (Molin et al. 2015, Baker et al. 2018, FDA 2020, FSANZ 2020).

Seafood is the major contributor to As in the diet since it has the highest total As concentrations; however, most of the present As forms are organic, which are less toxic than inorganic equivalents (Hong et al. 2014, Molin et al. 2015, Baker et al. 2018, WHO 2018). Both fish and seafood are generally low in iAs, usually < 0.2 mg As/kg dry weight, however a few exceptions do exist. Some examples are the marine algae *Hizikia fusiforme,* which can have extremely high iAs concentration (arsenate > 60 mg/kg), and blue mussels which can present relatively high levels of the inorganic forms of the metalloid (0.001 to 4.5 mg/kg). Limits are established for fish and crustacea iAs content, being 2mg/kg (FDA 2020, FSANZ 2020). Organoarsenicals predominate with levels between 1 and 100 mg/kg w. w. (Molin et al. 2015). Among organoarsenicals, fin fish or crustaceans contain high levels of arsenobetaine, which is relatively nontoxic and excreted intact in urine; seaweed and marine algae contain arsenosugars, which may also be present in bivalves such as clams, mussels, oysters and scallops and other marine food feeding on them (Baker et al. 2018); arsenolipids are present in cod liver oil, capelin and tuna (Molin et al. 2015). Not much is known about the biotransformation of organoarsenicals, though recent observations indicate that the bioconversion of seafood' arsenosugars and arsenolipids results in urinary excretion of DMA, which also possibly produce reactive iAs^{3+} intermediates; DMA^{5+} have been associated with cellular toxicity and genotoxicity, but their impact on human health is still unclear (Molin et al. 2015, Baker et al. 2018). Arsenocholine is another organoarsenical present in marine organisms. Experimental studies indicate that a major part of an orally administered dose of arsenocholine is absorbed from the gastrointestinal tract, and rapidly excreted in urine with biotransformation (Kaise et al. 1992, Molin et al. 2015).

Consumer products

The natural iAs content of tobacco taken up from soil along with tobacco plants that have been treated with lead arsenate insecticides, expose smokers to As (WHO 2018, PAHO/WHO 2021); approximately 0.25 µg of As is ingested after smoking one cigarette. Cigarette smoking *per se* is a main risk factor for lung cancer, cigarette smoking and ingesting As in drinking water has also proved to have a synergistic effect (Hong et al. 2014).

Additionally, the presence of metals in cosmetic products, which includes the metalloid As, is particularly dangerous for human health and prohibited in the countries of the European Union (EU); some other countries provide guidelines for limiting their use and tests to control their levels (Hong et al. 2014, Borowska and Brzóska, 2015). Many kinds of colour cosmetics produced and used in various parts of the world contain As, with values of 0.006–0.31 mg/kg already seen in lipsticks, lip glosses and skin-whitening creams used in Nigeria. Other products such as make-up, face creams, hair bleaches, shampoos, lacquers and brilliantine were found to have As mean concentrations of 0.107 mg/kg (Borowska and Brzóska 2015). Excessive exposure to As may also occur after use of some traditional remedies or ayurvedic medication from several Asian countries (Baker et al. 2018).

Toxicity

Signs or symptoms of As toxicity depend on the route of exposure, and whether this exposure is acute, subacute or chronic (Baker et al. 2018). The toxicity brought about by As is also dependent on the biological species, age, sex, individual sensitivity, genetics and nutritional factors (Hong et al. 2014). But mostly, the physicochemical properties of many As species are important determinants for their potential toxic effects and therefore, the use of analytical methods which allows to speciate As is required for a reliable toxicological estimation (Hata et al. 2003, Hughes 2006). As mentioned, it is traditionally accepted that iAs species are more toxic than organoarsenicals, and that iAs^{3+} is two to ten times more toxic than iAs^{5+} (Hong et al. 2014, Baker et al. 2018). On absorption into organisms, the reducing environment in tissues will result in conversion of iAs^{5+} to iAs^{3+}, which can be methylated to organic forms (substantially less toxic than iAs^{3+}), MMA and DMA (Hata et al. 2003). Among organoarsenicals, arsenobetaine is the major arsenical in most species, which humans cannot metabolize but is considered to have negligible toxicity (Hata et al. 2003, Hong et al. 2014, Erickson et al. 2019).

Acute toxicity

The immediate symptoms of acute As poisoning are vomiting, abdominal pain and diarrhoea; these signs are followed by numbness and tingling of the extremities, muscle cramping and death, in extreme cases (WHO 2018). For humans the lethal dose of orally administered As trioxide has been estimated as 70 to 180 mg and 200 to 300 mg (1 to 5 mg/kg b.w.); outstandingly, survival has been reported after doses as high as 10 g (DeSesso et al. 1998).

Chronic toxicity

The effects of chronic poisoning are very different from those found in acute poisoning and occur mostly due to environmental or occupational exposure, having a more insidious onset. The most specific overt sign of chronic iAs ingestion is skin or dermal effects, usually hyperpigmentation and hyperkeratosis with bilateral thickening of the palms and soles (Baker et al. 2018). This occurs after a minimum exposure of approximately 5 yr, these could progress from skin lesions to nonmelanoma skin cancers including squamous cell carcinoma, basal cell carcinoma or Bowen disease (Baker et al. 2018, WHO 2018).

Lung cancer has been associated with inhalation exposures by workers in smelting and pesticide , chronically exposed to As, whereas incident bladder or prostate cancer in exposures to high concentrations of As through drinking water has been verified (Hong et al. 2014, Baker et al. 2018). Very few studies investigated the association between As exposure and leukaemia, but some reported increased incidence of the disease in areas where drinking water was contaminated with the metalloid (Hong et al. 2014). More limited evidence has associated As exposure through drinking water with other cancers, of the kidney, angiosarcoma of the liver and other liver cancers (Baker et al. 2018). Exposure to As is also associated with infant mortality and impacts on child health; exposures *in utero* and in early childhood has been linked to increased mortality in young adults due to multiple cancers, lung disease, heart attacks and kidney failure (WHO 2018). Exposure to elevated levels of As *in utero* was also associated with elevated bladder and lung cancer rates in adulthood, despite As exposures that ended as much as 40 yr earlier (Baker et al. 2018). The International Agency for Research on Cancer (IARC) classifies iAs as carcinogenic to humans (Group 1) (usually taking more than 10 yr to develop) and consider that As in drinking-water is carcinogenic to humans (Chen et al. 2004, Baker et al. 2018, WHO 2018, ACS 2021). Both MMA and DMA are classified as possible human carcinogens (IARC Group 2B), whereas arsenobetaine and other organoarsenicals have not been categorized as carcinogens (Group 3) (Hong et al. 2014, Baker et al. 2018). So far, the precise mechanisms that acute or chronic exposure to As conducts to induce cancer are not yet understood (Hong et al. 2014).

The negative impacts of As exposure on cognitive development, intelligence and memory are also demonstrated (WHO 2018). When accumulated in the body during childhood could induce neurobehavioral abnormalities during puberty, and neurobehavioral changes as an adult. Neuritis caused by As is another recognized complication of As toxicity, which affect the sensory functions of the peripheral nerves (Hong et al. 2014).

Many other adverse health effects can be associated with long-term ingestion of iAs: diabetes, pulmonary disease, renal system effects, enlarged liver, non-cirrhotic portal hypertension, bone marrow suppression (leukopenia and anaemia), hypertension and cardiovascular disease (As-induced myocardial infarction is a significant cause of excess mortality) (Baker et al. 2018, WHO 2018). The obliterative peripheral vascular disease known as Blackfoot disease, is also seen in populations exposed to iAs in drinking water from wells in southwestern Taiwan (Baker et al. 2018). A meta-analysis of adverse pregnancy outcomes following exposure to high levels of As in water (50 µg/L or greater) showed additionally, an increased risk of spontaneous abortion and stillbirth (Baker et al. 2018).

Biomarkers

As exposures are still a contemporary and worldwide health problem for which the assessment of human exposures, susceptibility and effects, through biomarkers (BM), is an undoubtable need. It should be noted when analyzing As exposure, it is useful to distinguish between As species, since they differ in their origin and toxicity (Orloff et al. 2009)

Biomarkers of exposure

BMs of exposure to As include the analysis of urine, blood, hair or nails, which are indicative of systemic absorption after exposure; functional BMs for the metalloid has also been proposed (Hughes 2006).

Arsenic levels

Blood As is cleared from the blood within a few hours after it is absorbed, which makes the analysis of blood for As levels the best suited for recent high-dose exposures (ATSDR 2007, Hong et al. 2014). Even though blood may have steady-state levels in chronic exposure to iAs, some authors believe that blood As levels may not be a reliable BM of As exposure as is cleared very rapidly; this is particularly true for low levels of iAs (ATSDR 2007). There is a poor relationship between As levels in drinking water and blood as the experimental exposure to 200 μg/L of iAs in water, already resulted in blood As levels increasing d to 8 μg/L (different from total urinary As, which increased to 261 μg/L) (Hong et al. 2014). Background As blood levels in humans range from 0.5 to 2 μg/L (ATSDR 2007).

Urine The As absorbed through the respiratory or digestive systems is converted within a few hours into more than five different metabolites. They are then primarily expelled through the urine with a half-life of approximately 4 d in humans; for this reason, urinary As is considered to reflect respiratory and digestive exposures (Hong et al. 2014). Hence, the most common and useful general indicator of As exposure is the analysis of total As in urine. Since after absorption iAs is methylated in the body to MMA and DMA, the sum of urinary iAs^{3+}, iAs^{5+}, MMA, DMA and organic forms such as arsenobetaine, are often reported by laboratories as the total As level (Baker et al. 2018). However, the separation and quantification of trivalent and pentavalent iAs, as well as of MMA and DMA, is believed to be the most accurate indicator (Farmer and Johnson 1990, ATSDR 2007, Hughes 2006, Hong et a. 2014). A study of total urine As and speciated As led to finding that MMA, DMA and arsenobetaine were the major contributors to the total urine As level, implying that exposures to iAs can be overestimated when only total As is determined (Baker et al. 2018). A well-known condition is when seafood ingestion is suspected to have occurred before the analysis; due to the high orgnoarsenicals content in those foods, subjects should refrain from ingesting seafood 2 to 3 d before collection of urine if only total urinary As is to be determined (ATSDR 2007).

Other concerns regarding urinary As analysis are sample collection times (i.e., if 24 hr, spot or first morning void urines). The collection of urine for 24 hr is sometimes difficult because of quality assurance logistics; as urinary As does not appear to vary over time, spot collection or first morning void may be used (ATSDR 2007, Rivera-Núñez et al. 2010). Additional concerns are whether to adjust for urinary concentration, adjusting for creatinine to account for the dilution or concentration of the urine. Many laboratories report 24-hr urine As level as both μg/L (not corrected for creatinine) and μg/g creatinine (corrected for creatinine). Adjusting As concentration for creatinine levels is less important in 24-hr urine specimens, as urine concentration is typically a 24-hr average and not as variable as could occur with a spot specimen; for spot urine As samples, creatinine correction is probably be most helpful when comparing serial As levels over time in a single individual patient. However, in other cases, as urine creatinine concentration is significantly associated with age, sex, race/ ethnicity and body mass index, As levels reported as μg/g creatinine may reflect changes to these demographic factors instead of actual differences in As exposure. Creatinine adjustment of urinary As in children or malnourished individuals, which eliminate relatively less creatinine, may also yield a value that seems extremely high (Baker et al. 2018).

Background levels of urinary As range from 5 to 50 μg/L, with reports of upper 95th percentiles exhibiting values of 65.4 μg/L (or 50.2 μg/g creatinine) (ATSDR 2007, Hong et al. 2014, Baker et al. 2018). Confounding factors such as sex, age, smoking and dietary factors must be considered in exposure assessments (ATSDR 2007, Hong et al. 2014). In addition, laboratory reports of minimally elevated urine As levels cannot be interpreted by just comparing results to a "normal range", to determine whether As toxicity is present (particularly in the case of chronic toxicity) as it is necessary to look for signs of actual As toxicity. Even so in acutely symptomatic As toxicity total urine As is typically greater than 1,000 μg/L (Baker et al. 2018).

The levels of urinary iAs have been correlated with external doses found in soil, air or water, although for soil these correlations were less consistent and only found for high levels of iAs. For air, urinary concentrations of 50–200 μg/L in workers of smelting were already determined to be equivalent to airborne levels of 1-7-53 μg/m³; in another study a nonlinear association of higher air As concentrations with urinary As was also attributed to factors such as the use of respirators by workers (Farmer and Johnson 1990, ATSDR 2007). The American Conference of Governmental Industrial Hygienists (ACGIH) recommend a biologic exposure determinant value of 35 μg/L in urine, for occupational exposure to iAs. This guideline for potential workplace health hazards includes excreted iAs plus its methylated metabolites (ATSDR 2007). For drinking water, a study in Korea reported average urinary As concentration of 7.10 μg/L (males, 7.63 μg/L; females, 6.75 μg/L), with a similar work mentioning values of 8.47 μg/L. Such levels were much lower than the ones found in a high-risk area for As exposure through drinking water in Bangladesh, where the average total urine As concentration were 20.77 μg/L) (ATSDR 2007).

Hair and nails Collection of hair and nails provides the advantage of being less invasive than collecting blood, making more subjects willing to participate in As exposure studies (Hughes 2006). The levels of As in hair and nails are referred as

BMs of exposure due to the accumulation of the metalloid in these tissues, thought to be due to the binding of iAs^{3+} to sulphydryl groups in keratin (Hong et al. 2014). As hair and nails grow slowly, the analysis of As concentrations in these matrices may give an indication of past As exposure.

The predominant form of As in hair is iAs followed by small amounts of dimethylated As; arsenobetaine does not accumulate in hair and thus consumption of seafood should not be a complicating factor in As hair analysis (Hong et al. 2014). Moreover, significant correlation between As in hair and total urinary As is reported (Katz 2019). Background As levels in hair are < 1 μg/g, which increase in populations exposed to elevated levels of iAs in air, soil or drinking water. An increase of 10 μg/L of As in drinking water is reported as corresponding to an augment of 0.1 μg/g in hair As (Hong et al. 2014).

Background As levels in nails range from < 1.5 to 7.7 μg/g and, in the same way as for hair, populations exposed to elevated levels of iAs from air, soil or drinking water have increased levels of As in their nails (Karagas et al. 2000, Hong et al. 2014). There are significant correlations between As in drinking water (> 1 μg/L) and toenails As levels, with a 10-fold increase of As in well water corresponding to a toenail As augment of about 2-fold (Karagas et al. 2000).

Some limitations exist regarding both hair and nail As levels as BMs, because As from external sources may bind to these matrices which can complicate exposure analysis. For someone who consumes and bathes in water or is in contact with soil with elevated levels of iAs, As from these external sources are most likely be detected in the samples. Even though washing procedures in the laboratory have developed, the possibility exists that some As coming from internal sources may be also removed (Hughes 2006).

Other biomarkers

More than 200 enzyme systems can be affected by As poisoning, with some of them having the potential to use as BMs of As exposure. Haematological system alteration is one of the consequences of As toxicity, which is among the most promising BMs that range the of effects which As causes to the group of enzymes responsible for heme biosynthesis and degradation (Fig. 1). As can inhibit two enzymes of the heme biosynthesis pathway, coproporphyrinogen oxidase and heme synthase, and that these events can lead to an accumulation of some heme precursors, which are uroporphyrins and coproporphyrins (Ng et al. 2005, Kumari et al. 2018). Due to ensuing changes in their urinary excretion, the urinary levels of these porphyrins are proposed as BMs for As exposure. Increased ratios of coproporphyrin/uroporphyrin and coproporphyrin-III/coproporphyrin-I are demonstrated experimentally, and in humans, for subjects exposed to burning As-bearing coal and for individuals chronically exposed to As via drinking water (Ng et al. 2005).

Skin hyperpigmentation and palmoplantar hyperkeratosis were also proposed for long-term (many years) internal dose BMs (An et al. 2008). Significant associations between As in well water and the prevalence of both skin alterations among residents of an As-exposed area have already been determined. The authors suggested that these As-induced skin lesions could be considered a long-term BM of cumulative exposure to As (Chen et al. 2005).

An exposure marker for biologically effective dose of As is suggested taking into consideration that ingested or inhaled iAs is methylated into MMA and DMA. A study examining the association between As methylation and skin cancer risk, revealed that the risk for developing the disease was around three-fold for those who had a high percentage of urinary MMA in total iAs metabolites, as compared with those who had a low percentage (Chen et al. 2005). In this way, the urinary percentage of MMA in total metabolites of iAs could be considered as an exposure marker for biologically effective dose (An et al. 2008). Other researchers studied the urinary As methylation index, defined as the ratio of DMA to MMA, finding that subjects who had a high cumulative As exposure and low methylation index, had a substantially increased risk of skin cancer and bladder cancer, peripheral vascular diseases, hypertension and atherosclerosis (Chen et al. 2005, Hsieh et al. 2011).

Biomarkers of susceptibility

BMs of susceptibility to As-induced health hazards include genetic polymorphisms of enzymes involved in xenobiotic metabolism, DNA repair and serum levels of carotenoids (An et al. 2008).

Enzymes of xenobiotic metabolism

Genetic polymorphisms for several of the enzymes involved in As metabolism have been proposed as being in part responsible for the individual variation in sensitivity to As health effects (Hughes 2006, Baker et al. 2018). Glutathione S-transferase (GST) P1 catalyze the formation of a bond between a thiol in an important antioxidant substrate, glutathione (GSH), and iAs^{3+}. For people exposed to As the prevalence of carotid atherosclerosis is significantly associated with the genetic polymorphism of GST P1; the risk of developing the disease is around two-fold for those exhibiting variant genotypes (VV/WV) of GST P1 compared with those with wild genotype (WW). In fact, epidemiological studies have demonstrated a correlation between environmental or occupational As exposure and a risk of vascular diseases related to atherosclerosis (Simeonova and Luster 2004). It is also known that in As-induced skin cancer patients, the risk of the disease is significantly associated with the combination of three GST genotypes; those who have at least one null or variant genotype of GST M1, T1 or P1 have skin cancer risk five-fold higher, compared with those exhibiting wild genotypes of all three GSTs (Chen et al. 2005).

In turn microsomal epoxide hydrolase (EPHX1) is another enzyme that plays a role in the metabolism of environmental pollutants, detoxification and bioactivation, functioning to regulate the oxidation status of a wide range of xenobiotic and substrates (Farin et al. 2001, Taha et al. 2021). While gene polymorphisms of EPHX1 could be associated with variations in enzyme activity, the risk of As-induced skin cancer was found to be significantly associated with the activity of EPHX1; the relative risk of skin cancer is around three-fold higher for genotypes TC/CC compared with persons with the TT genotype (Chen et al. 2005, Taha et al. 2021).

DNA repair enzymes

Genetic polymorphisms of certain DNA repair enzymes, which work in concert to protect the genome of the cell from carcinogenic exposures, are additional potential BMs of susceptibility to As. The DNA repair Xeroderma Pig-mentosum complementation group D (XPD) is a Nucleotide Excision Repair (NER) helicase, providing NER the major way for the removal of bulky DNA lesions; several important Single Nucleotide Polymorphisms (SNP) have been identified in the XPD locus (McCarty et al. 2007). Since it was hypothesized that As inhibits the repair of UV-induced DNA damage through inhibition of the NER path , some studies were conducted in As-induced skin cancer patients. A significant association was found between genetic polymorphisms of the XPD exon 6 and the relative risk of developing skin cancer, which was around two-fold higher for AA/AC genotypes, compared with CC genotype of XPD. Similar studies suggested that differences in NER phenotypes could modulate genetic susceptibility to skin lesions associated with increased risk for skin cancer (Chen et al. 2005, McCarty et al. 2007).

Carotene

Different carotenoid concentrations in the body could occur due to differences in absorption, assimilation and metabolism of carotenoids, which may be attributed to genetic factors; specific genetic variants that influence circulating carotenoid concentrations have been identified (Farook et al. 2017). Increased risk of As-induced skin cancer seems to be significantly associated with decreased serum-carotene levels, in the same way as a significant reverse dose-response relationship with As-induced ischemic heart disease was observed for serum levels of carotene (Chen et al. 2005).

Biomarkers of effect

BM candidates for early biological effects induced by iAs include molecular markers such as blood levels of reactive oxidants, antioxidant capacity and genetic expression of inflammatory molecules, as well as cytogenetic changes and altered structures and function of target organs.

Molecular biomarkers

People living in an As-contaminated area are known to exhibit blood levels of As positively associated with plasma levels of reactive oxidants, and negatively associated with antioxidant capability (Chen et al. 2005). Regarding gene expression of inflammatory molecules, investigations were performed on molecular targets of inflammatory routes possibly involved in As-associated atherosclerosis, aimed to identify genes with differential expression in As-exposed apparently healthy individuals. Upregulations in gene expression of a great number of inflammatory molecules were found in circulating lymphocytes; those encompassed interleukin-1h, interleukin-6, chemokine C–C motif ligand 2/monocyte chemotactic protein-1 (CCL2/MCP1), chemokine C–X–C motif ligand 1/growth-related oncogenea, chemokine C–X–C motif ligand 2/growth-related oncogeneh, CD14 antigen and matrix metalloproteinase 1 (Wu et al. 2003).

Another BM candidate is the induction of Heme Oxygenase (HO) (Hughes 2006). Arsenate and arsenite are good inducers of rat hepatic and renal HO, while MMA or DMA, are not. This could suggest that HO enzyme induction should theoretically be an excellent BM of biologically active arsenite concentration (Kitchin et al. 1999). In the case of HO1, the induction in human lymphoblastoid cells by arsenite occurs in a dose-related manner, but HO1 is also induced by cadmium or mercuric chloride. Therefore, circulating lymphocyte HO1 levels may be useful only in assessing the biological activity of As exposure *in vivo* under correctly controlled conditions of simultaneous urinalysis for As, cadmium and mercury (Menzel et al. 1998).

Cytogenetic biomarkers

Cytogenetic BMs such as Chromosomal Aberrations (CA), Sister Chromatid Exchanges (SCE) or micronuclei formation (MN) are traditionally used for biomonitoring genotoxic effects in humans. These cytogenetic endpoints are extensively used to identify the genotoxic potentials of toxic compounds and for cancer risk assessment.

It is acknowledged that As provokes chromosome breaks, and accordingly increased frequency of CA was found in the lymphocytes of psoriatic patients who were treated with As (Chen et al. 2009, Roy et al. 2018). In an As-polluted area, increased cancer risk was also significantly associated with elevated frequency of CA in peripheral lymphocytes; such a risk ranged from five to 12 for those who had the chromosome-type aberration than those without it (Chen et al. 2005).

SCE is a process whereby, during DNA replication, two sister chromatids break and rejoin with one another, physically exchanging regions of the parental strands in the duplicated chromosomes (Wilson and Thompson 2007). Following the exposure to DNA-damaging agents, their frequency increases substantially and so, they have frequently been used as an indicator of genotoxic effects (Eastmond 2014). This modification is reported in the lymphocytes of patients treated with As in the form of potassium arsenite, indicating it is a significant genotoxicity of the metalloid. Several other reports describe similar effects, such as in patients affected with As-induced Bowen's disease due to long-term exposure to As from drinking well water (Roy et al. 2018).

MNs are small and independent nuclei located in the cytoplasm appearing from the loss of one or more chromosomes due to mitotic errors or from chromosomal breakage due to the mis repair of DNA damage. While the genotoxic potential of As is currently acknowledged, curiously MMA^{3+} and DMA^{3+} are more genotoxic than iAs^{3+}. Several studies were also conducted to investigate whether MN could be used as a BM for As. Increased MN was detected among exposed individuals, using MN assays conducted with lymphocytes, although there was lack of correlation between the exposure level and the increase in MN; similar augments were found using other cells, such as buccal and urothelial cells. (Chen et al. 2005). In the light of the evidence collected so far, it is being proposed that MN can be used as an effective BM for the biological monitoring of As-exposed populations (Dong et al. 2019).

Altered structures or functions of target organs

Long-term exposure to ingested As has been documented to induce atherosclerotic diseases including peripheral vascular disease, ischemic heart disease and cerebral infarction in a dose-response relationship (Chen et al. 2005). In this view, associations between ingested As and electrocardiogram abnormality have also been detected. Three indices of chronic As poisoning (duration of consuming well water, average As level and cumulative As exposure) were found to be all dose-dependently associated with QT (time taken for ventricular depolarization and repolarization) prolongation and increased QT dispersion on electrocardiograms, which are parameters associated with increased all-cause and cardiovascular mortality (Chen et al. 2005).

Final Remarks

This chapter described that different chemicals species of As are present in the environment, and that pronounced differences exist in their potential for toxicity. Chronic exposures to most toxic forms (iAs) through drinking water is a problem for which there is a current and global apprehension. Among numerous As-induced health hazards, cancer incidences are largely associated with chronic intake of drinking water contaminated with metalloid. Several means exist to assess As exposures as being the most reliable indicators, total urinary As measures; changes in urinary excretion of heme precursors or urinary As methylation index that constitute additional ways of assessment. Several genetic polymorphisms modulating susceptibilities to As-adverse effects or BM candidates for early biological effects have been studied and proposed so far.

Major concerns are likely to prevent further exposure to As in affected communities. Such measures include the provision of a safe water supply for drinking, food preparation and irrigation of food crops. Several options to reduce levels of As in drinking-water should additionally be implemented in-field (WHO 2018).

References

[ACS] American Cancer Society, Arsenic and Cancer Risk, 2021. New York, USA.

An, Y., L. Yin, S. Wang, Z. Wang et al. 2008. An overview on biomarkers of arsenic-induced health hazardsan. Wei Sheng Yan Jiu. 37(2): 237–241.

[ATSDR] Agency for Toxic Substances and Disease Registry. 2007. Toxicological Profile for Arsenic. Agency for Toxic Substances and Disease Registry, U.S. Department of Health and Human Services, Public Health Service, Atlanta, USA.

Baker, B.A., V.A. Cassano and C. Murray. 2018. Arsenic exposure, assessment, toxicity, diagnosis, and management guidance for occupational and environmental physicians. J. Occup. Environ. Med. 60(12): e634–e639.

Borowska, S. and M.M. Brzóska. 2015. Metals in cosmetics: implications for human health. J. Appl. Toxicol. 35(6): 551–572.

Chen, C.J., L.I. Hsu, C.H. Wang, W.L. Shih, Y.H. Hsu, M.P. Tseng et al. 2005. Biomarkers of exposure, effect, and susceptibility of arsenic-induced health hazards in Taiwan. Toxicol. Appl. Pharmacol. 206(2): 198–206.

Chen, C.L., L.I. Hsu, H.Y. Chiou, Y.M. Hsueh, S.Y. Chen, M.M. Wu et al. 2004. Ingested arsenic, cigarette smoking, and lung cancer risk: a follow-up study in arseniasis-endemic areas in Taiwan. JAMA. 292(24): 2984–90.

Chen, Y., F. Parvez, M. Gamble M., T. Islam, A. Ahmed, M. Argos et al. 2009. Arsenic exposure at low-to-moderate levels and skin lesions, arsenic metabolism, neurological functions, and biomarkers for respiratory and cardiovascular diseases: Review of recent findings from the Health Effects of Arsenic Longitudinal Study (HEALS) in Bangladesh. Toxicol. Appl. Pharmacol. 239(2): 184–192.

Chung, J.Y., S.D. Yu and Y.S. Hong. 2014. Environmental source of arsenic exposure. J. Prev. Med. Public Health. 47(5): 253–257.

DeSesso, J.M., C.F. Jacobson, A.R. Scialli, C.H Farr and J.F. Holson. 1998. An assessment of the developmental toxicity of inorganic arsenic. Reprod. Toxicol. 12(4): 385–433.

Dong, J., J.-Q. Wang, Q. Qian, G.-C. Li, D.-Q. Yang and C. Jiang. 2019. Micronucleus assay for monitoring the genotoxic effects of arsenic in human populations: A systematic review of the literature and meta-analysis. Mutat. Res. Rev. Mutat. Res. 780: 1–10.

Eastmond, D.A. 2014. Sister chromatid exchanges. pp. 276–277. *In*: P. Wexler [ed.]. Encyclopedia of Toxicology. Academic Press, Cambridge, USA.

Erickson, R.J., D.R. Mount, T.L. Highland, J.R. Hockett, D.J. Hoff, C.T. Jenson et al. 2019. The effects of arsenic speciation on accumulation and toxicity of dietborne arsenic exposures to rainbow trout. Aquat. Toxicol. 210: 227–241.

[FAO/WHO] Food and Agriculture Organization/World Health Organization. 1999 Position paper on arsenic (Prepared by Denmark). Joint FAO/WHO standards programme Codex committee on food additives and contaminants, Rome, Italy.

Farin, F.M., P. Janssen, S. Quigley, D. Abbott, C. Hassett, T. Smith-Weller et al. 2001. Genetic polymorphisms of microsomal and soluble epoxide hydrolase and the risk of Parkinson's disease. Pharmacogenetics 11(8): 703–708.

Farmer, J.G. and L.R. Johnson. 1990. Assessment of occupational exposure to inorganic arsenic based on urinary concentrations and speciation of arsenic. Br. J. Ind. Med. 47(5): 342–348.

Farook, V.S., L. Reddivari, S. Mummidi, S. Puppala, R. Arya, J.C. Lopez-Alvarenga et al. 2017. Genetics of serum carotenoid concentrations and their correlation with obesity-related traits in Mexican American children. Am. J. Clin. Nutr. 106: 52–58.

[FDA] Food and Drug Administration, Arsenic in Food and Dietary Supplements, 2020. FDA, New Hampshire, USA.

[FSANZ] Food Standards Australia New Zealand, Arsenic, 2020. Wellington, New Zealand.

Hata, N., I. Kasahara, S. Taguchi and K. Goto. 2003. Arsenic, properties and determination pp. 304–311. *In*: B. Caballero [ed.]. Encyclopedia of Food Sciences and Nutrition. Academic Press, Cambridge, USA.

Hong, Y.S., K.H. Song and J.Y. Chung. 2014. Health effects of chronic arsenic exposure. J. Prev. Med. Public Health. 47(5): 245–252.

Hsieh, Y.C., L.M. Lien, W.T. Chung, F.I. Hsieh, P.F. Hsieh, M.M. Wu et al. 2011. Significantly increased risk of carotid atherosclerosis with arsenic exposure and polymorphisms in arsenic metabolism genes. Environ. Res. 111(6): 804–10.

Hughes, M.F. 2006. Biomarkers of exposure: a case study with inorganic arsenic. Environ. Health Perspect. 114(11): 1790–1796.

Kaise, T., Y. Horiguchi, S. Fukui, K. Shiomi, M. Chino and T. Kikuchi. 1992. Acute toxicity and metabolism of arsenocholine in mice. Appl. Organomet. Chem. 6(4): 369–373.

Karagas, M.R., T.D. Tosteson, J. Blum, B. Klaue, J.E. Weiss, V. Stannard et al. 2000. Measurement of low levels of arsenic exposure: a comparison of water and toenail concentrations. Am. J. Epidemiol. 152(1): 84–90.

Katz, S.A. 2019. On the use of hair analysis for assessing arsenic intoxication. Int. J. Environ. Res. Public Health. 16(6): 977.

Kitchin, K.T., L.M. Del Razo, J.L. Brown, W.L. Anderson and E.M. Kenyon. 1999. An integrated pharmacokinetic and pharmacodynamic study of arsenite action. 1. Heme oxygenase induction in rats. Teratog. Carcinog. Mutagen. 19(6): 385–402.

Kumari, A. 2018. Heme synthesis. pp. 33–36. *In*: A. Kumari [ed.]. Sweet Biochemistry. Academic Press, Cambridge, USA.

McCarty, K.M., T.J. Smith, W. Zhou, E. Gonzalez, Q. Quamruzzaman, M. Rahman et al. 2007. Polymorphisms in XPD (Asp312Asn and Lys751Gln) genes, sunburn and arsenic-related skin lesions. Carcinogenesis 28(8): 1697–1702.

Menzel, D.B., R.E. Rasmussen, E. Lee, D.M. Meacher, B. Said, H. Hamadeh et al. 1998. Human lymphocyte heme oxygenase 1 as a response biomarker to inorganic arsenic. Biochem. Biophys. Res. Commun. 250(3): 653–656.

Molin, M., S.M. Ulven, H.M. Meltzer and J. Alexander. 2015. Arsenic in the human food chain, biotransformation and toxicology-Review focusing on seafood arsenic. J. Trace Elem. Med. Biol. 31: 249–59.

Ng, J.C., J.P. Wang, B. Zheng, C. Zhai, R. Maddalena, F. Liu et al. 2005. Urinary porphyrins as biomarkers for arsenic exposure among susceptible populations in Guizhou province, China. Toxicol. Appl. Pharmacol. 206(2): 176–84.

[NIOSH] National Institute for Occupational Safety and Health, Arsenic, 2019. Centers for Disease Control and Prevention (CDC), Atlanta, USA.

Orloff, K., K. Mistry and S. Metcalf. 2009. Biomonitoring for environmental exposures to arsenic. J. Toxicol. Environ. Health B Crit. Rev. 12(7): 509–524.

[PAHO/WHO] Pan American Health Organization/World Health Organization, Arsenic, 2021. WHO, Geneva, Switzerland.

Reid, M.S., K.S. Hoy, J.R.M. Schofield, J.S. Uppal, Y. Lin, X. Lu et al. 2020. Arsenic speciation analysis: A review with an emphasis on chromatographic separations. Trends Anal. Chem. 123: 115770.

Rivera-Núñez, Z., J.R. Meliker, A.M. Linder and J.O. Nriagu. 2010. Reliability of spot urine samples in assessing arsenic exposure. Int. J. Hyg. Environ. Health. 213(4): 259–64.

Roy, J.S., D. Chatterjee, N. Das and A.K. Giri. 2018. Substantial evidences indicate that inorganic arsenic is a genotoxic carcinogen: A review. Toxicol. Res. 34(4): 311–324.

Simeonova, P.P. and M.I. Luster. 2004. Arsenic and atherosclerosis. Toxicol. Appl. Pharmacol. 198(3): 444–9.

Taha, M.M., E.M. Shahy and H. Mahdy-abdallah. 2021. Alteration in antioxidant status in slow and fast alleles of EPHX1 gene polymorphisms among wood workers. Environ. Sci. Poll. Res. 36: 49678–49684.

[WHO] World Health Organization, Arsenic. 2018. WHO, Geneva, Switzerland.

Wilson, D.M. 3rd and L.H. Thompson. 2007. Molecular mechanisms of sister-chromatid exchange. Mutat. Res. 616(1-2): 11–23.

Wu, M.M., H.Y. Chiou, I.C. Ho, C.J. Chen and T.C. Lee. 2003. Gene expression of inflammatory molecules in circulating lymphocytes from arsenic-exposed human subjects. Environ. Health Perspect 111(11): 1429–1438.

CHAPTER 6

Cadmium

Sources of Exposure

In its pure form, cadmium (Cd) is a silvery white, malleable metal with a bluish hue. The metal is found naturally on the Earth's crust and is relatively rare, ranking 67th in abundance among the 90 naturally occurring elements on Earth. Cd is primarily found in zinc-containing ores but may also be detected in lead and copper ores; the primary mineral form of the metal is greenockite or Cd sulfide (DTM 2021). The metal can be naturally released to the environment in a number of ways, such as by volcanic activity (both on land and in the deep sea), weathering and erosion and river transport (WHO 2019, DTM 2021). Human activity through mining, smelting and refining metal ores, particularly zinc, lead and copper, have played a significant role in creating concentrated sources of Cd. Nowadays their levels have greatly increased in environmental media, which has been relevant to populations exposure (WHO 2019, DTM 2021). Infact Cd is a "modern metal", since commercial Cd production started only at the beginning of the 20th century in applications such as electroplating in auto industries, production of pigments as Cd sulfate or Cd selenide, as a stabilizer for polyvinyl plastic and in rechargeable nickel-Cd batteries. The disposal and recycling of electronic and electrical waste (e-waste) has also been identified as a potentially important source of Cd emissions (WHO 2011, 2019). No efficient recycling way exists so far, for these Cd-containing products (Rahimzadeh et al. 2017).

Air

Cd is present in ambient air in the form of particles, relevant sources being fossil fuel burning, waste incineration and steel production; non-ferrous smelting and refining represent very important sources of emissions (HC 2008, DTM 2021). The guideline for Cd in air is 5 ng/m³ (annual average); examples of measured annual average concentrations of the metal are 1–3 ng/m³ in Germany, 0.7–2 ng/m³ in the Netherlands, and in Canada ambient outdoor levels ranging from 0.02 to 14.89 ng/m³ (median 0.04 ng/m³) and indoor concentrations of 0.005 to 1.30 ng/m³ (median 0.03 ng/m³) (HC 2008, WHO 2011, 2019). Such results reveal that non-occupational exposure to Cd from air is generally low, and it is estimated that for the general population Cd intakes from air are unlikely to exceed 0.8 µg/day and are not expected to pose hazards to health (ATSDR 2008, HC 2008, WHO 2011). In another way the levels of Cd in air are generally higher in the vicinity of metallurgical plants, with the annual average levels in Belgium being reported as 10–60 ng/m^3 (WHO 2011).

Workplace

Occupational exposures occur mainly through inhalation, that take place during mining and work with Cd containing ores, during manufacture of products containing Cd (e.g., paints), and during work, such as plating, soldering and welding (ATSDR 2008). Cd air levels in these conditions can be thousands times greater than in the general environment, as established by the Occupational Safety and Health Agency (OSHA) a Permissible Exposure Limit (PEL) of Cd fume or Cd oxide in the workplace of 0.1 mg/m^3, to control workers' exposure (ATSDR 2008, HC 2008). Minimal Risk Levels (MRL) of $3x10^{-5}$ mg/m^3 has also been obtained for acute-duration inhalation exposure to Cd ($\leq$ 14 d). No intermediate-duration inhalation MRL is decided, but an MRL of $1x10^{-5}$ mg/m^3 was determined for chronic-duration inhalation exposure ($\geq$ 1 yr) (ATSDR 2012).

Soils

Although the levels of Cd are generally low in soils, they vary with geology and soil types; there are reported values which are below the detection limit of the used analytical method (8.1 mg/kg). In some areas the atmospheric deposition of Cd exceeds its elimination, resulting in a gradual increase of Cd soil levels, which ultimately affects crops. In addition, the application of municipal sewage sludge to agricultural soil can be a significant source of Cd (WHO 2019). A critical Cd limit in light textured soils is considered as 5.33 mg/kg, in medium textured soils 6.33 mg/kg and in heavy textured soils 9.29 mg/kg (Sukarjo and Purbalisa 2019).

Water

Cd levels in drinking water can vary greatly depending on the geological formations surrounding the source water and on environmental factors affecting Cd mobility (HC 2008). As in soil , the solubility of Cd in water (via suspended or sediment-bound metals) is influenced to a large degree by acidity (HC 2008, WHO 2011). Anthropogenic sources include discharges from industrial facilities or sewage treatment plants or the use of phosphate fertilizers. Materials used in distribution and household plumbing systems may contain Cd as an impurity in zinc galvanized pipes, thus exhibiting an additional source of exposure to Cd via drinking water through the deterioration of pipes (HC 2008). The levels of Cd can be higher in areas supplied with soft water of low pH, since these factors could turn the water more corrosive in plumbing systems (WHO 2011). Concentrations of the metal in unpolluted natural waters are usually below 1 µg/L, whereas values from shallow wells in Sweden where the soil had been acidified may reach 5 µg/L. In Saudi Arabia mean concentrations of 1–26 µg/L were found in samples of potable water, some of which were taken from private wells or old corroded pipes; a maximum value of 100 µg/L had already been recorded in Rio Rimao, Peru (ATSDR 2012). Normally, Cd intake from drinking-water is usually less than 2 µg/d (ATSDR 2008). Different limit values for Cd in drinking water are decided by different agencies around the world. According with the Guidelines for Canadian Drinking Water Quality, a Maximum Acceptable Concentration (MAC) for Cd is 0.007 mg/L, while the US Environmental Protection Agency (EPA) or the European Union (EU) established a Maximum Contaminant Level (MCL) of 0.005 mg/L (HC 2008). Meanwhile, the World Health Organization (WHO) published a drinking-water quality guideline of 0.003 mg/L, the lowest settled value recommended by the Australian National Health and Medical Research Council (NHMRC), which is 0.002 mg/L (HC 2008, WHO 2019).

Food

Food is the main source of non-occupational exposure to Cd for the non-smoking general population, even though humans absorb only a small proportion of ingested Cd (1–10%) (ATSDR 2008, WHO 2011, 2019, HFS 2021). Cd reaches into food as once in soil, water can be taken up by certain crops and aquatic organisms, accumulating in the food chain. Other concerns appear from the fact that to achieve high quality crops, a large amount of Cd-containing phosphate fertilizers and sludge waste are applied; crops with increased concentrations of the metal may be consumed by animals grazing meat from animals grazing on these contaminated pastures (Grant et al. 1999, WHO 2011, Roberts 2014, Rahimzadeh et al. 2017). Dietary Cd daily intakes from food are estimated as 10–35 µg, with some authors mention that a typical dietary intake is about 30-50 µg/d ; in contaminated areas in Japan, daily intake was already determined in the range 150–250 µg (ATSDR 2008, WHO 2011).

Food groups that contribute the most for dietary Cd exposure are cereals and cereal products, vegetables, nuts and pulses, starchy roots or potatoes and meat and meat products. Due to a high consumption of cereals, nuts, oilseeds and pulses, vegetarians have a higher dietary Cd intake (EC 2021). A very appropriate source of Cd through the diet is rice, which absorbs the metal from the soil. A lifetime of eating this Cd-contaminated rice, namely in Japan, can lead to a serious kidney and bone disorder called "Itai-Itai" disease (ATSDR 2008). About 25 µg/kg w.w. of Cd can be found in cereals, while average Cd levels in fish are reported as 20 µg/kg w.w., and as the liver and kidneys of animals have concentrate Cd, levels of 10–100 and 100–1000 µg/kg, respectively, are described for these organs; the highest levels are found in shellfish (200–1000 µg/kg) (WHO 2011). The permitted level of Cd in rice is 0.2 mg/kg based on the Food and Agriculture Organization/World Health Organization (FAO/WHO) rules. The Hong Kong Centre for Food Safety establishes a limit value of 0.1 mg/kg in vegetables and cereals, 2 mg/kg for fish, crab meat, oysters, prawns and shrimps, and 0.2 mg/kg for meat (HK CFS 2017, Rahimzadeh et al. 2017).

The Joint FAO/WHO Expert Committee on Food Additives (JECFA) also concluded that the total intake of Cd should not exceed 1 µg/kg b.w./d, and that a

Provisional Tolerable Weekly Intake (PTWI) should be set as 7 µg/kg b.w. Due to the Cd's exceptionally biological long half-life (10-30 yr), this value was withdrawn, and a provisional tolerable monthly intake (PTMI) of 25 µg/kg b.w. was established (HK CFS 2017, WHO 2011, 2019). Later in in 2012 the Agency for Toxic Substances and Disease Registry (ATSDR) created a minimum risk level (MRL) of 5 µg/kg bw/d to protect children from developmental harm (Neltler 2019). Other values exist such as a MRLs of 5×10^{-4} mg/kg b.w./d for intermediate-duration oral exposure (15–364 d), and 1×10^{-4} mg/kg b.w./d for chronic-duration oral exposure ($\geq$ one yr) (ATSDR 2012).

Consumer products

Tobacco smoking is an important source of exposure to Cd, since the tobacco plant takes up Cd fervently from the environment. There is about 2.0 µg of Cd in a cigarette, of which nearly 2–10% is transferred to cigarette smoke. Smokers typically have Cd blood (and body burdens) more than double than those of nonsmokers, and it is also known that cigarette smoking can cause significant increases in the concentration of Cd in the kidney, which is the main target organ for Cd toxicity (ATSDR 2008, Siedel 2011, WHO 2019, Rahimzadeh et al. 2017). An additional motive for apprehension is the fact that cigarette smoking increases Cd concentrations inside houses (WHO 2011).

Inexpensive jewelry, toys and plastics can constitute significant sources of exposure to Cd, especially for children (WHO 2019). Many countries have moved to restrict or ban Cd in such products, namely in Canada, where regulation for Cd in children's jewelry was developed given the potential of children's exposure from ingestion of Cd-containing jewelry (HC 2008, WHO 2019).

Toxicity

Cd is a metal of considerable toxicity with a destructive impact on most organ systems, depending on these toxic impacts on the body burden of Cd. The metal induces tissue injury through creating oxidative stress, epigenetic changes in DNA expression, inhibition or upregulation of transport pathways as well as competitive interference with the physiologic action of zinc or magnesium, inhibition of heme synthesis and impairment of mitochondrial function potentially inducing apoptosis (Bernhoft 2013). Oxidative stress mechanisms are showed in Fig. 4.

Acute toxicity

Cd compounds have a moderate acute oral toxicity the oral LD50 values for mice and rats being from 60 to over 5000 mg/kg b.w. The major effects of Cd acute toxicity are desquamation of epithelium of the gastrointestinal tract, necrosis of the gastric and intestinal mucosa and dystrophic changes of liver, heart and kidneys (WHO 2011). Unfortunately, unpredictable accidental exposure to Cd still occurs. Workers during rescue and recovery activities at the World Trade Center after the September

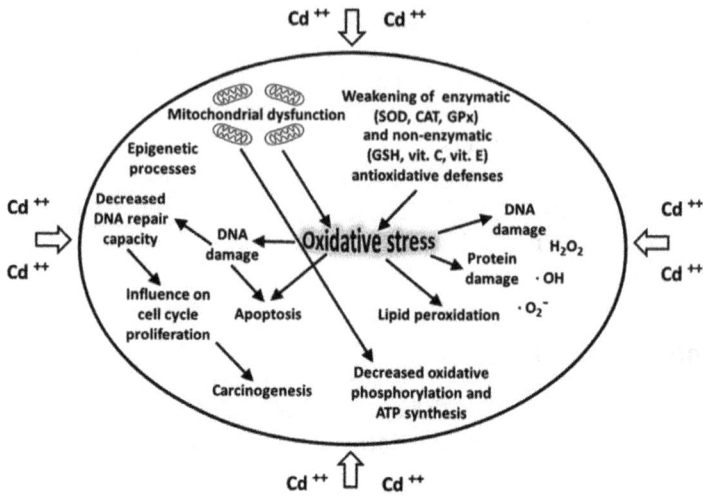

Figure 4. Cadmium-induced oxidative stress mechanisms (Genchi et al. 2020).

11 attacks, when using a torch cutter were overexposed to Cd. A fuel oil leak due to the accidental sinking of the oil tanker, Prestige, was cleaned up by French and Spanish troops and local personnel; cytogenetic and endocrine effects were found in those exposed to the fuel oil which contained toxic metals, which included Cd. Acute lung injury and respiratory distress syndrome occurred among employees, who were accidentally exposed to metal fumes containing Cd, while working in the local silver jewelry industry and in a galvanized steel industry (Miura 2009).

High inhalation exposure to Cd oxide fumes results in acute pneumonitis with pulmonary oedema, which may be lethal (WHO 2019). It has also been seen that acute high-level human inhalation exposure, typically through flame cutting or brazing Cd-containing materials, is associated with diffuse alveolar damage (Rahimzadeh et al. 2017).

Chronic toxicity

Chronic exposure to Cd leads to mainly toxic effects on the kidneys, in the skeletal system and in the respiratory system; Cd is also classified as a human carcinogen (WHO 2019).

Effects on the kidneys

Both occupational and environmental exposures to Cd have been implicated in renal dysfunction (Rahimzadeh et al. 2017). The kidney is the critical target organ following long-term exposure to Cd, as the metal accumulates primarily here. Such an accumulation may lead to renal tubular dysfunction, resulting in increased excretion of low-molecular-weight proteins in the urine. While a modest increase in urinary excretion of these proteins is generally reversible, an increase of more than an order of magnitude indicates irreversible tubular dysfunction, which progresses to overt nephropathy (Rahimzadeh et al. 2017, WHO 2019). Cellular damage can affect functional integrity resulting in complications such as glucosuria, aminoaciduria, hyperphosphaturia, hypercalciuria, polyuria and decreased buffering capacity (Rahimzadeh et al. 2017).

Effects in the bones

Cd can directly interact with bone cells, diminish mineralization, inhibit procollagen C-proteinases and collagen production and interfere with vitamin D3 metabolism. Clinical findings associated with Cd-induced osteoporosis include pain, physical impairment and decreased quality of life. Besides, decreased bone density leads to increased risk for bone fractures as severe skeletal decalcification may be observed as well. The earlier mentioned "Itai-itai" disease is the most severe form of chronic Cd intemperance, being characterized by osteomalacia with severe bone pain and is also associated with renal tubular dysfunction. The first recognition of the disease occurred in Japan affecting mainly women residing in rice farming areas irrigated with contaminated water (Rahimzadeh et al. 2017). In this area, the Cd levels in unpolished rice samples ranged from 0.00 to 5.20 mg/kg, with a mean value of 0.37 mg/kg, which was 2.5 times higher than the values found for samples from non-polluted areas (Aoshima 2016, Nishijo et al. 2017).

Effects in the lungs

Extensive data has established that Cd exposure from cigarette smoking and occupational sources, causes chronic lung toxicities (Hu et al. 2019, WHO 2019). Long-term high-level occupational exposure is associated with chronic obstructive pulmonary disease. It is also possible that interactions between the natural Cd content of cigarettes and environmental Cd by obstructive lung disease caused by smoking (Moitra et al. 2013, WHO 2019). In experimental animal models repeated Cd inhalation causes emphysema, while human data has shown a possible involvement of Cd in the same disorder (Moitra et al. 2013, Rahimzadeh et al. 2017). Much less is known about the impact of low-level chronic Cd exposure in nonsmokers, mainly obtained from dietary intake (Hu et al. 2019).

Effects in the nervous system

Cd can be up taken from the nasal mucosa or olfactory routes into the peripheral and central neurons where the production of free radicals may increase and cellular defenses against oxidation could decrease, causing cellular damage and lipid peroxidation in the brain. Olfactory dysfunction, neurobehavioral defects in attention, disorders in psychomotor activity and memory are associated with Cd exposure; Cd poisoning may also lead to neurodegenerative disorders, such as Parkinson, Alzheimer and Huntington's diseases (Leal et al. 2012, Rahimzadeh et al. 2017). Mechanisms underlying Cd neurotoxicity are not completely understood, but effects on neurotransmitters, oxidative damage, interaction with other metals (such as cobalt and zinc), increased blood brain barrier permeability and epigenetic modification may all be underlying mechanisms (Wang and Du 2013).

Carcinogenicity

The International Agency for Research on Cancer (IARC) has classified Cd and Cd compounds as carcinogenic to humans (Group 1), meaning that there is sufficient evidence for their carcinogenicity in humans (WHO 2019). The metal is a lung carcinogen and an inducer of prostatic or renal cancers. Some reports suggested that Cd may also be involved in malignancies of the liver, hematopoietic system, bladder, pancreas and stomach, and that may also be a risk factor for breast cancer. Cellular and molecular mechanisms implicating Cd carcinogenicity include the activation of proto-oncogenes, inactivation of tumor suppressor genes, disruption of cell adhesion and inhibition of DNA repair. Cd exposure can additionally affect cell proliferation, differentiation, apoptosis, cell signaling and other cellular activities, which could directly or indirectly bear on carcinogenesis (Rahimzadeh et al. 2017).

Other effects

Other systems affected by Cd include the reproductive system, with induced decrease in density, volume and number of sperms, as well as diminished libido, fertility and serum testosterone. In females the function of the ovary and development of oocytes may also be inhibited (Rahimzadeh et al. 2017). Cd could act like an estrogen, mimicking the action of 17β-estradiol (E2) by forming high-affinity complexes with Estrogen Receptors (ER) (Miura 2009). The rate of spontaneous abortion may be increased and the rate of live births decrease due to exposure to Cd (Rahimzadeh et al. 2017).

The cardiovascular system *in vitro* studies have indicated the involvement of Cd in endothelial dysfunction as well as in carotid intima-media thickness. Moreover, the formation of atherosclerotic plaques is promoted *in vivo*. Epidemiologic studies had shown the association of Cd exposure with the risk of high blood pressure, while results exist supporting the hypothesis of Cd involvement in cardiovascular disease and myocardial infarction (Tellez-Plaza et al. 2013, Rahimzadeh et al. 2017).

Biomarkers

Cd exposures vary widely and there is very little information on whether during the last four decades such exposures are increasing, decreasing or unchanged. What is certainly known is that biomarker (BM) based risk assessment of exposures to Cd, including low-level exposures (when considering the long half-life of the metal in the body), are vital to provide a basis for preventive action (Nordberg et al. 2018).

Biomarkers of exposure

The levels of Cd in blood, urine, hair and nails samples are often determined in laboratory tests, with blood or urine Cd levels being the most frequently used BM of exposure (ATSDR 2012, Rahimzadeh et al. 2017). However, while possibly it might seem that the most direct way to monitor levels of Cd exposure would be to simply measure their blood or urinary concentrations, this question is largely complicated by the unique toxicokinetics of Cd in the body, where the tendency of Cd to be sequestered in organs such as the liver and kidneys is especially problematic (Prozialeck and Edwards 2010). Hence, monitoring and interpretation of these data are not as simple as one might expect (Prozialeck and Edwards 2010).

Cadmium levels in blood

Once Cd is absorbed into the bloodstream, whether from the lungs or from the gastrointestinal tract, it tends to concentrate in blood cells (mainly erythrocytes, but also leucocytes) and only a small percentage ($< 10\%$) remains in the plasma. For this reason, the monitoring of blood samples for levels of Cd exposure typically involves the analysis of the whole blood (Prozialeck and Edwards 2010, Borné et al. 2019). The half-life of Cd in blood is short (3 to 4 mon), and thus blood Cd levels are indicative of recent exposure rather than providing information regarding the total body burden of Cd (Prozialeck and Edwards 2010, ATSDR 2012, Rahimzadeh et al. 2017). Other authors have a different opinion contending that blood Cd may also include a contribution from long-term body burden; environmental studies advocate this position (Adams and Newcomb 2014, CDC 2017).

Confounder factors have been identified, being acknowledged that blood Cd concentrations increase with age (in the same way as noted for urinary Cd levels). Men have significantly higher blood Cd levels than women, whereas the urinary excretion rates of Cd were higher in women than in men (Sirivarasai et al. 2002, HC 2008). The blood levels of Cd in non-exposed populations are typically less than 0.5 µg/L, with blood levels higher than 1.0 µg/L considered as generally indicative of Cd exposure; levels higher than 5 µg/L are considered hazardous (Prozialeck and Edwards 2010).

Cadmium levels in urine

It is generally considered that Cd levels in single spot urine samples are indicative of long-term exposure (ATSDR 2012, Akerstrom et al. 2014, Vacchi-Suzzi et al. 2016). Cd accumulates in the kidneys and is slowly released into the urine, usually proportionally to the levels found in the organ (Vacchi-Suzzi et al. 2016). Hence in typical environmental exposure, urinary Cd also reflect Cd concentrations in kidneys (CDC 2017). Significant correlations between estimated Cd exposure and urinary Cd levels have been determined in populations exposed to environmental contamination (Vacchi-Suzzi et al. 2016). However, this BM presents some limitations such as the variability within individuals. This is attributed to many factors, such as natural physiological variations, the choice of sampling strategy (e.g., the time of day) and the method used to adjust the urine Cd concentrations for diuresis; there is also some variability in urinary Cd excretion between individuals, according to differences in absorption and age; however, the increase of urinary excretion of Cd with age is interpreted as an evidence that urine Cd reflects the accumulation of the metal in the body (Akerstrom et al. 2014, Wang et al. 2017). Additionally, gastrointestinal Cd absorption is dependent on iron status and so young women with low iron stores might have increased Cd absorption (Akerstrom et al. 2014).

In the absence of unusually high environmental or occupational sources, urine concentration of Cd is < 2 µg/g creatinine in western populations (Vacchi-Suzzi et al. 2016). Other authors refer values below 0.5 µg/g creatinine in non-exposed populations, and mention that urinary Cd concentrations equal or greater than 0.5 µg/g creatinine are associated with renal damage, whereas concentrations higher than 2.0 µg/g of creatinine may be translated into extensive damage (Prozialeck and Edwards 2010, Rahimzadeh et al. 2017). In a different manner values between 5.5 and 6.6 µg/g creatinine were also proposed to evaluate the threshold value of urinary Cd for renal dysfunction. These values were suggested with the basis on relationships unconfounded by protein degradation, diuresis and renal effects associated with chronic smoking, in an occupationally exposed population (Chaumont et al. 2011). Accordingly, a study with workers moderately exposed to Cd provides evidence that irreversible effects on renal tubular function may occur at levels somewhat exceeding 5 µg/g creatinine (Verschoor et al. 1987). However there are several aspects in monitoring urinary (and blood) Cd which have been problematic and controversial; one of them is the definition of critical levels to identify Cd exposure and the onset of proximal tubule injury (Prozialeck and Edwards 2010). With low or even moderate levels of exposure, any Cd that is filtered at the glomerulus is almost completely reabsorbed by epithelial cells of the proximal tubule and hence little or no Cd is excreted in the urine. It is only when the body burden of Cd is fairly large and/or kidney injury begins to appear that urinary excretion of Cd increases significantly (Järup and Alfvén 2004, Prozialeck and Edwards 2010). Renal function must be adjunctively considered when interpreting urinary Cd values, since the values will increase with renal tubular damage (HC 2008).

Ultimately, one of the more controversial aspects of monitoring the urinary levels of Cd involve the use of chelating agents such as ethylenediamine tetra acetic acid (EDTA) to enhance the urinary excretion of Cd, a process referred to as "provoked urine excretion tests". These chelators are administered 24-hr prior to the collection of urine for the determination of urinary Cd levels. It is not clear how this increase in excretion relates to the total burden of Cd in critical organs such as the kidney, since the EDTA is cell membrane impermeant and do not mobilize Cd from intracellular stores. Thus, the increase in urinary Cd excretion following EDTA administration most likely represents the removal of Cd from easily accessible pools in body fluids and from cell surfaces, each of which represent only a small amount of the total body burden of the metal (Prozialeck and Edwards 2010).

Cadmium levels in hair and nails

Since Cd accumulates in the body for a long time and concentration can gradually increase several years after exposure, the levels of Cd in hair are thought as potential BMs of exposure (Rahimzadeh et al. 2017). However, while some studies have considered that Cd hair levels effectively reflect a long-term environmental exposure, the results of others involving industrially exposed individuals led to concluding that hair Cd levels are not a good index of body burden (Ellis et al. 1981, Razi et al. 2012). Conflicting results exist with regard to hair Cd concentrations in smokers versus non-smokers. A nationwide German environmental survey found little correlation with Cd in hair and active cigarette smoking (although it was the major predictor for blood and urine Cd concentrations). Factors such as outdoor activities, seasonality and Cd in tap water were more correlated with hair concentrations, but these results emphasize the role of exogenous deposition of Cd into hair. Even animal studies show inconsistent results with respect to any correlation between Cd in hair and Cd target organ, the kidney (Siedel, 2011).

"Normal" values exist even though they differ among countries: e.g., in Italy it is 0.03 mg/kg, in England 0.11 mg/kg and in Japan 0.05 mg/kg (Rahimzadeh et al. 2017).

Amounts of Cd in nails are reported as 1.11 ± 0.83 µg/g (Rahimzadeh et al. 2017). Considering that the levels of potentially toxic elements in fingernails can reflect sub-chronic exposure over a period of 5 to 6 mon, studies using nail samples to measure Cd levels in preschool children were conducted however, leading to negative results (Oliveira et al. 2021).

Cadmium levels in saliva

A mean level of Cd in the saliva with a tolerable standard limit is considered less than 0.55 µg/L (Rahimzadeh et al. 2017). But again, controversial results also exist for these BM; Cd concentrations were already found in higher concentrations in the saliva than in blood, with saliva levels markedly elevated in occupationally exposed subjects. Meanwhile, smokers had more Cd in their blood than non-smokers, but this was not the case for saliva (Gervais et al. 1981). Despite concluding that the measurement of salivary Cd could reflect recent exposure to the metal, some researchers emphasized that considerable care should be taken in collecting samples, as a risk of contamination during sampling is apparent with procedures frequently used for saliva collection; for this reason, the applicability of such measurements for biological monitoring was limited (White et al. 1992).

Biomarkers of susceptibility

For several decades efforts have been made to use blood metallothionein (MT) gene transcription levels as a BM of metal exposure (Miura 2009). By sequestrating metal ions, these cysteine-rich proteins, are considered to play a primary role in metal detoxification. MTs involvement in the bodily response to Cd exposure has been well-documented, with numerous studies demonstrating Cd-induced MT expression both *in vitro* and *in vivo* (McNeill et al. 2019). As some subjects have lower MTs levels regardless of increased Cd, these led to postulate if they would have abnormal MT synthesis and could be more sensitive to Cd-induced renal dysfunction. Further genetic polymorphisms analysis allowed to find a genotype involving limited ability for Cd to induce MT synthesis. Workers with the MT-2A G/G genotype (and also perhaps the A/G type) who work in industries that handle Cd, seems to be prone to developing various biological dysfunctions when exposed to the metal. It was even suggested that if these workers were exposed to Cd concentrations below regulation values, symptoms of long-term Cd accumulation could gradually start to appear; therefore, these persons should be carefully monitored (Miura 2009). Similar findings were attained regarding the potential validity of MT gene expression in peripheral blood lymphocytes serving as BM of susceptibility to Cd-induced renal dysfunction. In this work, MT mRNA levels measured from residents living in a Cd-contaminated area were found to be larger with the increase of urinary or blood Cd levels (Lu et al. 2005).

Biomarkers of effect

As referred to earlier, due to the Cd's tendency to accumulate in epithelial cells of the kidney proximal tubules, the kidney is usually the primary critical target organ of Cd toxicity in the body (Prozialeck and Edwards 2010). Cd-induced tubular dysfunction can result in the inability to generate appropriately concentrated urine in response to a physiologic stimulus (Gillham 2007). Moreover, this condition may develop in a dose-dependent manner according to the internal dose of Cd, as already assessed on the basis of Cd levels in the kidneys, urine or in blood (Bernard 2004). The kidneys are, in effect, a sentinel of Cd exposure (Prozialeck and Edwards 2010). Traditional urinary markers of Cd nephrotoxicity include Cd and Cd-binding proteins such as MTs and low molecular weight proteins, (Prozialeck and Edwards 2010).

Toxic actions of cadmium in the kidneys

Metallothioneins Even though the Cd-MT complex is nontoxic to most organs, it can be filtered at the glomerulus and be taken up by the epithelial cells of the proximal tubule (Prozialeck and Edwards 2012). For these reasons, the urinary excretion of Cd and MT have been used in conjunct as markers of Cd exposure and Cd-induced proximal tubule injury (Prozialeck and Edwards 2010). Additionally, several low molecular weight proteins present in the plasma, are small enough to be easily filtered at the glomerulus. Under normal circumstances, these filtered proteins are efficiently reabsorbed by the proximal tubule and are not excreted to any great extent in the urine (Maack et al. 1979, Prozialeck and Edwards 2010). There is a general agreement that one of the most sensitive and specific indicators of Cd-induced renal dysfunction is decreased tubular reabsorption of low molecular weight proteins, leading to the so-called tubular proteinuria. Cd-induced microproteinuria is usually considered irreversible except at the incipient stage of the intoxication, where partial or complete reversibility has been found in some studies.

Low molecular weight proteins Among low molecular weight proteins, β2-microblobulin and α1-microglobulin are the most used for screening renal damage in populations at risk (Bernard 2004). However, despite β2-microglobin being most widely used as a standard marker for monitoring the early stages of Cd exposure and toxicity, it is questionable whether this effect parameter is sufficient to detect reversible renal damage (Verschoor et al. 1987, Prozialeck and Edwards 2010,). Even so urinary levels of β2-microglobulin greater than 1,000 μg/g creatinine are considered to indicate irreversibility of renal effects, as this level is typically associated with urinary Cd concentrations greater than 5 μg/g creatinine (Nomiyama et al. 1992, Prozialeck and Edwards 2010). Meanwhile α1-microglobulin is possibly a better BM than β2-microglobulin, since among non-smoking women of various ages, β2-microglobulin showed a less close correlation with urinary Cd levels than α1- macroglobulin, even after correcting for urine density. It is also described that α1-microblobulin-uria prevalence increase with Cd exposure by 3.3 times (Moriguchi et al. 2004).

Kidney enzymes Some other authors extensively used markers of Cd-induced proximal tubule injury are enzymes expressed in proximal tubule epithelial cells. These are namely, N-Acetyl-β-D-Glucosamidase (NAG) and later, α-Glutathione-S-Transferase (α-GST) was also studied in this context. The appearance of these enzymes in urine is classically thought to result from the leakage of intracellular contents when necrotic proximal tubule epithelial cells lose their membrane integrity and/or slough off into the urine (Prozialeck and Edwards 2010). NAG is a lysosomal brush-border enzyme present mainly in the proximal tubule cells and is the most active glycosidase in proximal tubule cell lysosomes. As is too large to be filtered through glomeruli, its urinary excretion reflects increased lysosomal activity and tubule cell injury (Khan et al. 2003). NAG has clearly proved to be especially useful in human monitoring for Cd effects, as consubstantiated by a study in a Chinese population residing in a polluted area; this area was contaminated by industrial wastewater from a nearby smeltering plant that discharged Cd-polluted waste into a river used for the irrigation of rice fields (Jin et al. 1999). There is a limit due to the fact that NAG exists as multiple isoforms, the B form being the most abundant in proximal tubule epithelial cells and regarded as the more sensitive and reliable marker of Cd-induced injury; assays that do not differentiate isoforms may not yield useful results (Prozialeck and Edwards 2010).

In turn, α-Glutathione S-transferases (GST) belongs to the GST family of ubiquitous enzymes which take part in the detoxification of free radicals. The human kidney contains the α and π forms in relatively high amounts in renal tubules, with the α-GST isoform mostly localized in the proximal tubule and the π-GST in the distal tubule. α-GST can detoxify lipid peroxidation end-products such as 4-Hydroxynonenal (4-HNE) via the formation of glutathione (GSH) conjugates, thereby limiting their cytotoxicity and providing a physiological response to the injury of tubular cells (McMahon et al. 2010). α-GST is exclusively released into urine during renal injury and its urinary levels are considered an especially useful early marker of Cd-induced kidney injury (McMahon et al. 2010, Prozialeck and Edwards 2010). Reports exist that the level of this BM is a sensitive indicator of kidney injury in workers who had been exposed to Cd and Pb. Moreover, α-GST was shown experimentally to be a more sensitive marker of kidney injury than NAG on Cd administration (Prozialeck and Edwards 2010).

Kidney injury molecule It may be also possible to detect early stages of Cd toxicity through using the kidney injury molecule-1 (Kim-1) levels (Prozialeck and Edwards 2012). Kim-1 is a type 1 membrane glycoprotein found in renal proximal tubule epithelial cells, whose expression is induced in a number of renal diseases, whereas in a healthy kidney tissue it is virtually undetectable. When the kidney is damaged, Kim-1 is expressed on the apical membrane being released in the urine. Kim-1 has been proved to correlate with urinary Cd *in vivo*, and Pennemans et al. (2011) also showed that urinary Kim-1 levels were positively correlated with urinary Cd concentration in humans in an elderly population after long-term low-dose exposure to Cd; in the same study other classical markers did not show any relationship .

Early Cadmium effects

All BMs described so far reflected specific toxic actions of Cd in the kidneys. However, Cd may produce very early systemic effects which may contribute to the pathophysiology of kidney injury; markers of these systemic effects are very important since they have the potential of provide a means of assessing risks of toxicity before the actual onset of the injury to the kidney. It has been noted that low level Cd exposure is associated with alterations in glucose levels and glucose metabolism that precede the onset of the injury to the kidney (Prozialeck and Edwards 2010). *In vivo* and epidemiological research suggest a positive association between Cd exposure and the incidence of type 2 diabetes, but the association remains controversial despite the knowledge that Cd accumulates in the pancreas and exerts diabetogenic effects in animals in both acute or sub-chronic exposure models (Schwartz et al. 2003, Prozialeck and Edwards 2010, Jacquet et al. 2018). There is a well-established link between diabetes and the development of the kidney disease, since over time high levels of sugar in the blood damage millions of tiny filtering units within each kidney; this eventually leads to kidney failure. Many people with diabetes also develop high blood pressure, which can also damage the kidneys (NIDDK 2017). Regarding Cd, the metal within the kidney alters the expression of two key enzymes of glucose metabolism, glucose 6-phosphate dehydrogenase and glyceraldehyde 3-phosphate dehydrogenase. Therefore, the ability of Cd to affect levels and metabolism of glucose could have significant implications for the pathophysiology of Cd-induced kidney injury. The fact that the Cd-induced changes in glucose and insulin levels appear before changes in indicators of renal dysfunction, such as urinary $\beta2$ microglobulin or NAG, indicates that measurements of serum glucose and insulin may be more sensitive markers of Cd exposure than traditional effect BMs of renal dysfunction (Prozialeck and Edwards 2010).

Final Remarks

In this chapter Cd was presented as a "modern metal " used in more or less novel applications, being the newest, e-waste and for which no efficient recycling way exists so far. Food contamination with Cd is the major exposure for non-smokers and non-occupational exposed persons. It was also described how Cd is a metal of considerable toxicity with a destructive impact on most organ systems, and that even at low levels can induce toxicity. A concern for risk assessment entities is the fact that the metal has a very long half-life (10–30 yr) in the body (Schaefer et al. 2020). Cd total body burden can be easily accessed through urinary Cd measurements, whereas blood Cd levels can indicate recent exposure. People susceptible to Cd toxicity include those with genotypes expressing limited ability for Cd to induce MT synthesis, which can potentially serve as a susceptibility BM. Since the kidneys are the primary critical target organ for Cd toxicity, various effect markers of Cd nephrotoxic effects have been developed. Among the large use in controlling Cd-induced toxicities are markers which can provide means to assess toxicity before the onset of kidney damages, such as glucose metabolism molecules currently explored as BMs of early effects. Additionally, developing mitigation strategies to reduce exposures to Cd are certainly necessary.

References

Adams, S. and P. Newcomb. 2014. Cadmium blood and urine concentrations as measures of exposure: NHAES 1999–2010. J. Expo. Sci. Environ. Epidemiol. 24: 163–170.

Akerstrom, M., L. Barregard, T. Lundh and G. Sallsten. 2014. Variability of urinary cadmium excretion in spot urine samples, first morning voids, and 24 h urine in a healthy non-smoking population: Implications for study design. J. Expo. Sci. Environ. Epidemiol. 24: 171–179.

Aoshima, K. 2016. Itai-itai disease: Renal tubular osteomalacia induced by environmental exposure to cadmium—historical review and perspectives. Soil Sci. Plant Nutr. 62(4): 319–326.

[ATSDR] Agency for Toxic Substances and Disease Registry, Cadmium Toxicity. How Are People Exposed to Cadmium? 2008. U.S. Department of Health and Human Services, Public Health Service, Atlanta, USA.

[ATSDR] Agency for Toxic Substances and Disease Registry, ToxGuideTM for Cadmium, AS# 7440-43-9, 2012. U.S. Department of Health and Human Services, Public Health Service, Atlanta, USA.

Bernard, A. 2004. Renal dysfunction induced by cadmium: biomarkers of critical effects. Biometals. 17: 519–523.

Bernhoft, R.A. 2013. Cadmium toxicity and treatment. Sci. World J. 2013: 394652.

Borné, Y., B. Fagerberg, G. Sallsten, B. Hedblad, M. Persson, O. Melander et al. 2019. Biomarkers of blood cadmium and incidence of cardiovascular events in non-smokers: results from a population-based proteomics study. Clin. Proteom. 16: 21.

[CDC] Centers for Disease Control and Prevention Biomonitoring Summary, Cadmium, CAS No. 7440-43-9, 2017. CDC, Atlanta, USA.

Chaumont, A., F. De Winter, X. Dumont, V. Haufroid and A. Bernard. 2011. The threshold level of urinary cadmium associated with increased urinary excretion of retinol-binding protein and β2-microglobulin: A re-assessment in a large cohort of nickel-cadmium battery workers. Occup. Environ. Med. 68: 257–264.

[DTM] Dartmouth Toxic Metals Superfound Research Program, The Facts on Cadmium. 2021. National Institute of Environmental Health Sciences (NIEHS), Research Triangle Park, USA.

[EC] European Commission. 2021. Cadmium in food. European Commission, Brussels, Belgium.

Ellis, K.J., S. Yasumura and S.H. Cohn. 1981. Hair cadmium content: Is it a biological indicator of the body burden of cadmium for the occupationally exposed worker? Am. J. Ind. Med. 2(4): 323–330.

Genchi, G., M.S. Sinicropi, G. Lauria, A. Carocci and A. Catalano. 2020. The Effects of Cadmium Toxicity. Int. J. Environ. Res. Public Health. 17: 3782.

Gervais, L., Y. Lacasse, J. Brodeur and A. P'an. 1981. Presence of cadmium in the saliva of adult male workers. Toxicol. Lett. 8(1-2): 63–6.

Gillham, M. 2007. Acute renal failure and renal support. pp. 481–494. *In*: D. Sidebotham, A. Mckee and M.J.H. Levy [eds.]. Cardiothoracic Critical Care. Butterworth-Heinemann, Oxford, UK.

Grant, C.A., L.D. Bailey, M.J. Mclaughlin and B.R. Singh. 1999. Management factors which influence cadmium concentrations in crops. pp. 151–198. *In*: M.J. McLaughlin and B.R. Singh [eds.]. Cadmium in Soils and Plants. Developments in Plant and Soil Sciences. Springer, Dordrecht, Netherlands.

[HC] Health Canada, Guidelines for Canadian Drinking Water Quality: Guideline Technical Document – Cadmium. 2008. HC, Ottawa, Canada.

[HFS] Directorate-General for Health and Food Safety, Cadmium in Food, 2021. European Commission (EC), Brussels, Belgium.

[HK CFS] Hong Kong Centre for Food Safety, Food Contaminants, Cadmium in Food, 2017. HK CFS, Hong Kong, China.

Hu, X., K.-H. Kim, Y. Lee, J. Fernandes, M.R. Smith, Y.-J. Jung et al. 2019. Environmental cadmium enhances lung injury by respiratory syncytial virus infection. Am. J. Pathol. 189(8): 1513–1525.

Jacquet, A., J. Arnaud, I. Hininger-Favier, F. Hazane-Puch, K. Couturier, M. Lénon et al. 2018. Impact of chronic and low cadmium exposure of rats: Sex specific disruption of glucose metabolism. Chemosphere 207: 764–773.

Järup, L. and T. Alfvén. 2004. Low level cadmium exposure, renal and bone effects—the OSCAR study. Biometals. 17(5): 505–509.

Jin, T., G. Nordberg, X. Wu, T. Ye, Q. Kong, Z. Wangd et al. 1999. Urinary N-Acetyl-β-D-glucosaminidase isoenzymes as biomarker of renal dysfunction caused by cadmium in a general population. Environ. Res. 81(2): 167–173.

Khan, K.N.M., G.C. Hard and C.L. Alden. 2013. Kidney. *In*: W.M. Haschek, C.G. Rousseaux, Matthew A. Wallig [eds.]. Haschek and Rousseaux's Handbook of Toxicologic Pathology. Academic Press, Cambridge, USA.

Leal, R.B., D.K. Rieger, T.V. Peres, M.W. Lopes and C.A.S. Gonçalves. 2012. Cadmium neurotoxicity and its role in brain disorders. pp. 751–766. *In*: Y. Li and J. Zhang. [eds.]. Metal Ion in Stroke. Springer Series in Translational Stroke Research. Springer, New York, USA.

Lu, J., T. Jin, G. Nordberg and M. Nordberg. 2005. Metallothionein gene expression in peripheral lymphocytes and renal dysfunction in a population environmentally exposed to cadmium. Toxicol. Appl. Pharmacol. 206(2): 150–156.

Maack, T., V. Johnson, S.T. Kau, J. Figueiredo and D. Sigulem. 1979. Renal filtration, transport, and metabolism of low-molecular-weight proteins: A review. Kidney Int. 16(3): 251–270.

McMahon, B.A., J.L. Koyner and P.T. Murray. 2010. Urinary glutathione S-transferases in the pathogenesis and diagnostic evaluation of acute kidney injury following cardiac surgery: A critical review. Curr. Opin. Crit. Care 16(6): 550–555.

McNeill, R.V., A.S. Mason, M.E. Hodson, J. Catto and J. Southgate. 2019. Specificity of the Metallothionein-1 Response by Cadmium-Exposed Normal Human Urothelial Cells. Int. J. Mol. Sci. 20(6): 1344.

Miura, N. 2009. Individual susceptibility to cadmium toxicity and metallothionein gene polymorphisms: with references of current status of occupational cadmium exposure. Ind Health. 47(5): 487–94.

Moitra, S., P.D. Blanc and S. Sahu. 2013. Adverse respiratory effects associated with cadmium exposure in small-scale jewellery workshops in India. Thorax. 68: 565–570.

Moriguchi, J., T. Ezaki, T. Tsukahara, K. Furuki, Y. Fukui, S. Okamoto et al. 2004. α1-Microglobulin as a promising marker of cadmium-induced tubular dysfunction, possibly better than β2-microglobulin. Toxicol. Lett. 148(1–2): 11–20.

Neltler, T. 2019. Too much cadmium and lead in kids' food according to estimates by FDA. https://blogs.edf.org/health/2019/05/07/cadmium-and-lead-kids-food-fda-study/.

[NIDDK] National Institute of Diabetes and Digestive and Kidney Diseases, Diabetic Kidney Disease, 2017. National Institutes of Health (NIH), Bethesda, USA.

Nishijo, M., H. Nakagawa, Y. Suwazono, K. Nogawa and T. Kido. 2017. Causes of death in patients with Itai-itai disease suffering from severe chronic cadmium poisoning: a nested case-control analysis of a follow-up study in Japan. BMJ Open 7(7): e015694.

Nomiyama, K., S.J. Liu and H. Nomiyama. 1992. Critical levels of blood and urinary cadmium, urinary beta 2-microglobulin and retinol-binding protein for monitoring cadmium health effects. IARC Sci. Publ. 118: 325–340.

Nordberg, G.F., A. Bernard, G.L. Diamond, J.H. Duffus, P. Illing, M. Nordberg et al. 2018. Risk assessment of effects of cadmium on human health (IUPAC Technical Report). Pure Appl. Chem. 90(4): 755–808.

Oliveira, A.S., E.A.C. Costa, E.C. Pereira, M.A.S. Freitas, B.M. Freire B.L. Batista et al. 2021. The applicability of fingernail lead and cadmium levels as subchronic exposure biomarkers for preschool children. Sci. Total Environ. 758: 143583.

Pennemans, V., L. M. De Winter, E. Munters, T. S. Nawrot, E. Van Kerkhove, J.-M. Rigo et al. 2011. The association between urinary kidney injury molecule 1 and urinary cadmium in elderly during long-term, low-dose cadmium exposure: a pilot study. Environ. Health. 10: 77.

Prozialeck, W.C. and J.R. Edwards. 2010. Early biomarkers of cadmium exposure and nephrotoxicity. Biometals. 23(5): 793–809.

Prozialeck, W.C. and J.R. Edwards. 2012. Mechanisms of cadmium-induced proximal tubule injury: new insights with implications for biomonitoring and therapeutic interventions. J. Pharmacol. Exp. Ther. 343(1): 2–12.

Rahimzadeh, M.R., M.R. Rahimzadeh, S. Kazemi and A.A. Moghadamnia. 2017. Cadmium toxicity and treatment: An update. Caspian J. Int. Med. 8(3): 135–145.

Razi, A.C.H., K.O. Akin, K. Harmanci, O. Ozdemir, A. Abaci, S. Hizli et al. 2012. Relationship between hair cadmium levels, indoor ETS exposure and wheezing frequency in children. Allergol. Immunopathol. 40(1): 51–59.

Roberts, T.L. 2014. Cadmium and Phosphorous Fertilizers: The Issues and the Science. Procedia Eng. 83: 52–59.

Schaefer, H.R., S. Dennis and S. Fitzpatrick. 2020. Cadmium: Mitigation strategies to reduce dietary exposure. J. Food Sci. 85(2): 260–267.

Schwartz, G.G., D. Il'yasova and A. Ivanova. 2003. Urinary cadmium, impaired fasting glucose, and diabetes in the NHANES III. Diabetes Care. 26(2): 468–470.

Siedel, S., Hair Analysis Panel Discussion: Section: Appendix C. 2011. Content source: Agency for Toxic Substances and Disease Registry (ATSDR). U.S. Department of Health and Human Services, Public Health Service, Atlanta, USA.

Sirivarasai, J., S. Kaojaren, W. Wananukul and P. Srisomerang. 2002. Non-occupational determinants of cadmium and lead in blood and urine among a general population in Thailand. Southeast Asian J. Trop. Med. Public Health. 33(1): 180–187.

Sukarjo, I.Z. and W. Purbalisa. 2019. The critical limit of cadmium in three types of soil texture with shallot as an indicator plant. AIP Conference Proceedings. 2120: 040012.

Tellez-Plaza, M., E. Guallar, B.V. Howard, J.G. Umans, K.A. Francesconi, W. Goessler et al. 2013. Cadmium exposure and incident cardiovascular disease. Epidemiology 24(3): 421–429.

Vacchi-Suzzi, C., D. Kruse, J. Harrington, K. Levine and J.R. Meliker. 2016. Is urinary cadmium a biomarker of long-term exposure in humans? A review. Curr. Environ. Health Rep. 3(4): 450–458.

Verschoor, M., R. Herber, J. van Hemmen, A. Wibowo and R. Zielhuis. 1987. Renal function of workers with low-level cadmium exposure. Scand. J. Work Environ. Health. 13(3): 232–238.

Wang, B. and Y. Du. 2013. Cadmium and its neurotoxic effects. Oxid. Med. Cell. Longev. 2013: 898034.

Wang, H., X. Dumont, V. Haufroid and A. Bernard. 2017. The physiological determinants of low-level urine cadmium: An assessment in a cross-sectional study among schoolchildren. Environ. Health. 16(1): 99.

White, M.A., S.A. O'Hagan, A.L. Wright and H.K. Wilson. 1992. The measurement of salivary cadmium by electrothermal atomic absorption spectrophotometry and its use as a biological indicator of occupational exposure. J. Expo. Anal. Environ. Epidemiol. 2(2): 195–206.

[WHO] World Health Organization, Cadmium in Drinking-water, Background document for development of WHO Guidelines for Drinking-water Quality. 2011. WHO, Geneva, Switzerland.

[WHO] World Health Organization, Preventing disease through healthy environments. Exposure to cadmium: a major public health concern. 2019. WHO, Geneva, Switzerland.

CHAPTER 7

Chromium

Sources of Exposure

Chromium (Cr) is a naturally occurring element in animals, plants, rocks, soil, and volcanic dust and gases, it is the sixth most abundant element, on the Earth's crust and seawater (EPA 1992, Dioni et al. 2017). It is present predominantly in two valence states. One of them is trivalent Cr (Cr^{3+}) which occurs naturally and is an essential nutrient to normal glucose, protein and fat metabolism; the other one is hexavalent Cr (Cr^{6+}), which along with the less frequent metallic Cr [(Cr 0)] is mostly produced by industrial processes, rarely found in nature (EPA 1992, Ventura et al. 2021). Cr compounds, in either Cr^{3+} or Cr^{6+} forms, are used for chrome plating, manufacture of dyes and pigments, leather and wood preservation and treatment of cooling tower water. Smaller amounts are used in drilling muds, textiles and toners for copying machines (EPA 1992). The several uses of Cr have resulted in its widespread release into the environment, and nowadays Cr is present in all environmental parts, including water, air and soil, at different concentrations (Dioni et al. 2017, Ertani et al. 2017). The general population is exposed to Cr, generally Cr^{3+}, by inhaling air that contains chemical, drinking water and eating food (EPA 2000).

Air

Air emissions of Cr are predominantly in the trivalent form, as small particles or aerosols. The most important industrial sources of Cr in the atmosphere are those related to ferrochrome production. Ore refining, welding, chemical and refractory processing, cement-producing plants, automobile brake lining and catalytic converters for automobiles, leather tanneries and chrome pigments also contribute to the atmospheric burden with the metal (EPA 1992, Ventura et al. 2021). Although information regarding concentrations of total and speciated Cr in the atmosphere is limited, some measurements were already carried out several thousands of km away from major land masses, showing concentrations ranging 0.07–1.1 ng/m³. In inhabited areas concentrations of 0.7 ng/m³ have been reported for Norway and 0.6 ng/m³ in Northwest Canada; values of 45–67 ng/m³ were estimated in Hawaii and 1–140 ng/m³ in Japan. In the Member States of the European Union remote areas can exhibit 0–3 ng/m³ Cr concentrations in air, urban areas 4–70 ng/m³ and industrial areas 5–200 ng/m^3 (WHO 2000). It is estimated from other studies that median concentrations of Cr in ambient air are < 20 ng/m^3, but indoor air in areas with cigarette smoking can have levels 10–400 times higher than outdoor air (ATSDR 2012). It is also known that people who live in the vicinity of Cr waste disposal sites or Cr manufacturing and processing plants, have a greater probability of elevated Cr exposure than the general population; such exposures are generally to mixed Cr^{3+} and Cr^{6+} (EPA 1992).

Workplace

Occupational exposure can be two orders of magnitude higher than the exposure of the general population and occur mainly in the following settings: welding and other types of "hot work" on stainless steel and other metals containing Cr, use of pigments, spray paints and coatings and operating chrome plating baths (Alvarez et al. 2021, OSHA 2021). The main exposure routes in these contexts are the inhalation of dust and mist or fumes and dermal contact (ATSDR 2012, Ventura et al. 2021). No acute-duration inhalation Minimal Risk Levels (MRLs) ($\leq$ 14 d) are obtained for Cr^{3+} or Cr^{6+}, but several intermediate-duration inhalation MRLs (15–364 d) have been acquired: 5×10^{-6} mg/m^3 for Cr^{6+} aerosols and mists, 3×10^{-4} mg/m^3 for Cr^{6+} particulates, 5×10^{-3} mg/m^3 for insoluble Cr^{3+} particulates and 1×10^{-4} mg/m^3 for soluble Cr^{3+} particulates. Chronic-duration inhalation exposure ($\geq$ 1 yr) to aerosols and mists with Cr^{6+} have an MRL of 5×10^{-6} mg/m^3 (ATSDR 2012). The American Conference of Governmental Industrial Hygienists (ACGIH) considers a Threshold Limit Value (TLV) value for inhalable Cr^{6+} compounds of 0.0002 mg/m^3; TLV the concentrations in air of various chemical substances represent the conditions under which it is estimated that most workers can be repeatedly exposed to these substances, day after day, during a working life, without showing any adverse effect due to exposure. A Short-Term Exposure Limit (STEL) of 0.0005 mg/m^3 for inhalable Cr^{6+} compounds, is also regulated by the Occupational Safety and Health Administration (OSHA), as well as a Permissible Exposure Limit (PEL) of 0.005 mg/m^3. The National Institute for Occupational Safety and Health (NIOSH) sets a Recommended Exposure Limit (REL) of 0.0002 mg/m^3 (Alvarez et al. 2021). Apart from other agencies the European Scientific Committee on Occupational Exposure Limits (SCOEL) and the German Research Council Foundation have classified the compounds of Cr^{6+} as carcinogens of category 1 (whose carcinogenicity cannot be excluded) and due to that, no recommended limit value is given for specific occupational exposures (Alvarez et al. 2021).

Soil

The levels of Cr in soils may vary considerably according to the natural composition of rocks and sediments that compose them (Oliveira 2012). The metal is abundant in mafic and ultramafic igneous rocks, while much lower concentrations are present in acid igneous and sedimentary rocks. Under natural conditions soil concentrations may range between 10 and 50 mg/kg, having been recorded as a mean concentration of 37.0 mg/kg (Zampella et al. 2010, ATSDR 2012). Yet the levels of Cr in soil may increase due to dumping of Cr-bearing liquids and solid wastes of Cr by-products, ferrochromium slag or Cr plating baths (Wyszkowska 2002, Oliveira 2012). Consequently, contaminated agricultural soils can reach Cr concentrations varying up to values as high as 350 mg/kg. Most Cr is found in the trivalent form since Cr^{6+} tends to be reduced to Cr^{3+}, particularly in soils with a high organic matter content, and complexed with mineral structures as oxides. However, if pH conditions are favourable and MnO_2 is present, Cr^{3+} can be oxidized to the toxic hexavalent form, even though this transformation is quite rare (Ertani et al. 2017). In a different manner, when the reducing capacity of the soil is overcome, Cr^{6+} may persist in soils or sediments for years, especially if the soils are sandy or present low levels of organic matter (Oliveira 2012). Cr^{6+} is known to have a harmful effect on soil microorganisms depressing their biological activity (Wyszkowska 2002).

Water

The primary source of exposure to Cr for the general population is drinking water (and food); thus, the oral route is the predominant route of exposure (ATSDR 2012, Sun et al. 2015). Cr may enter natural waters by weathering of Cr-containing rocks, from leaching of soils and by direct discharge from industrial operations (WHO 2000, Oliveira 2012, Ertani et al. 2017). The solubility of Cr^{3+} depends on the pH of the water: under neutral to basic pH, Cr^{3+} precipitates, and conversely under acidic pH tend to solubilize. The forms of Cr^{6+} chromate and dichromate are very soluble under all pH conditions but can precipitate with divalent cations (Oliveira 2012). Cr levels in uncontaminated water are very low, presenting an average concentration of 0.3 ug/L in ocean water, ranging 0.2–1 ug/L in rainwater and 1–10 ug/L in rivers and lakes (Sun et al. 2015). Tap-water in typical cities of the USA have Cr concentrations ranging from 0.4 to 8 µg/L, while in effluents in the vicinity of Cr industries levels ranging from 2 to 5 g/L can be determined (Oliveira 2012, WHO 2000). Elevated levels of Cr^{6+} found in drinking water of more than 30 USA cities, presented apprehensions regarding possible health effects due to exposure from this source, for the general population. Limit values were established, such as the current drinking water standard established by the US Environmental Protection Agency (EPA) for total Cr, which is 0.1 mg/L; however, there is no specific drinking water standard for Cr^{6+}. Other agencies, such as the Food Safety Commission of Japan (FSCJ) considered that the quantitative risk assessment of Cr^{6+} through drinking water was difficult to conduct, based on results available from epidemiological studies on non-occupational and occupational exposures. Despite such difficulties, the FSCJ specified that the tolerable daily intake (TDI) of Cr^{6+} would be 1.1 µg/kg b.w./d , after applying an uncertainty factor of 100 to a benchmark dose modelling ascribed on diffuse epithelial hyperplasia in the duodenum of male mice observed in a 2-yr oral exposure study. The agency estimated that the daily intake of Cr^{6+} from consumption of mineral water and tap water for mean and high intakes, was 0.04 and 0.29 µg/kg b.w./d , respectively. Since both values were lower than the TDI, it was concluded that the risk of health effects from Cr^{6+}, at current exposures, through the consumption of mineral or tap water was extremely low (FSCJ 2019).

Food

Cr is present in many foods, including meats, grain products, fruits, vegetables, nuts, spices, brewer's yeast, beer and wine; their levels are very low, and as Cr^{3+} (Dioni et al. 2017, NIH 2021). However, according to an EFSA report (2014) there is less data concerning the presence of Cr^{6+} in food (Vincent and Lukaski 2018, FSCJ 2019). Known concentrations of Cr range from < 10 to 1,300 ug/kg, with the highest amounts present in meat, fish, fruits and vegetables (Sun et al. 2015). In fact, Cr amounts in these foods vary widely depending on the local soil and water conditions, and on agricultural and manufacturing processes used to produce them (NIH 2021). The Cr content of British commercial alcoholic beverages was reported to be slightly higher than that of wines produced in the USA, being 0.45 mg/L for wine, 0.30 mg/L for beer and 0.135 mg/L for spirits (WHO 2000).

Additional food contamination can be obtained on Cr transfer from stainless steel equipment during food processing, and from pots and pans during cooking (NIH 2021). It seems that most dietary Cr is procured from these sources, which is a quite relevant fear for people living in developed nations (Vincent and Lukaski 2018).

Most multivitamin/mineral supplements contain Cr, typically 35–120 μg and supplements containing only Cr are also available , usually providing 200 μg to 500 μg, although some contain up to 1,000 μg (NIH 2021).

The National Health and Nutrition Examination Survey (NHANES) provides dietary intake data for many nutrients, but not for Cr (NIH 2021). As with drinking water, it is mentioned that the daily Cr intake from food is difficult to assess because studies conducted so far have used methods that are not easily comparable. However the Cr intake from typical North American diets was estimated in a study to be 60–90 μg/d (WHO 2000). Another research listing well-balanced diets designed by nutritionists found that the mean Cr content per 2,000 kcal was about 27 μg, ranging from 17 to 47 μg. In 2018 a dietary intake assessment in northern Italy showed that the median Cr intake was about 57 μg/d , from a typical Italian diet. With regards to food supplements and according to an analysis of the NHANES III (1988–1994) data, the median supplemental intake of Cr was about 23 μg/d among those taking supplements containing the metal (NIH 2021).

No acute-duration oral MRL ($\leq$ 14 d) has been collected for Cr^{6+}; an MRL of 5×10^{-3} mg/kg b.w./d was collected for intermediate-duration oral exposure to Cr^{6+} (15–364 d), while the MRL for chronic-duration oral exposure to Cr^{6+} ($\geq$ 1 yr) is 9×10^{-4} mg/kg b.w./d (ATSDR 2012). Concerning Cr^{3+} no acute- intermediate- or chronic duration oral MRLs has been decided, however the recommended Daily Dietary Intake (DDI) for healthy adults from the age of 14 to 50 yr of 24–35 μg, and 20–30 μg after 50 yr of age (ATSDR 2012, Dioni et al. 2017). In as no adverse effects have been convincingly associated with excess intake of Cr^{3+} from food or food supplements, no Upper Tolerable Limit (UL) has been established. However, this does not mean that no toxic effects might be associated with high intakes of Cr^{3+} (Vincent and Lukaski 2018).

Consumer products

Relevant exposures to Cr can occur through tobacco smoke, including electronic cigarette vaping (Ventura et al. 2021). The average Cr found in cigarettes may range from 0.96 to 3.85 and 0.32 to 0.80 µg/cigarette, for tobacco and ashes, respectively (Lisboa et al. 2020). The metal in the hexavalent state is mostly up taken from the soil and translocated to aerial parts of the tobacco plant; once Cr^{6+} reaches the leaves, it gets reduced to Cr^{3+} by binding to specific ligands and sequestered into leaf vacuoles of the tobacco plant where is stored as Cr^{3+}, which as mentioned is the non-toxic form. However, tobacco leaves contain manganese which further oxidizes Cr^{3+} to the highly toxic Cr^{6+} form. Moreover, while smoking tobacco, the Cr^{3+} present in it may get oxidized to Cr^{6+} due to combustion, as it involves oxygen and high temperature. It is also known that when tobacco smoke is inhaled through the mouth or nose, and gets mixed with moisture thereby forming Cr^{6+}. Due to these reasons, some authors emphatically defend the probable involvement of Cr^{6+} rich tobacco in oral cancer (Samal et al. 2020). In e-cigarettes, Cr is a component of heating coils, and some studies reveal positive associations of Cr aerosol concentrations with higher BM levels indicating an increased metal internal dose. These authors recommended the establishment of metal level standards to prevent involuntary metal exposure among e-cigarette users (Aherrera et al. 2017).

Cr can be found in several other consumer products such as wood treated with copper dichromate, leather tanned with chromic sulphate and metal-on-metal hip replacements (EPA 1992, ATSDR 2012). The possibility that traces of Cr (among other metals) could give rise to allergic contact dermatitis via those products was reviewed. However, it was concluded that the risk was negligible and that the risk of elicitation in pre sensitized individuals was acceptably low if the level of contamination was kept to very low levels. Less than 5 mg/L of Cr is considered an acceptable standard for consumer products, and to minimize the risk for very sensitive individuals the ultimate target should be not more than 1 mg/L (Basketter et al. 2003).

Toxicity

A significant amount of research has evaluated the impact of Cr in the environment and in the health of humans . It should be noted however, that due to the significant toxicities between Cr^{3+} and Cr^{6+}, these two distinct valence states must be examined to assess potential risks to exposure (Dioni et al. 2017). It was already mentioned that Cr^{3+} is an essential nutrient for mammals including humans. Its lack in the diet can cause serious cardiac problems, metabolic dysfunctions and diabetes; on the other hand, its excessive presence in the body has dangerous health effects. The most relevant apprehensions pertain to the hexavalent forms, which are 10–100 times more toxic than trivalent Cd because they tend to act as strong oxidants; Cr^{6+} is one of the 14 most harmful substances for the health of living organisms (Ertani et al. 2017). Despite the body possessing several systems for reducing Cr^{6+} to Cr^{3+}, it should be noted that such Cr^{6+} detoxification process leads to increased levels of Cr^{3+}, that could result in excessive levels (EPA 1992).

Acute toxicity

Acute poisoning with Cr is likely to occur through the oral route, usually being accidental or intentional (suicide), whereas occupational or environmental chronic poisoning is mainly from inhalation or skin contact.

Acute Cr^{6+} poisonings are often fatal regardless of the therapy used, the average oral lethal dose in humans is 1–3 g (ATSDR 2013). Depending on the dose, signs of acute intoxications can include intense gastrointestinal irritation or ulceration and corrosion, epigastric pain, nausea, vomiting, diarrhoea, vertigo, fever, muscle cramps, haemorrhagic diathesis, toxic nephritis, renal failure, intravascular haemolysis, circulatory collapse, liver damage, acute multisystem organ failure, coma and even death. In the respiratory system, shortness of breath, coughing and wheezing is also reported from a case of acute exposure to Cr^{6+} (EPA 1992).

Chronic toxicity

Chronic human exposure to high levels of Cr^{6+} by inhalation or ingestion, may cause effects on the liver, kidneys, gastrointestinal and immune systems, genetic mutations and changes in the blood. Studies on rats have shown that, following inhalation exposure, the lung and kidneys have the highest levels of Cr (EPA 1992, ATSDR 2013, Ertani et al. 2017). Epidemiological studies have also suggested that Cr status may be linked with cardiovascular diseases (Dioni et al. 2017).

Dermal exposure to Cr^{6+} may cause contact dermatitis, sensitivity and ulceration of the skin (EPA 1992). Repeated skin contact with Cr dusts can lead to incapacitating eczematous dermatitis with oedema. When a solution of chromate contacts the skin, it can produce penetrating lesions known as chrome holes or chrome ulcers, particularly in areas where a break in the epidermis is already present. These usually occur on the fingers, knuckles and forearms. The characteristic Cr sore begins as a papule, forming an ulcer with raised hard edges. Ulcers can penetrate deep into the soft tissue or become the sites of secondary infection but are not known to lead to malignancy. Chromate dusts can also produce irritation of the conjunctiva and mucous membranes, nasal ulcers and perforations, keratitis, gingivitis and periodontitis (ATSDR 2013).

Human data have led the International Agency for Research on Cancer (IARC) to establish inhaled Cr^{6+} as a human carcinogen, resulting in exposures to increased risk of lung cancer, which is the most serious long-term effect induced by the metal (EPA 1992, ATSDR 2013, Sun et al. 2015, Kim et al. 2018). The same agency considers Cr^{3+} and metallic Cr as not classifiable as to their carcinogenicity to humans (ATSDR 2012). Several carcinogenicity mechanisms are described for Cr^{6+}, which itself does not bind to DNA or other macromolecules in cells. Instead, their metabolic intermediates Cr^{5+}, Cr^{4+} and the final product Cr^{3+}, are highly reactive and readily form Cr-DNA adducts, which may give rise to DNA single- and double-strand breaks (Sun et al. 2015, Ventura et al. 2021). Reactive Oxygen Species (ROS) are also formed, which are considered to mediate Cr^{6+}-induced changes in cell signalling and homeostasis leading to cell death by apoptosis; concomitantly ROS accumulation generates oxidative stress and contributes to chronic inflammation, metabolic reprogramming and genetic instability, ultimately leading to tumour development (Ventura et al. 2021). Numerous epidemiological studies report a high incidence of lung cancer among workers exposed occupationally to Cr^{6+} by inhalation (Sun et al. 2015).

As the respiratory tract is the main target organ for Cr^{6+} toxicity for both acute short-term and chronic long-term inhalation exposures, apart from the carcinogenic potential, prolonged exposure can result in bronchitis, rhinitis or sinusitis or the formation of nasal mucosal polyps (EPA 1992, ATSDR 2013); perforations and ulcerations of the septum, decreased pulmonary function, pneumonia and other respiratory effects have also been noted from chronic exposure (EPA 1992, Ertani et al. 2017). Limited human studies on the carcinogenic effects of ingested Cr^{6+} are available, although increased stomach cancer mortality for residents in an area where the drinking water was largely contaminated with Cr^{6+} (> 0.5 mg/L) is reported. In another area with water contamination problems, increased rates of liver, lung, kidneys and genitourinary organs cancers were also observed (Sun et al. 2015).

Cr^{3+} is mainly regarded as an essential element and Cr^{3+} supplementation is generally considered safe for humans, the use Cr^{3+} supplements can be seen as leading to improvements in glucose metabolism in type 2 diabetes. Yet, there is growing concern over the possible genotoxicity of these compounds. This may occur because of the presence of partial hydrolysis products of Cr^{3+} in these products are capable of binding to biological macromolecules and alter their functions, and as highly reactive $Cr^{(6+/5+/4+)}$ species and organic radicals can be formed in reactions of Cr^{3+} with biological oxidants (Sawicka et al. 2021).

Biomarkers

Several human biomonitoring studies have presented data on exposure to Cr, in particular to Cr^{6+} using biomarkers (BMs) as tools. These are essential to support improved inferences of regulations and policy making.

Biomarkers of exposure

Exposure to higher-than-normal levels of Cr may result in increased Cr levels in blood, urine, hair, nails, expired air, as Cr elevations in blood and urine are accepted as the most reliable BMs of exposure (Qu et al. 2008, ATSDR 2012).

Chromium levels in blood

Cr concentrations in blood (and urine) are mostly used to measure the degree of acute Cr exposure among workers in industrial sites (Son et al. 2018). Blood measurements may present some possibilities over the determination of urine as they allow to distinguish the sources and the types of exposure (Cr^{6+} versus other forms of Cr) (Qu et al. 2008, ATSDR 2013). This can be especially helpful if urine Cr levels are elevated and one wants to know if this indicates a toxic (e.g., Cr^{6+}) exposure or an essentially benign (e.g., Cr^{3+}) exposure. Since Cr rapidly clears from the blood, measurements in this tissue relate only to recent exposure, but Cr in erythrocytes has been recognized as a BM to index an integrated Cr^{6+} exposure over a lifespan of these cells (approximately 120 d) (Nomiyama et al. 1980, ATSDR 2013). Results from research on workers from a chromate factory, also proposed that erythrocytic Cr levels could serve as sensitive and reliable BMs for long-term exposure to Cr^{6+} (Qu et al. 2008).

For the general population in the absence of known exposure whole blood Cr concentrations are in the range of 2.0 to 3.0 µg/100 mL, the mean levels in serum being 0.10 to 0.16 µg/L; lower levels occur in rural areas, while higher levels occur in large urban centres (Nomiyama et al. 1980, ATSDR 2013).

Chromium levels in urine

When assuming no source of excessive exposure, urinary Cr values are typically less than 10 µg/L for a 24-hr period, in the general population (ATSDR 2012, 2013). Japanese residents in geographic areas without known Cr pollution, exhibited mean Cr levels of 0.41 µg/L and values less than 0.8 µg/L for all age and sex groups (Nomiyama et al. 1980). Urinary biomonitoring in occupational contexts has also been successfully used to assess high-level inhalation exposures to Cr^{6+} (Buisson and Cramer 2012, Paustenbach et al. 1997). A urinary Cr concentration of 40 to 50 µg/L, immediately after a work shift reflects exposure to air levels of 50 µg/m³ of soluble Cr^{6+} compounds, a concentration which was already associated with nasal perforations (ATSDR 2013).

Nevertheless, several factors limit the use of urinary Cr levels to obtain exposure. There are individual variations in metabolism and rapid depletion is the cause for urinary Cd reflecting absorption over the earlier 1 or 2 d only (ATSDR 2013). In addition, Cr levels in 2-hr urine samples do not correlate with those in 24 hr urine (Nomiyama et al. 1980). Evaluating low-level environmental exposures is also a problem as what is indicated in urine is usually the value of Cr^{3+} (as will be further explained) (Alvarez et al. 2021). In particular urinary Cr levels may only represent the total Cr excretion in urine not reflecting the real exposure or uptake of Cr^{6+}, due to several factors which govern the distribution and excretion of Cr^{6+} in the body (Qu et al. 2008). In the bloodstream Cr^{6+} enters red blood cells, but Cr^{3+} does not; inside cells Cr^{6+} is rapidly reduced into Cr^{3+} and trapped inside them, preventing the absorbed Cr^{6+} from being passed in urine, except for a portion which has been reduced in the plasma before entering the cells. Concomitantly Cr^{3+} in blood, whether absorbed or resulting from Cr^{6+} reduction in the plasma, can bind to plasma proteins being Cr^{3+} rapidly passed in urine. Due to these reasons urinary Cr seems to only provide information of total extracellular Cr^{3+} burden without any indication of Cr^{6+} uptake (Qu et al. 2008, ATSDR 2013).

Another constrain is the lack of availability of analytical techniques which allows the analysis of Cr levels for very low concentrations of exposure, which is important since there is no safe exposure to carcinogens (Paustenbach et al. 1997, Buisson and Cramer 2012). Normally biological samples contain concentrations of Cr lower than 1 µg/L and these values are well below the limits of detection for many analytical systems.

Chromium levels in hair or nails

Physiological concentrations of Cr in hair are up to 1,000 times higher than those in serum or urine (Randall and Gibson 1989). Even so some agencies do not defend the use of these biological samples, asserting that hair analysis is of little use in evaluating Cr exposure as it is impossible to distinguish Cr bound within the hair during protein synthesis from Cr deposited on the hair from dust, water or other external sources (such as shampoos or hair dyes) (ATSDR 2013, Son et al. 2018). Projects developed by Sazakli et al. (2014) showed that although positive associations could be found between the exposure dose and Cr levels in hair, no deviation from "normal Cr concentration" was evident in subjects exposed to up to 90 µg/L through drinking water for a long time. In addition, while occupational exposure to Cr seems to play a role in hair levels, it works only for individuals currently employed in the production line of factories with Cr exposure. Factory workers from the past or current workers in factory departments other than the production line, did not show such an association (ATSDR 2013). On the other hand, it is suggested that hair Cr concentrations may be used as indices of industrial exposure to Cr^{3+}, since tannery workers exposed to Cr^{3+} had significantly higher median Cr concentrations in hair compared with unexposed controls (Randall and Gibson 1989). Populations with no known Cr exposure have hair levels ranging from 50 to 100 mg/L (ATSDR 2013). In other reports reference values for Cr in hair were 0.03–1.20 and 0.001–4.56 µg/g; in a study from Greece, the 25th and 75th percentiles of Cr in hair were found to be 0.33 and 0.84 µg/g, respectively (Sazakli et al. 2014).

Augmented Cr levels in nails were already considered as being interpreted as a sign of increased health risk, for workers occupied in the production of special steels, mineral pigments and Cr plating (Georgieva and Tsalev 2014). This has been refuted by researchers defending that when Cr is measured in fingernails, concentrations are more likely to be contaminated by external materials, such as nail polishes and hand cream; rather, toenail Cr concentrations are usually less contaminated and, therefore, may be the most useful for assessing long-term Cr exposure (Son et al. 2018). One study described that toenail Cr concentrations were associated with the incidence of metabolic syndrome in young American adults, for which baseline median toenail Cr concentrations ranged from 0.2 (first quartile) to 3.5 µg/g (fourth quartile). European studies reported median toenail Cr concentrations of 0.5 µg/g for adults in Ireland, and 1.30 µg/g among men living in 10 European regions and in Iran (Son et al. 2018).

Other biomarkers

Exhaled breath condensate is a promising means of developing BMs of inhaled pneumotoxic metallic elements, providing complementary information to traditional blood and urine biomonitoring. Cr levels in this biological sample has been proposed as a BMs of exposure, given the correlations found with the concentrations of the metal in air and with Cr levels in the pulmonary tissue of patients with non-small cell lung cancer. Additional studies showed a close relationship between both Cr and Cr^{6+} concentrations in exhaled air and erythrocytic Cr levels, in an occupational context (Goldoni et al. 2010).

Efforts has also been developed to use DNA-Protein Cross-Links (DPC) measurements as BMs of Cr exposure, since lymphocytic DPC correlate largely with Cr levels in erythrocytes, which as mentioned earlier are considered indicative of Cr^{6+} exposure (Zhitkovich et al. 1998).

Biomarkers of susceptibility

Human susceptibility to Cr-induced pathologies may depend on individual characteristics, such as enzymatic polymorphisms, carriers, endogenous reducing systems, adduct formation and stability and efficiency of DNA repair mechanisms, among other factors (Urbano et al. 2012, Pavesi and Moreira 2020). Genomic and epigenomic variations seem to justify distinct responses generated by similar Cr exposure in different subjects (Mignini et al. 2004, Martinez-Zamudio and Ha 2011, Pavesi and Moreira 2020).

Genetic biomarkers

Polymorphisms are recognized as genetic factors creating an individualized environment for Cr toxicity. Associating erythrocytic Cr levels and polymorphisms in genes encoding proteins that carry anions or proteins in endoplasmic reticulum membranes, such as the erythroid anion exchange protein (EPB3) gene, is an example; higher erythrocytic Cr levels are reported in subjects with the EPB3 wild-type gene, when compared to variant alleles (Pavesi and Moreira, 2020).

Cr^{3+} formation from the intracellular reduction of Cr^{6+} is necessary for decreasing Cr^{6+} body burden, Cr reducing enzymatic systems present genetic variability that may affect the kinetics and efficiency of those enzymatic processes (Pavesi and Moreira 2020). This is the case of NAD(P)H: quinone oxidoreductase (NQO1), also known as DT-diaphorase, for which polymorphic forms exists among human populations. A transition of base C to T in the 609 codon of NQO1 results in no detectable NQO1 activity, which clearly affect the rate of Cr^{6+} reduction (Sharma et al. 2012).

Another example is Glutathione S-Transferase (GST), involved in the metabolic detoxification of various environmental toxics. In an appreciable percentage of the human population, two of the most relevant GST isoenzymes, GSTM1 (mu) and GSTT1 (theta), are non-functional (due to deletion of a portion of the coding gene); these deficiencies have been suggested to play an important role in cancer susceptibility. With respect to susceptibilities of Cr toxicity, variations in the polymorphic status of the GSTM1 gene was shown as probably influencing dermal outcomes among residents from Cr^{6+} contaminated areas and among Cr-exposed workers (Sharma et al. 2012, Pavesi and Moreira 2020).

The enzyme 8-oxoguanine glycosylase I (hOGG1), which is expressed as one of the DNA base excision repair genes, is another case. hOGG1 presents polymorphic forms, with one of these forms having an affected biological activity. This genotype is considered a potential BMs for individual susceptibility to lung cancer risk in occupational exposure to Cr^{6+}; a higher susceptibility to Cr^{6+} induced DNA damages, which could explain inter-individual variations observed in effect BMs in response to Cr^{6+} exposures (Qu et al. 2006, Pavesi and Moreira 2020).

Human glutathione peroxidase (GPX) is an antioxidant enzyme seen in blood vessels, its main role being to protect cells against oxidative damage (Buraczynska et al. 2017). The most common isoform of GPX is GPX1 and it was found that heterozygous alleles of either hGPX1 Pro197Leu or Pro198Leu polymorphisms made people susceptible to Cr^{6+} associated DNA Protein Crosslink (DPC); elevated levels of DPC were determined in peripheral blood lymphocytes among welders, chrome platers and leather tanners (Qu et al. 2006, Macfie et al. 2010). A likely impediment of DNA replication by Cr-induced DPC is suggested to provoke gross genetic rearrangements, mutations or S-phase specific DNA double-strand breaks (Macfie et al. 2010).

Epigenetic biomarkers

Epigenetic modifications provide a plausible link between the environment and alterations in gene expression that might lead to disease phenotypes. Epidemiological evidence increasingly suggests that environmental exposures early in development have a role in susceptibility to disease in later life, and that these environmental effects seem to be passed on through subsequent generations (Jirtle and Skinner 2007). Epigenetics refers to the reversible yet heritable changes in gene expression, independent of DNA sequence, caused by DNA hypo- or hypermethylation, histone tail post-translational modifications or microRNAs (miRNAs) expression (Chen et al. 2019). Epigenetic modifications altering the patterns of gene expression are critical factors for cancer development due to exposure to carcinogenic agents, such as Cr. Only. Extensive studies have shown that Cr^{6+} is capable altering gene expression and induce cancer development through multiple epigenetic mechanisms and that such modifications may occur through the action of ROS formed from Cr^{6+} reduction (Chen et al. 2019, Pavesi and Moreira 2020). It is also acknowledged that the mitochondrially encoded tRNA phenylalanine (MT-TF) and the mitochondrially encoded 12S RNA (MT-RNR1) genes in mitochondrial DNA (mtDNA), are hypomethylated in the presence of Cr. The methylation level is negatively associated to blood Cr concentration, suggesting it as an alternative BM for Cr exposure.

In addition, the NOD-, LRR- and pyrin domain-containing protein 3 (NLRP3) is a multimeric protein receptor that acts as a cytosolic innate immune signalling platform and a is key mediator of inflammation and immunity (Bai et al. 2021). NLRP3, whose activation is fine-tuned by miRNAs, is a critical component of the innate immune system mediating caspase-1 activation and the secretion of the proinflammatory cytokines interleukin (IL)-1β/IL-18 in response to cellular damage; aberrant activation of the NLRP3 inflammasome has been linked with several inflammatory disorders (Kelley et al. 2019). There is evidence that activation of the NLRP3 inflammasome and consequent increase in IL-1β, is also related to Cr^{6+}-induced toxicity mechanisms (Pavesi and Moreira 2020). Recent studies have shown that pterostilbene attenuates Cr^{6+} induced allergic contact dermatitis by inhibiting IL-1β-related NLRP3 inflammasome activation, which consubstantiate these evidences (Wang et al. 2018).

It is also described that Cr^{6+} silences the tumour suppressor gene MLH1 through promoting hypermethylation in human lung cell cancer (Sun et al. 2009, Pavesi and Moreira 2020). The capacity of chromate to modulate histone methylation and subsequently silence specific tumour suppressor genes may underlie Cr carcinogenicity (Sun et al. 2009). Accordingly, MLH1 gene promoter methylation was found to have increased in 63% of tumours of patients occupationally exposed to Cr^{6+}, when compared to unexposed subjects, MLH1 inactivation being strongly related to microsatellite instability.

MicroRNAs (miRNAs) are non-coding, small single-stranded RNAs with important regulatory roles in gene expression, being aberrant miRNA expression implicated in diabetes, insulin resistance and cardiovascular disease (CVD). Evidence exist that Cr^{6+} disrupts certain transcriptional pathways through direct deregulation of the miRNA expression profile; association of CVD with exposure to Cr was demonstrated earlier. Cr exposure is negatively associated with microR-146a in peripheral blood leukocytes, micro R-143 in epithelial cells and micro R-3940-5p plasma levels, among several other correlations (Pavesi and Moreira 2020). Changes in miR-451 and miR-486-3p expression, in association with urinary Cr are also reported (Dioni et al. 2017).

Biomarkers of effect

Although several human biomonitoring studies have presented data on exposure to Cr and Cr^{6+}, only a few include results on effect BMs, which were used only in occupational contexts. They is certainly a need to identify associations between exposure and adverse outcomes to assess public health implications (Ventura et al. 2021).

Traditional tests currently considered include a complete blood count, liver function tests [aspartate aminotransferase (AST) or serum glutamic-oxaloacetic transaminase (SGOT), alanine aminotransferase (ALT) or serum glutamic-pyruvic transaminase (SGPT), and bilirubin], Blood Urea Nitrogen (BUN), creatinine and urinalysis (ATSDR 2013, Siti et al. 2020).

Hexavalent chromium

The association of Cr^{6+} exposure with hepatotoxicity is well established, as well as the induction of apoptosis in human liver cells (HepG2) via redox imbalance; Cr hepatotoxicity exacerbate liver function and is involved in fibrosis/cirrhosis and carcinogenesis (Wang and Lin 2013, Das et al. 2015).

Haematological parameters and inflammation Since it is recognized that occupational exposure to Cr^{6+} may induce several haematological disorders, the haematological profile seems to be a potent indicator of Cr toxicity; decreased lymphocytes count and an increase in basophils number are described as associated to Cr exposures (Ventura et al. 2021, Ray 2016).

The C-reactive protein (hs-CRP) is an acute phase protein synthesized by the liver in response to interleukin-6 secretion by macrophages and T cells, being a nonspecific BM of inflammation (Kerner et al. 2005). Serum hs-CRP was already found to be significantly elevated in Cr^{6+}-exposed workers, while interestingly, other studies show that Cr^{3+} supplementation can significantly reduce the levels of this protein in serum, helping to improve inflammation (Ventura et al. 2021, Zhang et al. 2021). However, when considering inflammation, lymphocyte induced proliferation is a parameter used to evaluate the subject cell-mediated immune responsiveness (Nikbakht et al. 2019). While some studies did not show Cr-induced changes in this BM in stainless steel welders exposed to welding fumes containing Cr, levels were found to be augmented in shoe, hide and leather industry workers exposed to Cr^{6+} (Shrivastava et al. 2002, Ventura et al. 2021).

Renal parameters Controversially negative results are reported for the mentioned renal parameters. Even so, the concentration of certain proteins and enzymes in the urine of workers are recognized as indicating renal effects of Cr^{6+} (NIOSH 2013). To relate exposure to Cr (and to Cd) with renal tubular dysfunction, renal BMs such as β-2-microglobulin (B2-MG) and N-acetyl-β-D-glucosaminidase (NAG) are well-established, considered sensitive and are the most used; namely, hard-chrome platers exposed to airborne Cr^{6+} concentrations of 4.20 μg/m³ had significantly higher urinary NAG concentrations (4.9 IU/g creatinine) than other metal exposed workers (NIOSH 2013). As with Cd, KIM-1 could also serve as sensitive BMs of acute kidney damage in response to Cr^{6+} and might additionally serve as an early and sensitive indicator of proximal tubular damage in children (ATSDR 2013, Ventura et al. 2021). A dose-dependent association between urinary KIM-1 across Cr exposure was found in Mexican children which had Cr urinary values higher than the biological exposure index (Ventura et al. 2021).

Genotoxicity Cr^{6+} is known to be genotoxic and could induce DNA single- and Double-Strand Breaks (DSBs); unrepaired or erroneously repaired DBS may give rise to structural chromosome anomalies (Varga and Aplan 2005, Nickens et al. 2010). These anomalies can be assessed through chromosome aberrations or micronuclei (MN) in human peripheral lymphocytes or in buccal cells (in the case of MN) (Ventura et al. 2021). Workers with higher levels of Cr in urine, erythrocytes and lymphocytes, than unexposed controls, were demonstrated to have increased DNA strand breaks, as evaluated by comet assays (NIOSH 2013). After combining the analyses of chromosome aberrations, micronuclei and DNA damage in peripheral blood cells, a higher degree of genetic alterations in environmentally Cr^{6+}-exposed subjects (due to their living in the vicinity of tannery industries) was also detected (Ventura et al. 2021).

It seems that inside cells Cr^{6+} and Cr^{5+} interact with adenines and guanines of genomic DNA to cause bulky DNA adducts and oxidative DNA damage, which are poorly repaired (Ventura et al. 2021). Cr is generally believed to induce mainly mutagenic binary and ternary Cr^{3+}–deoxyguanosine (dG)-DNA adducts in human cells, with Cr^{6+}-induced bulky DNA adducts and oxidative DNA damage may contribute to mutagenesis of the p53 gene, leading to lung carcer (Arakawa et al. 2012). Cr^{6+}-induced extensive formation of DNA adducts and pronounced positivity in genotoxicity tests with high predictive values for carcinogenicity, are experimentally observed (Sawicka et al. 2021).

Other parameters BMs targeting DNA and lipid peroxidation, respectively 8-hydroxy-2'-deoxyguanosine (8-OHdG) and malondialdehyde (MDA), are used most frequently to associate exposure to Cr^{6+} with cancer development (Ventura et al. 2021). Positive correlations are reported between urinary 8-OHdG concentrations and both urinary Cr concentration and airborne Cr concentration in an electroplating workplace (NIOSH 2013).

In vivo cellular senescence phenotype is accompanied by elevated protein levels of apolipoprotein J/clusterin (ApoJ/CLU), which is a heterodimeric secreted glycoprotein regulated differentially in many severe physiological disturbance states, which include neurological diseases, and *in vivo* cancer progression. Blood and urine samples from shipyard industry welders (exposed to different levels of Cr^{6+} over a period of 5 mon) were analyzed to assess the relation of ApoJ/CLU serum levels with exposure to Cr^{6+}. In agreement with earlier *in vitro* data, reduction of Cr levels after a worksite intervention, resulted in lower levels of ApoJ/CLU serum levels (Alexopoulos et al. 2008).

Trivalent chromium

Despite being less investigated, occupational exposure to Cr^{3+} can induce oxidative stress and DNA damage, namely in tannery workers. Research demonstrated that blood Cr levels, DNA damage and superoxide dismutase (SOD) and MDA levels, were significantly higher is an exposed group of workers; their glutathione (GSH) concentrations were notably lower. Important correlations between blood Cr levels and DNA damage were also detected (Sawicka et al. 2021). Other studies with workers exposed to Cr^{3+} associated blood Cr levels with increased p53 expression (Elhosary et al. 2014).

Final Remarks

This chapter refers to two valence states of Cr with completely different features in terms of origin, the role in the human body and toxicity. Cr^{3+} is a natural essential nutrient whose deficiency may result in serious adverse health effects; benefits are even attributed to nutritional supplementation and general populations are typically exposed to Cr^{3+}. In excessive levels due to intake or occupational exposure, Cr^{3+} can induce toxicity. Insufficient studies have been conducted on Cr^{3+} overexposure (namely unsupervised consumption of nutritional supplements), their consequences or even on BMs which could provide such information. Much more data exists on Cr^{6+} due to the severity of the adverse health outcomes attributed to exposures to this chemical specie. Serious relevance are occupational exposures for which BMs of exposure, and in a much lesser extend BMs of susceptibility, have been explored; these BMs have been also been investigated for environmental exposures. In a different manner, BMs of effect are mostly used in occupational contexts pointing to the need of develop and apply them for long-term low-level exposures in the general population , namely when considering the strong carcinogenic potential of this dangerous metal specie.

References

Aherrera, A., P. Olmedo, M. Grau-Perez, S. Tanda, W. Goessler, S. Jarmul et al. 2017. The association of e-cigarette use with exposure to nickel and chromium: A preliminary study of non-invasive biomarkers. Environ. Res. 159: 313–320.

Alexopoulos, E.C., X. Cominos, I.P. Trougakos, M. Lourda, E.S. Gonos and V. Makropoulos. 2008. Biological monitoring of hexavalent chromium and serum levels of the senescence biomarker apolipoprotein J/Clusterin in welders. Bioinorg. Chem. Appl. 2008: 420578.

Alvarez, C.C., M.E. Bravo Gomez and A.H. Zavala. 2021. Hexavalent chromium: Regulation and health effects. J. Trace Elem. Med. Biol. 65: 126729.

Arakawa, H., M.W. Weng, W.C. Chen and M.S. Tang. 2012. Chromium (VI) induces both bulky DNA adducts and oxidative DNA damage at adenines and guanines in the p53 gene of human lung cells. Carcinogenesis. 33(10): 1993–2000.

[ATSDR] Agency for Toxic Substance and Disease Registry, ToxGuideTM for Chromium. CAS# 7440-47-3, 2012. U.S. Department of Health and Human Services, Public Health Service, Atlanta, USA.

[ATSDR] Agency for Toxic Substances and Disease Registry. 2013. Chromium Toxicity. Clinical Assessment—History, Signs and Symptoms. Agency for Toxic Substances and Disease Registry, U.S. Department of Health & Human Services, Atlanta, USA.

Bai, B., Y. Yang, S. Ji, S. Wang, X. Peng, C. Tian et al. 2021. MicroRNA-302c-3p inhibits endothelial cell pyroptosis via directly targeting NOD-, LRR- and pyrin domain-containing protein 3 in atherosclerosis. J. Cell Mol. Med. 25: 4373–4386.

Basketter, D.A., G. Angelini, A. Ingber, P.S. Kern and T. Menné. 2003. Nickel, chromium and cobalt in consumer products: Revisiting safe levels in the new millennium. Contact Derm. 49(1): 1–7.

Buisson, D. and B. Kramer. 2012. Chromium (III & VI) toxicity. It works well in practice, but how does it work in theory? Occupational Health Southern Africa. 18(1): 23–26.

Buraczynska, M., K. Buraczynska, M. Dragan and A. Ksiazek. 2017. Pro198Leu Polymorphism in the glutathione peroxidase 1 gene contributes to diabetic peripheral neuropathy in type 2 diabetes patients. Neuromol. Med. 19(1): 147–153.

Chen, Q.Y., A. Murphy, H. Sun and M. Costa. 2019. Molecular and epigenetic mechanisms of Cr (VI)-induced carcinogenesis. Toxicol. Appl. Pharmacol. 377: 114636.

Das, J., A. Sarkar and P.C. Sil. 2015. Hexavalent chromium induces apoptosis in human liver (HepG2) cells via redox imbalance. Toxicol. Rep. 2: 600–608.

Dioni, L., S. Sucato, V. Motta, S. Iodice, L. Angelici, C. Favero et al. 2017. Urinary chromium is associated with changes in leukocyte miRNA expression in obese subjects. Eur. J. Clin. Nutr. 71: 142–148.

[EFSA] European Food Safety Authority 2014. Chromium in food and drinking water. EFSA Journal. 12(3): 3595.

Elhosary, N., A. Maklad, E. Soliman, N. El-Ashmawy and M. Oreby. 2014. Evaluation of oxidative stress and DNA damage in cement and tannery workers in Egypt. Inhal. Toxicol. 26: 289–298.

[EPA] Environmental Protection Agency, Chromium Compounds, Hazard Summary. 1992. EPA, Washington, USA.

Ertani, A., A. Mietto, M. Borin and S. Nardi. 2017. Chromium in agricultural soils and crops: A review. Water Air Soil Pollut. 228: 190.

Feng, L., X. Guo, T. Li, C. Yao, H. Xia, Z. Jiang et al. 2020. Novel DNA methylation biomarkers for hexavalent chromium exposure: an epigenome-wide analysis. Epigenomics. 12(3): 221–233.

[FSCJ] Food Safety Commission of Japan. 2019. Hexavalent chromium (Contaminants). Food Saf. (Tokyo). 28; 7(2): 56–57.

Georgieva, R. and D.L. Tsalev. 2014. Chromium levels in erythrocytes, nails and urine as biomarkers of exposure—informational value and relevance. Bulg. J. Chem. 3: 21–31.

Goldoni, M., A. Caglieri, G. De Palma, O. Acampa, P. Gergelova, M. Corradi et al. 2010. Chromium in exhaled breath condensate (EBC), erythrocytes, plasma and urine in the biomonitoring of chrome-plating workers exposed to soluble Cr (VI). J. Environ. Monit. 12: 442–447.

Jirtle, R.L. and M.K. Skinner. 2007. Environmental epigenomics and disease susceptibility. Nat. Rev. Genet. 8(4): 253–262.

Kelley, N., D. Jeltema, Y. Duan and Y. He. 2019. The NLRP3 Inflammasome: An overview of mechanisms of activation and regulation. Int. J. Mol. Sci. 20(13): 3328.

Kerner, A., O. Avizohar, R. Sella, P. Bartha, O. Zinder, W. Markiewicz et al. 2005. Association Between Elevated Liver Enzymes and C-Reactive Protein-Possible Hepatic Contribution to Systemic Inflammation in the Metabolic Syndrome. Arterioscler. Thromb. Vasc. Biol. 25(1): 193–197.

Kim, J., S. Seo, Y. Kim and D.H. Kim. 2018. Review of carcinogenicity of hexavalent chrome and proposal of revising approval standards for an occupational cancer in Korea. Ann. Occup. Environ. Med. 30: 7.

Lisboa, T.P., A.M.S. Mimura, J.C.J. da Silva and R.A. de Sousa. 2020. Chromium levels in tobacco, filter and ash of illicit brands cigarettes marketed in Brazil. J. Anal. Toxicol. 44(5): 514–520.

Macfie, A., E. Hagan and A. Zhitkovich. 2010. Mechanism of DNA-protein cross-linking by chromium. Chem. Res. Toxicol. 23(2): 341–347.

Martinez-Zamudio, R. and H.C. Ha H. 2011. Environmental epigenetics in metal exposure. Epigenetics 6(7): 820–827.

Mignini, F., V. Streccioni, M. Baldo, M. Vitali, U. Indraccolo, G. Bernacchia et al. 2004. Individual susceptibility to hexavalent chromium of workers of shoe, hide, and leather industries. Immunological pattern of HLA-B8, DR3-positive subjects. Prev. Med. 39(4): 767–775.

Nickens, K.P., S.R. Patierno and S. Ceryak. 2010. Chromium genotoxicity: A double-edged sword. Chem.-Biol. Interact. 188(2): 276–288.

[NIH] National Institutes of Health. 2021. Chromium - Fact Sheet for Consumers. National Institutes of Health, Office of Dietary Supplements, Bethesda, USA.

Nikbakht, M., B. Pakbinand and G.N. Brujeni. 2019. Evaluation of a new lymphocyte proliferation assay based on cyclic voltammetry; an alternative method. Sci. Rep. 9: 4503.

[NIOSH] National Institute for Occupational Safety and Health, Occupational Exposure to Hexavalent Chromium, 2013. NIOSH, Washington, USA.

Nomiyama, H., M. Yotoriyama and K. Nomiyama. 1980. Normal chromium levels in urine and blood of Japanese subjects determined by direct flameless atomic absorption spectrophotometry, and valency of chromium in urine after exposure to hexavalent chromium. Am. Ind. Hyg. Assoc. J. 41(2): 98–102.

Oliveira, H. 2012. Chromium as an environmental pollutant: insights on induced plant toxicity. J. Bot. 2012: 375843.

[OSHA] Occupational Safety and Health Administration, Hexavalent Chromium, 2021. OSHA, Washington, USA.

Paustenbach, D.J., J.M. Lanko, M.M. Fredrick, B.L. Finley and D.M. Proctor. 1997. Urinary chromium as a biological marker of environmental exposure: what are the limitations? Regul. Toxicol. Pharmacol. 26(1 Pt 2): S23–34.

Pavesi, T. and J.C. Moreira. 2020. Mechanisms and individuality in chromium toxicity in humans. J. Apll. Toxicol. 40: 1183–1197.

Qu, Q., L. Xiaomei, R. Shore, F. An, G. Jia, L. Liu et al. 2006. Cr (VI) exposure: biomarkers and genetic susceptibility. Proc. Amer. Assoc. Cancer Res. 66(8): 108.

Qu, Q., X. Li, F. An, G. Jia, L. Liu, H. Watanabe-Meserve et al. 2008. CrVI exposure and biomarkers: Cr in erythrocytes in relation to exposure and polymorphisms of genes encoding anion transport proteins. Biomarkers. 13(5): 467–477.

Randall, J.A. and R.S. Gibson. 1989. Hair chromium as an index of chromium exposure of tannery workers. Br. J. Ind. Med. 46: 171–175.

Ray, R.R. 2016. Adverse hematological effects of hexavalent chromium: an overview. Interdiscip. Toxicol. 9(2): 55–65.

Samal, S., P. Debata and S.K. Swain. 2020. Role of chromium enriched tobacco in the occurrence of oral carcinogenesis. Int. J. Curr. Res. Rev. 12(18): 20–24.

Sawicka, E., K. Jurkowska1 and A. Piwowar. 2021. Chromium (III) and chromium (VI) as important players in the induction of genotoxicity—current view. AAEM. 28(1): 1–10.

Sazakli, E., C.M. Villanueva, M., Kogevinas, K. Maltezis, A. Mouzaki and M. Leotsinidis. 2014. Chromium in drinking water: Association with biomarkers of exposure and effect. Int. J. Environ. Res. Public Health. 11(10): 10125–10145.

Sharma, P., V. Bihari, S.K. Agarwal and S.K. Goel. 2012. Genetic predisposition for dermal problems in hexavalent chromium exposed population. J. Nucleic Acids. 2012: 968641.

Shrivastava, R., R.K. Upreti, P.K. Seth and U.C. Chaturvedi. 2002. Effects of chromium on the immune system. FEMS Immunol. Med. Microbiol. 34: 1–7.

Siti, A., A.R. Tualeka and Y.D. Ardyanto. 2020. The effect of chromium exposure on creatinine and bun level of tanners in leather industry in magetan. Indian J. Forensic Med. Toxicol. 14(3): 1115–1121.

Son, J., J.S. Morris and K. Park. 2018. Toenail chromium concentration and metabolic syndrome among korean adults. Int. J. Environ. Res. Public Health. 15(4): 682.

Sun, H., J. Brocato and M. Costa. 2015. Oral chromium exposure and toxicity. Curr. Environ. Health Rep. 2(3): 295–303.

Sun, H., X. Zhou, H. Chen, Q. Li and M. Costa. 2009. Modulation of histone methylation and MLH1 gene silencing by hexavalent chromium. Toxicol. Appl. Pharmacol. 237(3): 258–66.

Urbano, A.M., L.M.R. Ferreira and M.C. Alpoim. 2012. Molecular and cellular mechanisms of hexavalent chromium-induced lung cancer: An updated perspective. Curr. Drug Metab. 13: 284–305.

Varga, T. and P.D. Aplan. 2005. Chromosomal aberrations induced by double strand DNA breaks. DNA Rep. 4(9): 1038–1046.

Ventura, C., B. Costa Gomes, A. Oberemm, H. Louro, P. Huuskonen, V. Mustieles et al. 2021. Biomarkers of effect as determined in human biomonitoring studies on hexavalent chromium and cadmium in the period 2008–2020. Environ. Res. 197: 110998.

Vincent, J.B. and H.C. Lukaski. 2018. Chromium. Adv. Nutr. (Bethesda, Md.) 9(4): 505–506.

Wang B.-J., H.-W. Chiu, Y.-L. Lee, C.-Y. Li, Y.-J. Wang and Y.-H. Lee. 2018. Pterostilbene attenuates hexavalent chromium-induced allergic contact dermatitis by preventing cell apoptosis and inhibiting IL-1β-Related NLRP3 Inflammasome Activation. J. Clin. Med. 7(12): 489.

Wang, K. and B. Lin. 2013. Pathophysiological significance of hepatic apoptosis. ISRN Hepatology. 2013: 740149.

[WHO] World Health Organization Regional Office for Europe, Air quality guidelines for Europe, 2000. WHO, Geneva, Switzerland.

Wyszkowska, J. 2002. Soil contamination by chromium and its enzymatic activity and yielding. Pol. J. Environ. Stud. 11(1): 79–84.

Zampella, M., P. Adamo, L. Caner, S. Petit, D. Righi and F. Terribile. 2010. Chromium and copper in micromorphological features and clay fractions of volcanic soils with andic properties. Geoderma. 157(3-4): 185–195.

Zhang, X., L. Cui, B. Chen, Q. Xiong, Y. Zhan, J. Ye et al. 2021. Effect of chromium supplementation on hs-CRP, TNF-α and IL-6 as risk factor for cardiovascular diseases: A meta-analysis of randomized-controlled trials. Complement. Ther. Clin. Pract. 42: 101291.

Zhitkovich, A., V. Voitkun, T. KIuz and M. Costa. 1998. Utilization of DNA-protein cross-links as a biomarker of chromium exposure. Environ. Health Perspect. 106(4): 969–974.

CHAPTER 8

Lead

Source of Exposure

Lead (Pb) is a naturally occurring element found in small amounts on the Earth's crust, which can be released into air, soil and water through natural processes, such as soil erosion, volcanic eruptions, sea spray and bushfires (AD GAWE 2021). Since humans have used Pb in various applications for thousands of years and Pb does not breakdown over time, some of the practices of the past have led to serious environmental and human health problems. Thus, Pb is still a ubiquitous pollutant which can enter the environment both from past and current uses (AD GAWE 2021, EPA 2020, 2021b, WHO 2001).

Concerning the past applications , Pb-based house paint and leaded gasoline still constitute the major sources of environmental Pb at the present time (Charkiewicz and Backstrand 2020). Pb poisoning from deteriorating old paint is the primary source of elevated blood Pb levels in children, and in spite of motor-vehicle emissions being reduced by the phasing out of leaded gasoline, Pb is still used in general-aviation gasoline for piston-engine aircraft (Lin et al. 2020, TCEQ 2020, OSHA 2021).

Current emissions comprise Pb smelting, waste incinerators utilities, Pb-acid battery manufacturers and recycling (TCEQ 2020). Pb is still utilized in battery plates and equipment, in the production of sulphuric acid, cable covers, soldering materials, shields in atomic reactors, containers for radioactive materials, in paint and ceramics, in chemical and construction industries, glass and cement manufacturing and in the production of bearings and printing fonts (Charkiewicz and Backstrand 2020, TCEQ 2020, EPA 2021b). Pb can enter the human body via ingestion or inhalation, from Pb dust, in the workplace, from soil, water, food and dermal contact with products of everyday use (Charkiewicz and Backstrand 2020, EPA 2021b).

Air

Inhalation and ingestion of Pb settled on surfaces is the main way for human exposure to Pb originally released into the air, since most Pb in the air is in the form of fine particles (WHO 2001, Larssen and Hagen 2020, TCEQ 2020). Studies in China have shown that high Pb concentrations can be found in fine Particulate Matter (PM) with diameters less than 2.5 μm (PM2.5) and in inhalable particles with diameters less than 10 μm (PM10); both PM2.5 and PM10 can penetrate deep into the lungs (Shang and Sun 2018). The fraction of organic Pb (predominantly Pb alkyls) is generally below 10% of the total atmospheric Pb, the majority (> 90%) from leaded petrol emissions as inorganic particles (such as $PbBrCl$) (WHO 2001). It has been estimated that about 0.33×10^9 kg/yr of Pb is directly emitted into the atmosphere (AD GAWE 2021). Fortunately, air Pb concentrations in industrial and urban areas with high traffic density have decreased over the past year, subsequently leading to the decline of industrial emissions, reductions in the Pb content of petrol and increasing use of Pb-free petrol. A typical example is the annual mean of air Pb concentrations reported for an industrialized urban region in Germany; concentrations found in air were 0.81 to 1.37 μg/m^3 in 1974 and after 12 yr decreased to 0.17–0.19 μg/m^3. Pre-industrial levels of Pb in air from natural origins were about 0.6 ng/m^3, and today are usually below 0.15 μg/m^3 at non-urban sites. Mean Pb levels in several cities in Europe are now well below the European Union (EU) limit value, with levels above 0.3 μg/m^3 measured in some cities in Spain and Italy (WHO 2001, Larssen and Hagen 2020). Higher air Pb concentrations may be found in the vicinity of primary or secondary Pb smelters since larger particles predominate and settle at distances of only a few hundred m or 1–2 km; further away, the particle size distribution is indistinguishable from that of other urban sites (WHO 2001). The Environmental Protection Agency (EPA) establishes a standard value with the basis on a relationship between ambient air Pb and blood Pb concentrations, presuming that a blood Pb level of 0.15 μg/ mL (mean value for children) can be achieved at an ambient air Pb level of 1.5 μg/ m^3 (Larssen and Hagen 2020). The value settled in the EU is 0.5–1 μg/m^3, while the World Health Organization (WHO) sets the limit as 0.5μg/m^3 (annual average) (Larssen and Hagen 2020, Nag and Cummins 2021).

Workplace

In the work environment the main course of absorption of Pb and its compounds is through the respiratory system, although ingestion is also likely to occur (Charkiewicz and Backstrand 2020). Workers are exposed to Pb during production, use, maintenance, recycling and disposal of Pb materials and products in several industrial sectors involving construction, manufacturing, wholesale trade, transportation, remediation and even recreation. Exposures may take place during removal, renovation or demolition of structures painted with Pb pigments, when installing, maintaining or in the demolition of Pb pipes and fittings, Pb linings in tanks and radiation protection, in leaded glass work or in work involving soldering. Another source of exposure appears when workers take home Pb dust on their clothing and shoes (OSHA 2021). The EU has an 8-hr Time Weighted Average (TWA) value settled as 100 μg/m^3 for inorganic Pb, including Pb fumes and dusts with particle sizes below 10 μm (NRC 2012). Both the Occupational Safety and Health Administration (OSHA) and the National Institute for Occupational Safety and Health (NIOSH) sets a Permissible Exposure Limit (PEL) or Recommended Exposure Limit (REL), respectively, for Pb in workplace air of 50 μg/m^3 (8-hr time weighted average) (ATSDR 2019, CDC 2021). OSHA mandates periodic determination of blood Pb levels for those exposed to air concentrations at or above the action level of 30 µg/m^3 for more than 30 d in a year. In general industries the employer is required to remove an employee from excessive exposure, with maintenance of seniority and pay, until the employee's levels of Pb in blood falls below 40 µg/dL if a worker's one-time value reaches 60 µg/dL (ATSDR 2019). However, some studies suggest that the current OSHA PEL and the NIOSH REL may be too high to protect against certain health effects (CDC 2021).

Soil

Pb in soil can be distributed in a range of discrete mineral phases, with co-precipitated or sorbed Pb associated with soil minerals, clay and organic matter and dissolved Pb complexed with varied organic and inorganic ligands (Yan et al. 2017). Children and adults can be exposed to Pb in soil by playing, gardening, eating fruits and vegetables grown in contaminated areas, ingesting soil and touching hands to mouths (typically young children) (EPA 2020). Pb occurs naturally in the soil at concentrations between 10 to 50 mg/kg, but due to hundreds of years of human activities increased levels are found, especially in and around urban areas, and near older homes (Dudka and Miller 1999, EPA 2021b). Soil Pb levels in many urban areas exceed 200 mg/L and although Pb soil levels have generally decreased since 1970, they reduced much less near residential grounds . Soil samples near houses built before 1978, averaged 649 mg/kg of Pb which is more than three times the average level detected near streets (ATSDR 2020, Wade et al. 2021). Other areas where Pb levels in soil are augmented include places near roadways (because of air emissions from vehicles that used leaded gasoline in the past), near toxic waste sites and in other localities close to industrial facilities that release Pb into the environment (mining, smelting and refining industries) (EPA 2020, 2021). Additionally, when Pb is released to the air from industrial sources or spark-ignition engine aircraft, Pb may travel long distances before settling to the ground, where it usually sticks to soil particles (EPA 2021b). Maximum Permissible Levels (MPL) of Pb in soil have been recommended based on dose-response relationships between Pb in soil and blood Pb in children. An MPL of 250 mg/kg is recommended in areas without grass cover and repeatedly used by children below 5 yr of age, among whom mouthing objects is largely present; this level is believed to keep blood Pb levels under 2 µg/L in 99% of children, when soil is the main course through which they are exposed (Madhavan et al. 1998). EPA standards for Pb in bare soil on play areas is 400 mg/kg and 1200 mg/kg on non-play areas (ATSDR 2020).

Water

Pb may move from soil into ground water depending on the type of Pb compound and soil features (EPA 2021b). Despite exposure to Pb through drinking water is generally low (compared with exposure through food) different situations could occur when the metal enters drinking water, namely through corroded plumbing materials containing Pb (WHO 2001, EPA 2021a); this can happen particularly where the water has high acidity or low mineral content, corroding pipes and fixtures. These are the most common sources, which are more likely to be found in older cities and homes built before the 80s (Brown and Margolis 2012, EPA 2021a). Blood Pb levels in 6 yr old children living in houses with Pb pipes used for domestic drinking-water supply were found to be elevated by about 30%, relative to the ones living in houses without Pb pipes (WHO 2001). The average levels of Pb in blood declined following the removal of Pb from water pipes that were the current most common problems in homes without Pb service lines, brass or chrome-plated brass faucets and plumbing with Pb solder (Zahran et al. 2012, EPA 2021a).

Pb concentrations in groundwater and drinking-water vary from 1 µg/L to 60 µg/L. In most European countries, the levels of Pb in domestic tap water are relatively low, normally 20 µg/L (WHO 2001). Similarly, Pb levels in drinking water supplies do not represent a hazard in Canada since concentrations are normally below their maximum acceptable concentration which is 0.010 mg/L (HC 2009). WHO (2011) sets the Pb drinking water limits as 10µg/L, while in the USA threshold is 5µg/L (U.S. Code of Federal Regulations, 2019, Nag and Cummins 2021). US EPA is even more cautious considering a maximum contaminant level goal for Pb in drinking water as zero, since Pb is considered as harmful to human health even at low levels. According to the agency drinking water can make up 20%, or more, of a person's total exposure to the metal; moreover, infants who consume mostly mixed formula can receive 40 to 60% of their exposures from water (EPA 2021a).

Food

Most people receive the largest portion of their daily Pb intake via food (WHO 2001). Fortunately, according to a Food and Drug Administration (FDA) in a total diet study there was a steady fall in daily dietary intake during the period of 1979 to 1988; these decreases were from about 90 µg/d to below 10 µg/d for adult males and from about 30 µg/d to below 5 µg/d for infants; in addition, during the period 2003 and 2010, Pb levels in food decreased by 23% (WHO 2001, EFSA 2012). However, low levels of Pb continued to be detected in some foods due to the regular presence of Pb in the environment. Pb can enter food supplies, as the metal in the soil can be absorbed by fruits or vegetables through plant roots or direct foliar contamination; Pb in plants may also be ingested and absorbed by animals included in the human diet (WHO 2001, FDA 2020). Pb concentrations in different food items are highly variable. Cereal products (mostly), grains and vegetables (especially potatoes and leafy vegetables) are the most important contributors to Pb dietary exposure for the general European population (WHO 2001, FDA 2020). Groups with higher Pb exposures through the diet include high consumers of game meat and of game offal; dietary exposure for vegetarians is not different from that of the general adult population (DGHFS 2021). European average adult consumers have Pb dietary exposure ranges from 0.36 to 1.24 µg/kg b.w./d y, and high consumers up to 2.43 µg/kg b.w./d . Exposure of infants varies from 0.21 to 0.94 µg/kg bw/d , and in children ranges from 0.80 to 3.10 µg/kg b.w./d (average consumers), and up to

5.51 (high consumers) µg/kg b.w./d (EFSA 2010). As no safe level of Pb exposure has yet been identified for children, an Interine Reference Level (IRL) of 3 µg/d was reached, based on the correspondence to a reference value of 5 µg/dL blood Pb level (Flannery et al. 2020).

In the past can foods were important sources of exposure to Pb through food, until European legislative control measures were taken to remove Pb from cans in the 70s (EFSA 2012). In 1995 the FDA issued a final rule prohibiting the use of Pb solder in all canned food , including imported products (USDA 2019). Later (in 2006) the EU banned the use of Pb solders, enforcing the development of new joining techniques in the area of electronic and microsystem technology (EC 2012). Leaded crystal is another source of contamination of beverages as when the crystal comes in contact with the liquid (especially acidic beverages such as port wine, fruit juices and soft drinks) some Pb dissolves, in amounts depending on the length of time they are in contact with each other; Pb concentrations up to 20 mg/L were found in wines kept for 1 wk in crystal containers. Poorly fired ceramic ware with a Pb-based glaze can also release Pb into food, particularly acidic foods such as fruit juices and tomato sauces. In Canada releasable Pb from ceramic food ware is limited to 0.5, 1.0, 2.0, or 3.0 mg/L, depending on the type and size of the ceramic ware (HC 2009).

Consumer products

Consumer products are additional sources of exposure to Pb, the largest concern being young children as they tend to put their hands, toys, inexpensive children's jewellery or other objects into their mouths (Guney and Zagury 2014, CDC 2020). The metal may be found in painted toys, which while being banned in several countries, are still widely used in others; older toys before the ban or antique toys and collectibles passed down through generations also contain Pb in paint (CDC 2020).

Toxicity

Pb induced toxicity may largely be explained by interferences with different enzyme systems, as the metal can inactivate enzymes by binding to SH-groups or by displacing essential metal ions. Therefore, many organ systems are potential targets for Pb, and a wide range of biological effects of Pb may occur affecting the nervous, renal, cardiovascular, reproductive, hepatic, endocrine, immune and gastrointestinal systems, among others (WHO 2001, Charkiewicz and Backstrand 2020, TCEQ 2020). SIn 2004, 143,000 deaths and a loss of 8,977,000 disease-adjusted life years were attributed to Pb exposure worldwide, primarily from Pb-associated adult cardiovascular disease and mild intellectual disability in children (Brown and Margolis 2012). Children are more prone to the effects of Pb because their organs are in a developing stage, and particularly as in a child's developing brain, synapse formation is greatly affected in the cerebral cortex by this metal (Wani et al. 2015). This age group represents approximately 80% of the disease impact attributed to

Pb, with estimated 600,000 new cases of childhood intellectual disabilities resulting from blood Pb levels $\geq$ 10 µg/dL each year (Brown and Margolis 2012).

Acute toxicity

Symptoms of acute Pb poisoning include abdominal and muscle pains, fatigue, headache, nausea and vomiting, seizures and eventually, coma (AD GAWE 2021). Between April and May of 2015, 28 children died as a result of unregulated, rudimentary processing of Pb-rich gold ores in Nigeria; infants' exposure to excess Pb can lead to blindness (Lin et al. 2020). In acute cases it is imperative to begin a treatment immediately to minimize long-term damage, chelation therapy is the most used treatment. When there is extreme emergency, gastric lavage and accelerated intestinal transit are recommended (Al Khabbas et al. 2017, Tahtat et al. 2017).

Chronic toxicity

Symptoms of chronic Pb poisoning may involve fatigue, problems with sleep, irritability and shortened attention span, headaches, stupor, slurred speech, loss of appetite, learning disabilities, behavioural problems, poor coordination, impaired growth, anaemia, increased blood pressure, heart rate variability and fertility issues (Wani et al. 2015, AD GAWE 2021). The most common effects of chronic Pb found in the current population , are cardiovascular in adults (e.g., high blood pressure and heart disease) and neurological in children. Infants and young children are especially sensitive to even low levels of Pb, which may contribute to behavioural problems, learning deficits and lowered IQ (Wani et al. 2015, TCEQ 2020). The brain is the most sensitive organ to Pb exposure, and it has been revealed that different mechanisms can be disrupted by Pb neural systems (Wani et al. 2015). SPb substitutes for calcium and inappropriately trigger processes that rely on calmodulin, which, in turn, cause a wide range of mechanisms that disrupt synapse formation, axon dendritic extension and plasticity. Furthermore, Pb interferes with neurotransmitter release and neurotransmitter-related systems, particularly those of dopamine and gamma-aminobutyric acid (GABA) pathway decreasing GABA release. Pb also alters white matter by the expression of genes essential to myelin formation and alters the structure of myelin sheaths (Takeuchi et al. 2021).

The Department of Health and Human Services (HHS) has determined that Pb and Pb compounds can be anticipated to be human carcinogens, whereas EPA has classified Pb as a probable human carcinogen (ATSDR 2020).

Biomarkers

The appropriate selection and measurement of biomarkers (BM)s to control exposures, effects and monitor persons with higher susceptibility to Pb toxicity, is of critical importance for health care management purposes, public health decision making and primary prevention activities.

Biomarkers of exposure

Pb concentrations in a number of biological media, such as blood, urine, bone, saliva, tooth and hair, have been measured to serve as BM of exposure. These different markers have a different accuracy as surrogate measures of dose (WHO 2001, Klotz and Göen 2017).

Lead levels in blood

The main tool to detect elevated levels of body Pb is to measure the levels of the metal in blood samples, which is the method most used for screening and diagnostic purposes and for long-term biomonitoring (Barbosa Jr. et al. 2006, Wani et al. 2015).

Whole blood Pb concentration is the most widely used, since blood Pb is distributed mostly among erythrocytes, bound to the haemoglobin and its presence is less than 5% in the plasma (WHO 2001). Blood Pb shows at least two parts , one with a half-time of about 1 mon, and another with a half-time of decade(s) reflecting Pb released from the bone. Thus, Pb concentration in blood reflects a combination of exposures during the last month and the exposure several years in time (Bergdahl and Skerfving 2008). In exposed children bone-Pb contribution to the blood can be 90% or more, and typically there are reductions of blood-Pb levels of 30% after environmental remediation, when evaluated within several months after intervention. In children with blood Pb levels between 25 and 29 µg/dL, who were not treated with chelation drugs, the time required for blood Pb to decline to <10 µg/dL is about 2 yr. Therefore, there is a long time for mobilization and depletion of accumulated skeletal Pb in reserves and reduction in the absolute contribution to blood Pb levels from these supplies (Barbosa Jr. et al. 2006). The average whole blood Pb levels in people not exposed occupationally are 20.7 µg/L, while the average levels of 227.7 µg/L is for exposed workers (Sommar et al. 2014). Pb blood concentrations exceeding 400 µg/dL can cause severe signs of encephalopathy (a condition characterized by brain swelling) accompanied by increased pressure within the skull, delirium, coma, seizures and headache; such indications may inclusively appear in children with Pb blood levels of 70 µg/dL. In adults, abdominal colic, involving paroxysms of pain, can occur at concentrations higher than 80 µg/dL. Levels from 25 to 60 µg/dL may give rise to neuropsychiatric effects such as delayed reaction times, irritability and difficulty in concentrating, as well as slowing down motor nerve conduction and headache; anaemia may appear at Pb levels higher than 50 µg/dL (Wani et al. 2015, Nafti et al. 2020). However, the European Legislation sets Occupational Exposure Limits (OEL) for Pb in the blood of employees in battery plants at 70 µg/dL, and the EU Scientific and Social Committee for Occupational Exposure Limits (SCOEL) has proposed to decrease this limit to 30 µg/dL (Eurobat 2021). Medical management of exposure to Pb recommend that people should be kept away from the occupational exposure if a Pb level in the blood reaches or exceeds 30 µg/dL, or if two consecutive levels of Pb in blood measured over a 1-mon interval are ≥ 20 µg/dL. Medical surveillance of workers exposed to Pb is recommended every 3 mon for people with blood Pb levels between 10 and 19 µg/dL, and every 6 mon when blood Pb levels are less than 10 µg/dL.

For pregnant women it is recommended to avoid Pb exposure resulting in blood Pb levels greater than 5 µg/dL (Tahtat et al. 2017). The US Center for Disease Control and Prevention have set standard elevated blood Pb level for adults to be 10 µg/dL. For children this value is 5 µg/dL, in accordance with the National Toxicology Program which concludes that there is sufficient evidence for adverse health effects in children at these concentrations (ATSDR 2019, CDC 2021, EPA 2021b); IQ loss was observed in children with blood Pb levels below 10 µg/dL (WHO 2007).

From a physiological point of view the toxic effects of Pb are primarily associated with plasma Pb because this fraction is the most rapidly exchangeable one in the blood section (Barbosa Jr. et al. 2006). Thus, plasma Pb concentrations may better reflect the "active" fraction of Pb in blood and characterize the relationship between blood Pb and tissue accumulation (and effect), than whole blood Pb. But analytical problems appear since Pb concentrations in the plasma (and serum) are very low; a blood Pb concentration of 250 µg/L gives a concentration in the plasma of approximately 1 µg/L (WHO 2001, Bergdahl and Skerfving 2008). Plasma/serum Pb levels in non-exposed and exposed individuals range widely from 0.02 to 14.5 µg/L, again probably due to analytical instruments and/or methods for Pb determination limitations (Barbosa Jr. et al. 2006).

Research on associations between plasma Pb and toxicological outcomes is quite rare, and a significant gap in knowledge remains (Barbosa Jr. et al. 2006). Despite this a study comparing Pb workers with non-exposed subjects revealed higher levels of the metal in the plasma of the workers' group, the determined concentrations being respectively 0.57 and 0.09 µg/L, (Sommar et al. 2014).

Lead levels in urine or faeces

Several efforts have been done to use urine Pb levels as a surrogate for blood Pb. Pb concentrations in urine are generally lower by a factor of 10–100 as compared to blood (Bergdahl and Skerfving 2008). Both serial urine sampling and 24-hr urine may provide information on Pb body supplies, and spot urine samples when collected at approximately the same time of the day and provide valid information on Pb excretion rates (Gulson et al. 1998, Kim et al. 2014). Although controversial, urinary excretion of Pb after administration of a chelating agent has also been used as an index of total body burden; however, such a measurement does not reflect long-term accumulation (mainly occurring in the cortical bone) (Bergdahl and Skerfving 2008). Even so it may be a potential useful BM of internal exposure to Pb as it reflects the mobilized pool of Pb, which consists mainly of blood and soft tissue (Kim et al. 2014). Relationships between Pb concentrations in blood and urine are incorrectly understood. In a study, concentrations of urinary Pb were about 10% of that in whole blood but the correlations were not particularly sound (Gulson et al. 1998, Barbosa Jr. et al. 2006). On the contrary, Pb blood levels lower than 10 µg/dL were already associated with urinary Pb up to 0.55 µg/dL, as well as Pb in blood of 27.6 µg/dL corresponding to an amount of the metal in urine lower than 2.05 µg/dL. Urine is accepted to replace blood in occupational exposure to Pb assessments, but caution is advised in the case of environmental exposure where urinary Pb should be used just as an estimation of the metal content in blood (Moreira and Neves 2008). Better correlations between plasma and urinary Pb than between blood and urine Pb was found in Pb workers with low levels of Pb exposure (low levels occur similarly in environmental exposures), with urinary Pb concentration directly related to plasma Pb. This could be explained by the fact that urine Pb comes from plasma Pb that is filtered at the glomerular level, making urinary Pb a possible surrogate for the filterable fraction of Pb in the plasma, which is considered an interesting fraction (Barbosa Jr. et al. 2006, Bergdahl and Skerfving 2008). Concerning urinary values of Pb, exposed workers can exhibit average levels of 23.7 µg/L, while in a non-exposed group a mean concentration of 10.8 µg/L is documented (Sommar et al. 2014).

Faecal Pb reflects unabsorbed ingested Pb plus Pb that is eliminated via endogenous faecal (biliary) routes, providing an integrated measure of Pb exposure/ intake from all sources; these include dietary and environmental inside and outside the house . However, inter-individual variations in these physiologic processes may show up and the collection of complete faecal samples over multiple days may not be feasible (Barbosa Jr. et al. 2006). This approach is more useful in clinical cases where a large oral intake of Pb is suspected (Bergdahl and Skerfving 2008).

Lead levels in saliva

Pb in saliva is the direct excretion of the Pb fraction in diffusible plasma (i.e., the fraction not bound to proteins). Saliva Pb concentrations are closely related to Pb levels in this plasma fraction and intracellular Pb, reflecting the internal Pb level that can exert effects on human organs (Barbosa Jr. et al. 2006, Nriagu et al. 2006, Pasiga et al. 2019). Significant correlations have been reported between saliva Pb levels and both blood or plasma concentrations, with a study where a correlation of 0.72 was found between blood and salivary Pb levels (Koh et al. 2003, Barbosa Jr. et al. 2006). On the contrary, weak correlations were determined in another work, even after adjusting the health conditions and smoking habits of the subjects. The authors of this study considered the possibility of differential rates of excretion of Pb into saliva, at different levels of Pb concentrations in blood; in their opinion, measurements of saliva Pb levels for biological monitoring were not adequate (Koh et al. 2003). The Pb content in the saliva of unexposed children is usually below 0.15 µg/dL, but in reality the values of Pb reported in saliva vary from study to study (Barbosa Jr. et al. 2006).

Lead levels in hair or nails

Pb levels in the hair could be an adequate biological index for evaluating the history of a subject exposure, reflecting mean levels in the human body during a period of 2–5 mon (Pirsaraei 2007). Significant correlations between Pb-hair and blood suggests the efficiency and effectiveness of using hair for assessing occupational exposures; studies conducted in the 80s showed an exponential accumulation of the Pb content in hair simultaneously with an increase of the values in blood (Niculescu et al. 1983, (Nafti et al. 2020). Others revealed that hair Pb concentrations were significantly correlated with the design of the workplace and with the implementation of personal protection policy. However due to fluctuations in the results of some other studies, it is difficult to establish reference ranges (Barbosa Jr. et al. 2006, Nafti et al. 2020). Some authors mention that the normal range of hair Pb found in humans is 0-30 µg/g. In Poland the average values of Pb for people living in the country was reported as 2.39 µg/g, whereas people living in towns and cities presented average values of 4.17 µg/g (Pirsaraei 2007, Trojanowski et al. 2010). In Pakistan, values of 2.49 µg/g were determined for office workers and higher values were found for dyers and traffic constables (3.78 and 9.76 µg/g, respectively) (Batool et al. 2011). A great limitation of this BM is the disabilty to distinguish between endogenous Pb (i.e., absorbed into the blood and incorporated into the hair matrix) and exogenous Pb (i.e., derived from external contamination). For the laboratory washing step to treat this sample, no consensus exists on how to remove exogenous Pb, as well the length of the hair specimen to be collected, the amount or the position on the scalp. There are also significant variations in hair Pb concentrations according to age, gender, hair-colour and smoking (Barbosa Jr. et al. 2006).

Nail Pb levels are considered a reflection of long-term exposure, as this section remains isolated from other metabolic activities in the body and may indicate metal levels over a longer period than hair (12 to 18 mon) (Barbosa Jr. et al. 2006, Batool et al. 2011). Since toenails are less affected by exogenous environmental contamination than fingernails, they have been preferred for Pb exposure studies; moreover, toenails have a slower growth rate than fingernails, being 50% slower especially in winter (Barbosa Jr. et al. 2006). Others reject the use of both finger and toenails as BMs of exposure to Pb, due a high variability found in subjects serially sampled for over 6 mon (Gulson 1996). More recent studies resulted in similar conclusions, with poor correlations determined between Pb concentrations in both washed and non-washed nails and blood Pb levels (Olympio et al. 2020).

Lead levels in teeth or bones

Since teeth accumulate Pb over the long term, the concentration of teeth Pb is aparticularly attractive BM of past exposure. There is some evidence that teeth are even better than bones as an indicator of cumulative Pb exposure since losses from teeth are much slower and this is because teeth can exchange metal ions into the hydroxyapatite crystal structure and do not readily remobilize metals (different from the bone) (Barbosa Jr. et al. 2006, Arora and Hare 2015). Teeth can inclusively be used as a cumulative index of exposure since the prenatal period, when the tooth is formed, until the time of falling out (Bergdahl and Skerfving 2008). A great benefit is that children's deciduous teeth can be collected with relative ease and in a non-invasive manner. Reliable BMs are necessary to evaluate Pb exposures in this group, when considering that chronic Pb exposure from mouthing activity in early childhood may be camouflaged by "dilution" effects during periods of rapid skeletal growth in children; this may not be detected by a single blood-Pb measurement (Barbosa Jr. et al. 2006, Arora and Hare 2015). *In vivo* experiments demonstrated that Pb uptake in the first mandibular molar teeth of puppies was substantially greater than blood, the kidney, liver or brain samples, with Pb levels in teeth observed to be between 34 to 70 times greater than the concentration of the metal in blood. A strong link was also found between dental Pb levels and Pb concentrations in key organs, at doses between 10 and 40 mg/L during the prenatal and neonatal periods (Arora and Hare 2015). Another advantage is that since teeth are composed of several distinct tissues formed over a period of several years, and different parts of the tooth can bind Pb at different stages of the individual's life, each tooth section can yield actual information on the individual's prior exposure. Furthermore, the enamel of all primary teeth and parts of the enamel from some permanent teeth, are formed *in utero* and thus, can provide information on pre-natal exposure to Pb (which can be useful to study long-term health effects of *in utero* exposure) (Barbosa Jr. et al. 2006, Arora and Hare 2015). The dentine of the primary teeth provides evidence of exposure during the early childhood years, when hand-to-mouth activity is usually an important contributor to Pb body burden. In turn, enamel Pb levels may be also useful for indirectly estimating the Pb composition of the mother's bone (Barbosa Jr. et al. 2006). The analysis of deciduous teeth obtained from children living in rural areas showed that higher levels of Pb occur on the surface of the teeth, rapidly becoming less a few micrometres into the teeth, the ratios of the concentrations of Pb in enamel to dentine and to circumpulpal dentine 1:2:6; within dentine, the Pb levels were highest in the root dentine. It is also determined that the ratio ([Pb]tooth/[Pb]total set) decreases in the following order: first molars > central incisors > lateral incisors > canines > premolars > second molars > third molars. This order inversely correlates with the age of formation or emission of each tooth type (Purchase and Fergusson, 1986). Concentrations of deciduous teeth from children living in different areas of Turkey were determined as ranging from 1.30 to 1.77 µg/g, although in this study the highest levels were found in incisors (Karahalil et al. 2007).

Bone accounts for more than 94% of the adult body burden of Pb (70% in children) with a half-life of a year to decades; thus, bone Pb has been considered an indicator of cumulative Pb exposure and a source of body burden which can be mobilized into circulation (Barbosa Jr. et al. 2006, Specht et al. 2016). The importance of measuring bone Pb levels is supported by authors who argue that skeletal sources of Pb accumulated from past exposures should be considered along with the current ones (Barbosa Jr. et al. 2006). Non-invasive *in vivo* Cd-109 based K-shell X-Ray Fluorescence (KXRF) has been used to study bone Pb and results exist associating Pb levels in bone with blood concentrations (Börjesson and Mattsson 1995, Specht et al. 2016). These associations were also found additionally between Pb bone and serum, indicating the potential role of the skeleton as an important source of endogenous labile Pb, which may not be adequately discerned through measurement of blood Pb levels (Barbosa Jr. et al. 2006). Studies on children also revealed that correlations between KXRF bone Pb and blood had a higher slope and stronger r squared, than in adults. This could be due to children's higher bone resorption rate, indicating a more frequent transition of Pb between bone and blood and that in children, the majority of Pb in blood could come from the bone (Specht et al. 2016). It is also considered that differing bone types have differing bone-Pb mobilization characteristics, with trabecular bones, such as calcaneus and patella, having a faster turnover than cortical ones (e.g., tibia) (Bergdahl and Skerfving 2008). As suggested, smeltery workers with a tibia Pb concentration of 100 µg/g can expect a continuous endogenous contribution to blood Pb of 16 µg/dL; pregnant woman with a tibia Pb concentration of 50 μg/g could end up with a contribution of 8 μg/dL to blood Pb. Subjects not occupationally exposed typically display tibia Pb levels up to about 20 μg/g (Barbosa Jr. et al. 2006).

Biomarkers of susceptibility

Although the quantities of Pb to which individuals are exposed vary widely, susceptibility of an individual to the effects of a specific level of exposure is another highly important factor in development of Pb toxicity (Mahaffey 1974). Therefore, genetic polymorphisms and epigenetic changes have been explored as candidates of Pb susceptibility BMs.

Genetic biomarkers

Genetic factors can modify Pb toxicity or be protective from Pb poisoning. Differences in genes encoding molecules such as δ-aminolaevulinic acid dehydratase (ALAD), the hemochromatosis gene (HFE) and the Vitamin D Receptor (VDR) are among the ones considered as having potential to provide susceptibility for Pb-toxicity (Kim et al. 2014).

δ-aminolaevulinic acid dehydratase ALAD plays an important role in the susceptibility to Pb poisoning, and it has been reported how polymorphisms in the ALAD coding gene may affect the response of individuals to Pb toxicity (Shaik and Jamil 2008). Pb is a potent inhibitor of ALAD, coproporphyrinogen oxidase and ferrochelatase, enzymes which catalyze respectively, the second, sixth and final steps of the heme biosynthesis pathway. The metal has the greatest effect on ALAD, responsible for catalyzing the condensation of two molecules of 5-aminolevulinic acid (ALA) into one molecule of monopyrrole porphobilinogen (PBG); such inactivation from Pb occurs through the replacement of zinc from the ALAD active site (Scinicariello et al. 2007). ALAD inactivation has been implicated in the pathogenesis of Pb poisoning because the resulting accumulation of ALA causes a neuropathogenic effect, probably by acting as a γ-aminobutyric acid (GABA) receptor agonist in the nervous system (Percy et al. 1981, Scinicariello et al. 2007). ALAD is encoded in humans by a single gene, which exhibits two alleles: ALAD1 and ALAD2; the ALAD2 allele contains the substitution of a neutral asparagine for a positively charged lysine (van Bemmel et al. 2011, Kim et al. 2014). Differences are reported in the levels of heme precursors between the two types of ALAD genotypes, and this is probably due to different affinities of each ALAD isozyme to Pb. ALAD1 homozygotes have higher levels of ALA in comparison with ALAD2 carriers at high Pb exposures, which suggests that ALAD1 homozygotes might be more susceptible for disturbances in heme biosynthesis by Pb than ALAD2 carriers (Sakai 2000). Occupational studies showed subjects with the ALAD 1-2/2-2 genotype having higher blood Pb concentrations (80.5 μg/dL), when compared with subjects from the ALAD 1-1 genotype group (50.4 μg/dL) (Shaik and Jamil 2008). Caucasians have higher frequencies of the ALAD2 genotypes (approximately 18% ALAD 1-2 and 1% ALAD 2-2) than African and Asian people (Kim et al. 2014).

Glutathione-S-transferases Glutathione-S-transferases (GSTs) are known as phase II metabolic enzymes which play important roles in oxidative stress defence induced by various toxicants (particularly metals) (Kim et al. 2014). Pb is conjugated with glutathione non-enzymatically or enzymatically via GST (Lee et al. 2012). The GST superfamily comprises eight polymorphic genes, and of these, polymorphic variants of the GST-mu 1 (GSTM1), theta 1 (GSTT1) and pi (GSTP1) are the most reported globally; polymorphisms in GST genes vary with ethnicity. Deletions in GSTM1 and GSTT1 genes and polymorphism in GSTP1 contribute to interindividual differences in susceptibility to xenobiotic toxicity, with GSTP1 variant alleles and double-null genotypes of GSTM1 and GSTT1 associated with higher oxidative stress (Yohannes et al. 2021). These variations seem to play an important role in the susceptibility to the harmful effects of Pb. Pregnant mothers with combined GSTM1 and GSTT1 genetic variants show an inverse association of blood Pb and birth weight of their babies (Kim et al. 2014, Yohannes et al. 2021). People with the GSTP1-Val105 allele or GSTM1 deletion polymorphism also perform worse on cognitive assessments and show increased bone Pb levels (Kim et al. 2014). The GSTT1 positive type polymorphism might also be associated with Pb-related hypertension in Pb exposed male workers (Lee et al. 2012).

Hemochromatosis A candidate gene for susceptibility to Pb exposure is a gene which is known to be altered in hemochromatosis (HFE); HFE is an autosomal recessive genetic disease that produces an increase in the absorption of ingested iron (Wright et al. 2004). The HFE gene have been reported to be a modifier for Pb absorption and storage (Fan et al. 2014, Kim et al. 2014). The HFE-H63D genotype was shown to modify the association between Pb and iron metabolism, expressed by increased blood Pb levels associated with a higher body iron content (Fan et al. 2014). HFE-H63D variant carriers may be a potentially highly vulnerable sub-population if they are exposed to high Pb levels occupationally. Maternal HFE-H63D variant carriers also have a negative association between tibia Pb and birth weight (Fan et al. 2014, Kim et al. 2014).

Metallothioneins Metallothioneins (MTs), are proteins with numerous functions including toxic metal detoxification, including Pb, and maintenance of metal ion homeostasis that are responsible for the distribution of metals in the body (Yang et al. 2013, Kim et al. 2014, Fernandes et al. 2016). Four major isoforms of MT have been identified in mammals (MT-1 through MT-4), with the isoforms MT1 and MT2 expressed in almost all tissues. Genetic variations in genes that encode MTs are supposed to affect the Pb body burden (Kim et al. 2014). Pb is known as a renal toxin, variations of MT1 single polymorphisms (SNPs) can influence the levels of BMs of renal function and damage (urinary uric acid and N-acetyl-beta-d-glucosaminidase) in chronically Pb-exposed subjects; the SNP rs8052394 of the MT1A gene is also associated with increased metal accumulation (Yang et al. 2013, Kim et al. 2014, Singh et al. 2020). A study aimed to evaluate the genetic effects of MT2A rs10636 in blood Pb levels of occupationally exposed subjects, revealed that workers carrying at least one C allele had higher blood levels of Pb than those with the GG genotype (Fernandes et al. 2016).

Vitamin D receptor Polymorphisms in the Vitamin D Receptor (VDR) gene affect Pb toxicokinetics possibly because VDR is involved in calcium metabolism and Pb interacts with calcium; namely, Pb and calcium modify each other's absorption (Onalaja and Luz 2000, Pawlas et al. 2012). At least three genotypes of the VDR gene have been identified, resulting in the Restriction Fragment Length Polymorphisms (RFLPs): Taq I, Fok I and the most widely studied, BsmI. These RFLPs have been correlated with bone mineral density and circulating levels of osteocalcin. The polymorphism defined by the restriction BsmI results in three genotypes (BB, Bb and bb). The bb genotype (denoted bb when the restriction site is present) is associated with lower blood Pb concentrations (Onalaja and Claudio 2000, Pawlas et al. 2012). For FokI (rs2228570), the f genotype was associated with low levels of the metal in blood. It was also proved that BsmI and TaqI polymorphisms modified the relationship between IQ and Pb blood levels, leading to conclude that there is a fraction of the population for which due these gene genotypes, are particularly sensitive to Pb neurotoxicity (Pawlas et al. 2012).

Epigenetic biomarkers

There is evidence that environmental metal exposure results in epigenetic alterations, which may link heritable changes in gene expression with disease susceptibility. DNA methylation is the most studied epigenetic mechanism, corresponding to the addition of a methyl group to the 5-carbon position on the cytosine pyrimidine ring via DNA methyltransferases (DNMTs); this is mostly associated with gene silencing. Chronic exposure to Pb leads to an increase of reactive oxygen species (ROS), which can alter the function of DNMTs with consequent changes in DNA methylation (Kim et al. 2014). Pb exposure can increase the ALAD gene methylation and downregulate ALAD transcription.

The expression of other molecules can be also changed by Pb exposure, such as the Amyloid Precursor Protein (APP), which is one of the hallmarks of Alzheimer disease and collagen type 1 alpha-2 (COL1A2), an important component of connective tissues, as well as other compounds, whose alterations have been associated with increased risk of cancer (Kim et al. 2014).

Biomarkers of effect

Heme biosynthesis

A well-known critical effect of Pb is in the bone marrow, which appear mainly from the interaction of the metal with some enzymatic process involved in heme biosynthesis (Kim et al. 2014). Therefore, δ-aminolevulinic acid dehydratase (ALAD) and Erythrocyte Porphyrins (EPs) in the blood as well as δ-aminolaevulinic acid (ALA) in the urine and plasma and coproporphyrin in urine, have been be used to monitor Pb-induced alterations in this biochemical pathway (WHO 2001, Klotz and Göen 2017). In workers with low-level occupational Pb exposure ALAD can be used as an effect BM of low Pb blood level and in fact, the activity of ALAD (which is inhibited by Pb binding) is accepted as the most sensitive measurable biological index of Pb toxicity (Sanders et al. 2009, Yang et al. 2015). Zinc protoporphyrin (ZPP) concentrations are also widely used as BMs for Pb toxicity. Ferrochelatase is a crucial enzyme that catalyzes the insertion of iron into protoporphyrin IX in the final step of heme biosynthesis. As Pb can inhibit ferrochelatase, the pathway is interrupted and augmented ZPP occurs (Flora 2014). Other authors have considered that the effect of BMs alone is not sufficiently sensitive for an early detection of a health impairment caused by Pb. They recommended a diagnostic strategy for revealing Pb-induced effects that combines the determination of Pb in the whole blood with the analysis of ALA and coproporphyrin in urine or ALAD and ZPP in the blood (Klotz and Göen 2017).

Enzymes

5'-nucleotidases are a large functional group of enzymes that catalyze the dephosphorylation of various nucleoside 5'-monophosphates to their respective nucleosides and are present in the intestines, brain, heart, blood vessels, pancreas and liver (Chiarelli et al. 2006, Habib and Shaikh 2018). A major type of nucleosidases has been isolated from erythrocytes cells, as it is known that they are preferentially active towards pyrimidine nucleotides; these are called pyrimidine 5'-nucleotidases (Chiarelli et al. 2006). *In vitro* incubations of normal mature erythrocytes with Pb result in a significant inhibition of pyrimidine 5'-nucleotidase (Paglia et al. 1975, Rees et al. 2003, Kim et al. 2014). Subjects with chronic Pb intoxication secondary to industrial exposure also exhibited substantial and consistent impairment of erythrocyte pyrimidine-5'-nucleotidase activity. The inhibition of the enzyme in blood was already used to monitor workers at risk of Pb toxicity and considered to be a reliable adjunct to serum Pb levels (Paglia et al. 1975, Kim et al. 2014).

Nicotinamide Adenine Dinucleotide Synthetase (NADS) is another enzyme that catalyzes the final step in the Preiss-Handler pathway for NAD biosynthesis. The enzyme catalyzes the transfer of an amino group from glutamine to nicotinic acid adenine dinucleotide (NAAD) to form NAD in the presence of adenosine triphosphate (ATP), Mg^{2+} and K^+ (Zerez et al. 1990). The WHO refers to the decrease in NADS activity as one of the important effects of Pb on humans (Morita et al. 1997). A dose-effect relationship between Pb levels in blood and NADS activity is demonstrated in workers exposed to the metal, at Pb blood levels of 40 µg/dL. NADS activity can have even higher predictivity than ALAD. It is inclusively proposed that NADS measurements can be used for the evaluation of Pb effects at both low and high exposure contexts (Morita et al. 1997, Kim et al. 2014).

Brain-derived neurotrophic factor

The Brain-Derived Neurotrophic Factor (BDNF) is a member of the neurotrophin family which play critical roles in the nervous system development. The expression of BDNF is modulated by N-Methyl-D-Aspartate Receptor (NMDAR) via the Ca2þ signalling pathway, where the activated BDNF binds its receptors inducing multiple downstream signalling cascades (Yndestad et al. 2011, Zhao et al. 2020). The disruption of BDNF protection is an emerging mechanism for environmental neurotoxicants in the recent decade. The potential for the use of BDNF BMs in relation to Pb exposure in human biomonitoring studies is supported by toxicological data (Zhao et al. 2020, Gundacker et al. 2021). Pb may exert neurotoxicity by disrupting NMDAR-dependent BDNF signalling, inhibiting the NMDA receptor, thereby reducing the expression of BDNF and impairing the presynaptic BDNF-TrkB receptor signalling on presynaptic sites. Such events decrease the phosphorylation of synapsin-1, a vesicular protein important in vesicle-synaptic membrane interactions and imperative to vesicular release (Guariglia et al. 2016, Zhao et al. 2020). In agreement with these mechanistical evidences, the concentration, expression and release of BDNF were reported to be impaired during Pb exposure both *in vitro* and *in vivo*, and in one epidemiological study blood Pb levels were negatively associated with pre-school childrens' serum BDNF concentrations (Zhao et al. 2020). However, other researchers did not find this relation in girls; on the other hand, it is mentioned that other environmental chemicals (bisphenols, phthalates and polycyclic aromatic hydrocarbons) also interfere with BDNF signalling. In addition, epigenetic mechanisms (including DNA methylation) influence BDNF expression and regulation and thus, peripheral blood BDNF DNA methylation measurements could be advantageous over serum/plasma BDNF protein assessment, due to higher stability over time (Gundacker et al. 2021).

Cortisol and lipids

The stress hormone cortisol, synthesized from cholesterol, is the major glucocorticoid produced by the human adrenal cortex and the end product of the Hypothalamic-Pituitary-Adrenal (HPA) axis (Fortin et al. 2012, Braun et al. 2014, Gundacker et al. 2021) In this axis, the adrenocorticotropic hormone (ACTH) is produced in the pituitary gland, and its production stimulates the production and release of cortisol from the adrenal gland. Therefore, salivary cortisol is a measure which can be used as an index of HPA-axis functioning (Gundacker et al. 2021). Cortisol plays a major role in neurogenesis and is critical to brain development, especially in memory formation, the hippocampus having the highest concentration of glucocorticoid receptors (Braun et al. 2014). The hippocampus is a target of Pb toxicity, and lower blood cortisol levels from fasting children living in an area contaminated with Pb were associated with increased sensory integration difficulties, especially regarding touch, body awareness, balance and motion (Gundacker et al. 2021). Hence it is proposed that Pb exposure may be associated with alterations in HPA-axis function, where increased Pb exposure could alter the adrenal response to ACTH, resulting in a lower cortisol release (Fortin et al. 2012, Braun et al. 2014). Even though cortisol concentrations are highly variable due to circadian fluctuations and fast changes in response to stressors, it would be interesting to investigate the influence of Pb exposure on the response of the HPA-axis (Gundacker et al. 2021).

Extensive evidence exists of the association of serum lipid and lipoprotein levels with coronary artery disease, it has been recorded that both acute and chronic Pb poisoning cause impairment of heart and vessel function. In accordance, serum cholesterol and lipoprotein levels were found to be higher in subjects who were occupationally exposed to Pb, than in those who were not exposed. A dose-response relationship between blood Pb and serum cholesterol concentrations found in exposed subjects, suggested altered lipid metabolism related to the exposure. This could be explained by the fact that Pb depress the activity of cytochrome P-450 enzymes and may limit the biosynthesis of bile acids, which is the only significant route for elimination of cholesterol from the body (Kristal-Boneh et al. 1999).

As mentioned for other metals, Sister Chromatid exchanges (SCEs), High-SCE Frequency Cells (HFCs) and DNA-Protein Cross-links (DPCs) have also been shown to be reliable BMs for monitoring workers exposed to Pb (Sanders et al. 2009).

Final Remarks

In this chapter Pb was presented as a ubiquitous pollutant which even after being phased out of leaded gasoline, and could still enter the environment both from current and past uses; as Pb has been used by humans since thousand years and does not breakdown over time. In 2004, 143,000 deaths and a loss of 8,977,000 disease-adjusted life years were attributed to Pb exposure worldwide, primarily from Pb-associated adult cardiovascular disease and mild intellectual disability in children. Children are the most prone to Pb toxicity as their organs are in a developing stage, represent approximately 80% of the disease impact attributed to Pb. Markers of long-term exposure to Pb include bone or teeth Pb levels, and relatively short-term exposure (months) include blood, plasma, urine and cheatable Pb; it is seldom emphasized that Pb has no established BM of current (hours or days) exposure. As susceptibility BMs, differences in ALAD, hemochromatosis and vitamin D receptor coding genes are considered as having having the potential to provide susceptibility to Pb-toxicity. In turn, changes in heme precursor levels, such as ALA, coproporphyrin and ZPP are widely accepted as suitable to monitor Pb-induced alterations in heme synthesis biochemical pathway. Promising candidates for BMs of Pb effects have been explored, namely pyrimidine 5'-nucleotidases, nicotinamide adenine dinucleotide synthetase, the brain-derived neurotrophic factor and stress hormone cortisol. Pb constitutes an example of how even banning relevant sources to the environment (leaded gasoline) will not immediately solve pollution problems; Pb poisoning continues to exist, affecting mostly the neurodevelopment of children, who will become future adults. All efforts developed to control exposures and Pb-induced effects are not wasted, creating BMs vital tools to achieve these goals.

References

[AD GAWE] Australia Government Department of Agriculture Water and Environment, Lead, 2021. Canberra, Australia.

Al Khabbas, M.H., S.A. Ata, Kamal, I.A, M.F. Tutunjic and M.S. Mubarak. 2017. Synthesis and characterization of new 1-hydroxy-2-pyridinethione derivatives: Their lead complexes and efficacy in the treatment of acute lead poisoning in rats. J. Trace Elem. Med. Biol. 44: 209–217.

Arora, M. and D.J. Hare. 2015. Tooth lead levels as an estimate of lead body burden in rats following pre- and neonatal exposure. RSC Adv. 5(82): 67308.

[ATSDR] Agency for Toxic Substances and Disease Registry. 2020. Lead Tox Facts. Agency for Toxic Substances and Disease Registry, Division of Toxicology and Human Health Sciences, Atlanta, USA.

[Eurobat] Association of European Automotive and Industrial Battery Manufacturers 2021. Occupational Health & Safety. https://www.eurobat.org/environment-health-safety/occupational-health-safety EC.

[ATSDR] Agency for Toxic Substances and Disease Registry. 2020. Toxicological Profile for Lead. Agency for Toxic Substances and Disease Registry, U.S. Department of Health and Human Services, Atlanta, USA.

[ATSDR] Agency for Toxic Substances and Disease Registry, Lead Toxicity. What Are U.S. Standards for Lead Levels? 2019. U.S. Department of Health and Human Services, Public Health Service, Atlanta, USA.

Barbosa Jr, F., J.E. Tanus-Santos, R.F. Gerlach and P.J. Parsons. 2006. A critical review of biomarkers used for monitoring human exposure to lead: advantages, limitations and future needs. Ciênc. Saúde Coletiva. 113(12): 1669–74.

Batool, A.I., F.U. Rehman, N.H. Naveed, A. Shaheen and S. Irfan. 2011. Hairs as biomonitors of hazardous metals present in a work environment. Afr. J. Biotechnol. 10(18): 3602–3607.

Bergdahl, I.A. and S. Skerfving. 2008. Biomonitoring of lead exposure—Alternatives to blood. J. Toxicol. Environ. Health, Part A 71(18): 1235–1243.

Börjesson, J. and S. Mattsson. 1995. Toxicology; *in vivo* x-ray fluorescence for the assessment of heavy metal concentrations in man. Appl. Radiat. Isot. 46(6-7): 571–6.

Braun, J.M., R.J. Wright, A.C. Just, M.C. Power, M. Tamayo, Y. Ortiz, L. Schnaas et al. 2014. Relationships between lead biomarkers and diurnal salivary cortisol indices in pregnant women from Mexico City: A cross-sectional study. Environ. Health 13(1): 50.

Brown, M.J. and S. Margolis. 2012. Lead in drinking water and human blood lead levels in the United States. Centers for Disease Control and Prevention, Morbidity and Mortality Weekly Report. 61(04): 1–9.

[CDC] Centers for Disease Control and Prevention, Lead in Consumer Products. 2020. CDC, Atlanta, USA.

[CDC] Centers for Disease Control and Prevention, Lead: Exposure Limits, 2021. CDC, Atlanta, USA.

Charkiewicz, A.E. and J.R. Backstrand. 2020. Lead toxicity and pollution in Poland. Int. J. Environ. Res. Public Health. 17(12): 4385.

Chiarelli, R., E. Fermo, A. Zanella and G. Valentini. 2006. Hereditary erythrocyte pyrimidine 5'-nucleotidase deficiency: A biochemical, genetic and clinical overview. Hematology. 11(1): 67–72.

[DGHFS] Directorate-General for Health and Food Safety. 2021. Lead in food. European Commission, Directorate-General for Health and Food Safety, Brussels, Belgium.

Dudka, S. and W.P. Miller. 1999. Permissible concentrations of arsenic and lead in soils based on risk assessment. Wat. Air and Soil Poll. 113: 127–132.

[EC] European Commission, Lead-free joining for micro electronics and micro system technology devices, 2012. EC, Brussels, Belgium.

[EFSA] European Food Safety Authority, Panel on Contaminants in the Food. Scientific Opinion on Lead in Food. 2010. EFSA J. 8(4): 1570.

[EFSA] European Food Safety Authority. Lead dietary exposure in the European population. 2012. EFSA J. 10(7): 2831.

[EPA] Environmental Protection Agency, Lead in Soil, 2020. EPA, Washington, USA.

[EPA] Environmental Protection Agency, Ground Water and Drinking Water. Basic Information about Lead in Drinking Water 2021a. EPA, Washington, USA.

[EPA] Environmental Protection Agency, Learn about Lead, 2021b. EPA, Washington, USA.

Fan, G., G. Du, H. Li, F. Lin, Z. Sun, W. Yang et al. 2014. The effect of the hemochromatosis (HFE) genotype on lead load and iron metabolism among lead smelter workers. PLoS One. 9(7): e101537.

[FDA] Food and Drug Administration, Lead in Food, Foodwares, and Dietary Supplements. 2020. FDA. Maryland, USA.

Fernandes, K.C., A.C. Martins Jr, A.A. Oliveira, L.M. Antunes, I.M. Cólus, F. Barbosa Jr. et al. 2016. Polymorphism of metallothionein 2A modifies lead body burden in workers chronically exposed to the metal. Public. Health Genom. 19(1): 47–52.

Flannery, B.M., L.C. Dolan, D. Hoffman-Pennesi, A. Gavelek, O.E. Jones, R. Kanwal et al. 2020. U.S. Food and Drug Administration's interim reference levels for dietary lead exposure in children and women of childbearing age. Regul. Toxicol. Pharmacol. 110: 104516.

Flora, S.J.S. 2014. Metals. pp. 485–519. *In*: R.C. Gupta [ed.]. Biomarkers in Toxicology, Academic Press, Cambridge, USA.

Fortin, M.C., D.A. Cory-Slechta, P. Ohman-Strickland, C. Nwankwo, T.S. Yanger, A.C. Todd et al. 2012. Increased lead biomarker levels are associated with changes in hormonal response to stress in occupationally exposed male participants. Environ. Health Perspect. 120(2): 278–83.

Guariglia, S.R., K.H. Stansfield, J. McGlothan and T.R Guilarte. 2016. Chronic early life lead (Pb2+) exposure alters presynaptic vesicle pools in hippocampal synapses. BMC Pharmacol. Toxicol. 17(1): 56.

Gulson, B.L. 1996. Nails: Concern over their use in lead exposure assessment. Sci. Total Environ. 177: 323–327.

Gulson, B.L., M.A. Cameron, A.J. Smith, K.J. Mizon, M.J. Korsch, G. Vimpani et al. 1998. Blood lead–urine lead relationships in adults and children. Environ. Res. Section A78: 152–160.

Gundacker, C., M. Forsthuber, T. Szigeti, R. Kakucs, V. Mustieles, M.F. Fernandez et al. 2021. Lead (Pb) and neurodevelopment: A review on exposure and biomarkers of effect (BDNF, HDL) and susceptibility. Int. J. Hyg. Environ. Health. 238: 113855.

Guney, M.G. and G.J. Zagury. 2014. Children's exposure to harmful elements in toys and low-cost jewelry: Characterizing risks and developing a comprehensive approach. J. Haz. Mat. 271: 321–330.

Habib, S. and O.S. Shaikh. 2018. Approach to jaundice and abnormal liver function test results. pp. 99–116. In: A.J. Sanyal, T.D. Boyer, K.D. Lindor and N.A. Terrault [eds.]. Zakim and Boyer's Hepatology. Elsevier, Amsterdam. Netherlands.

[HC] Health Canada, Lead Information Package - Some Commonly Asked Questions About Lead and Human Health, 2009. HC, Ottawa, Canada.

Karahalila, B., B. Aykanata and N. Ertas. 2007. Dental lead levels in children from two different urban and suburban areas of Turkey. Int. J. Hyg. Environ. Health 210: 107–112.

Kim, J., Y. Lee and M. Yang. 2014. Environmental exposure to lead (Pb) and variations in its susceptibility. J. Environ. Sci. Health, Part C 32(2): 159–185.

Klotz, K. and T. Göen. 2017. Human biomonitoring of lead exposure. Met. Ions Life Sci. 17.

Koh, D., V. Ng, L.H. Chua, Y. Yang, H.Y. Ong and S.E. Chia. 2003. Can salivary lead be used for biological monitoring of lead exposed individuals? Occup. Environ. Med. 60: 696–698.

Kristal-Boneh, E., D. Coller, P. Froom, G. Harari and J. Ribak. 1999. The association between occupational lead exposure and serum cholesterol and lipoprotein levels. Am. J. Pub. Health 89(7): 1083–1087.

Larssen, S. and L.O. Hagen. 2020. Lead. In: Air quality in Europe. 1993 A pilot report. European Environment Agency (EEA), Copenhagen, Denmark.

Lee, B.K., B.K. Lee, S.J. Lee, J.S. Joo, K.-S. Cho and H.-J. Kim. 2012. Association of Glutathione S-Transferase genes (GSTM1 and GSTT1) polymorphisms with hypertension in lead-exposed workers. Mol. Cell. Toxicol. 8(2): 203–208.

Lin, K., W. Huang, R.B. Finkelman, J. Chen, S. Yi, X. Cui et al. 2020. Distribution, modes of occurrence, and main factors influencing lead enrichment in Chinese coals. Int. J. Coal. Sci. Technol. 7: 1–18.

Madhavan, S., K.D. Rosenman and T. Shehata. 1998. Lead in soil: Recommended maximum permissible levels. Environ. Res. 49(1): 136–42.

Mahaffey, K.R. 1974. Nutritional factors and susceptibility to lead toxicity. Environ. Health Perspect. 7: 107–112.

Moreira, M.F. and E.B. Neves. 2008. Uso do chumbo em urina como indicador de exposição e sua relação com chumbo no sangue [Use of urine lead level as an exposure indicator and its relationship to blood lead]. Cad. Saude Publica. 24(9): 2151–2159.

Morita, Y., T. Sakai, S. Araki, T. Araki and Y. Masuyama. 1997. Nicotinamide adenine dinucleotide synthetase activity in erythrocytes as a tool for the biological monitoring of lead exposure. Int. Arch. Occup. Environ. Health. 70(3): 195–198.

Nafti, M., M. Bani, D. Essid, I. Magroun, C. Hannachi, B. Hamrouni et al. 2020. Effectiveness of hair lead concentration as biological indicator of environmental and professional exposure. Jr. Med. Res. 3(2): 11–14.

Nag, R. and E. Cummins. 2021. Human health risk assessment of lead (Pb) through the environmental-food pathway. Sci. Total Environ. 810(1): 151168.

Niculescu, T., R. Dumitru, V. Botha, R. Alexandrescu and N. Manolescu. 1983. Relationship between the lead concentration in hair and occupational exposure. Br. J. Ind. Med. 40(1): 67–70.

[NRC] Committee on Potential Health Risks from Recurrent Lead Exposure of DOD Firing-Range Personnel; Committee on Toxicology; Board on Environmental Studies and Toxicology; Division on Earth and Life Studies; National Research Council. Occupational Standards and Guidelines for Lead. 2012. *In*: National Academy of Sciences. Potential Health Risks to DOD Firing-Range Personnel from Recurrent Lead Exposure. National Academies Press (US), Washington, USA.

Nriagu, J., B. Burt, A. Linder, A. Ismail and W. Sohn. 2006. Lead levels in blood and saliva in a low-income population of Detroit, Michigan. Int. J. Hyg. Environ. Health 209(2): 109–121.

Olympio, K., A. Ferreira, M. Rodrigues, M.S. Luz, L. Albuquerque, J. Barbosa et al. 2020. Are fingernail lead levels a reliable biomarker of lead internal dose? J. Trace Elem. Med. Biol. 62: 126576.

Onalaja, A.O. and C. Luz. 2000. Genetic susceptibility to lead poisoning. Environ. Health Perspect 108(1): 23–28.

[OSHA] Occupational Safety and Health Administration, Lead, 2021. OSHA, Washington, USA.

Paglia, D.N., W.N. Valentine and J.G. Dahlgren. 1975. Effects of low-level lead exposure on pyrimidine 5'-nucleotidase and other erythrocyte enzymes. Possible role of pyrimidine 5'-nucleotidase in the pathogenesis of lead-induced anemia. J. Clin. Invest. 56(5): 1164–1169.

Pasiga, B.D., R. Samad, R. Pratiwi and F.H. Akbar. 2019. Identification of lead exposure through saliva and the occurrence of gingival pigmentation at fuel station indonesian officers. Pesqui. Bras. Odontopediatria Clín. Integ. 19: e4266.

Pawlas, N., K. Broberg, E. Olewińska, A. Prokopowicz, S. Skerfving and K. Pawlas. 2012. Modification by the genes ALAD and VDR of lead-induced cognitive effects in children. Neurotoxicology 33(1): 37–43.

Percy, V.A., M.C. Lamm and J.J. Taljaard. 1981. delta-Aminolaevulinic acid uptake, toxicity, and effect on [14C] gamma-aminobutyric acid uptake into neurons and glia in culture. J. Neurochem. 36(1): 69–76.

Pirsaraei, S.R. 2007. Lead exposure and hair lead level of workers in a lead refinery industry in Iran. Indian J. Occup. Environ. Med. 11(1): 6–8.

Purchase, N.G. and J.E. Fergusson. 1986. Lead in teeth: The influence of the tooth type and the sample within a tooth on lead levels. Sci. Total Environ. 52(3): 239–250.

Rees, D.C., J.A. Duley and A.M. Marinaki. 2003. Pyrimidine 5' Nucleotidase deficiency. Br. J. Haematol. 120(3): 375–383.

Sakai, T. 2000. Biomarkers of lead exposure. Ind. Health. 38(2): 127–42.

Sanchez, O.F., J. Lee, N.Y.K. Hing, S.E. Kim, J.L. Freeman and C. Yuan. 2017. Lead (Pb) exposure reduces global DNA methylation level by non-competitive inhibition and alteration of dnmt expression. Metallomics 9(2): 149–160.

Sanders, T., Y. Liu, V. Buchner and P.B. Tchounwou. 2009. Neurotoxic effects and biomarkers of lead exposure: a review. Rev. Environ. Health, 24(1): 15–45.

Scinicariello, F., H.E. Murray, D.B. Moffett, H.G. Abadin, M.J. Sexton and B.A. Fowler. 2007. Lead and delta-aminolevulinic acid dehydratase polymorphism: where does it lead? A meta-analysis. Environ. Health Perspect. 115(1): 35–41.

Shaik, A.P. and K.A. Jamil. 2008. A study on the ALAD gene polymorphisms associated with lead exposure. Toxicol. Ind. Health. 24(7): 501–6.

Shang, Y. and Q. Sun. 2018. Particulate air pollution: Major research methods and applications in animal models. Environ. Dis. 3(3): 57–62.

Singh, P., P. Mitra, T. Goyal, P.V.K. Kumar, S. Sharma and P. Sharma. 2020. Effect of metallothionein 1A rs8052394 polymorphism on lead, cadmium, zinc, and aluminum levels in factory workers. Toxicol. Ind. Health. 36(10): 816–822.

Sommar, J.N., M. Hedmer, T. Lundh, L. Nilsson, S. Skerfving and I.A. Bergdahl. 2014. Investigation of lead concentrations in whole blood, plasma and urine as biomarkers for biological monitoring of lead exposure. J. Exp. Sci. Environ. Epidemiol. 24: 51–57.

Specht, A.J., Y. Lin, M. Weisskopf, C. Yan, H. Hu, J. Xu et al. 2016. XRF-measured bone lead (Pb) as a biomarker for Pb exposure and toxicity among children diagnosed with Pb poisoning. Biomarkers 21(4): 1–6.

Tahtat, D., M.N. Bouaicha, S. Benamer, A. Nacer-Khodja and M. Mahlous. 2017. Development of alginate gel beads with a potential use in the treatment against acute lead poisoning. Int. J. Biol. Macromol. 105(1): 1010–1016.

Takeuchi, H., Y. Taki, R. Nouchi, Y.R. Yokoama, Y. Kotozaki, S. Nakagawa, A. et al. 2021. Lead exposure is associated with functional and microstructural changes in the healthy human brain. Commun. Biol. 4: 912.

[TCEQ] Texas Commission on Environmental Quality, Air Pollution from Lead. 2020. TCEQ, Austin, USA.

Trojanowski, P., J. Trojanowski, J. Antonowicz and M. Bokiniec. 2010. Lead and cadmium content in human hair in central pomerania (northern poland). J. Elementol. 15(2): 363–384.

[USDA] United States Department of Agriculture, Do cans contain lead? 2019. USDA, Washington, USA.

van Bemmel, D.M., Y. Li, J. McLean, M.H. Chang, N.F. Dowling, B. Graubard et al. 2011. Blood lead levels, ALAD gene polymorphisms, and mortality. Epidemiology 22(2): 273–278.

Wade, A.M., D.D. Richter, C.B. Craft, N.Y. Bao, P.R. Heine, M.C. Osteen et al. 2021. Urban-soil pedogenesis drives contrasting legacies of lead from paint and gasoline in city soil. Environ. Sci. Technol. 55(12): 7981–7989.

Wani, A.L., A. Ara and J.A. Usmani. 2015. Lead toxicity: A review. Interdiscip. Toxicol. 8(2): 55–64.

[WHO] World Health Organization, Lead Air Quality Guidelines. 2001. WHO Regional Office for Europe, Copenhagen, Denmark.

[WHO] World Health Organization, Blood lead levels in children, Fact sheet. 2007. WHO, Geneva, Switzerland.

[WHO] World Health Organization. 2011. Lead in Drinking-water Background document for development of WHO Guidelines for Drinking-water Quality. WHO/FWC/WSH/16.53. World Health Organization, Geneva, Switzerland.

Wright, R.O., E.K. Silverman, J. Schwartz, S.W. Tsaih, J. Senter, D. Sparrow et al. 2004. Association between hemochromatosis genotype and lead exposure among elderly men: the normative aging study. Environ. Health Perspect 112(6): 746–750.

Yan, K., Z. Dong, M.A.A. Wijayawardena, Y. Liu, R. Naiduand and K. Semple. 2017. Measurement of soil lead bioavailability and influence of soil types and properties: A review. Chemosphere 184: 27–42.

Yang, C.C., H.I. Chen, Y.W. Chiu, C.H. Tsai and H.Y. Chuang. 2013. Metallothionein 1A polymorphisms may influence urine uric acid and N-acetyl-beta-D-glucosaminidase (NAG) excretion in chronic lead-exposed workers. Toxicology 306: 68–73.

Yang, H., H. Zhang, Q. Zhou, W. Gong, B. Zhu, W. Li et al. 2015. Study on relationships between biomarkers in workers with low-level occupational lead exposure. Zhonghua Lao Dong Wei Sheng Zhi Ye Bing Za Zhi. 33(6): 403–8.

Yndestad, A., J.W. Haukeland, T.B. Dahl, B. Halvorsen and P. Aukrust. 2011. Activin A in nonalcoholic fatty liver disease. pp. 323–342. *In*: G. Litwack [ed.]. Vitamins & Hormones. Academic Press, Cambridge, USA.

Yohannes, Y.B., S.M.M. Nakayama, J. Yabe, H. Toyomaki, A. Kataba, H. Nakata et al. 2021. Glutathione S-transferase gene polymorphisms in association with susceptibility to lead toxicity in lead- and cadmium-exposed children near an abandoned lead-zinc mining area in Kabwe, Zambia. Environ. Sci. Pollut. Res. 29(5): 6622–6632.

Zahran, S., M.A.S. Laidlaw, S.P. McElmurry, G.M. Filippelli and M. Taylor. 2012. Linking source and effect: Resuspended soil lead, air lead, and children's blood lead levels in Detroit, Michigan. Environ. Sci. Technol. 47: 2839–2845.

Zerez, C.R., M.D. Wong and K.R. Tanaka. 1990. Partial purification and properties of nicotinamide adenine dinucleotide synthetase from human erythrocytes: Evidence that enzyme activity is a sensitive indicator of lead exposure. Blood. 75(7): 1576–1582.

Zhao, J., Q. Zhang, B. Zhang, T. Xu, D. Yin, W. Gu et al. 2020. Developmental exposure to lead at environmentally relevant concentrations impaired neurobehavior and NMDAR-dependent BDNF signaling in zebrafish larvae. Environ. Poll. 257: 113627.

CHAPTER 9

Mercury

Sources of Exposure

Mercury (Hg) is generally found in nature in the form of ore , and is most prevalently as cinnabar (Hg sulphide) (EPA 2021). Most rocks, sediments, water and soils naturally contain small but varying amounts of Hg. The most abundant valence states are Hg^{+1} and Hg^{+2}, forming inorganic salts, usually mercuric chloride, mercurous chloride, mercuric nitrate, mercuric sulphide and mercuric sulphate. Hg also exists in organic forms, which include methylHg, methylmercuric chloride, dimethylHg and phenylmercuric acetate. Among these organic forms, the most frequently found in nature is methylHg. Hg methylation is primarily a result of anaerobic microbial activity in sediments, which is typically enhanced in environments with high concentrations of organic matter (WHO 2017, EPA 2021). There are significant behavioural differences among Hg species, as it is known that elemental Hg volatilizes easily and stays in the atmosphere for a long time, while ionic Hg deposits from the atmosphere readily and is very water soluble. Fish and mammals easily absorb methylHg when they ingest it via the food chain. (EPA 2021). Due to anthropogenic activity Hg emissions have grown about three times and in highly industrialized regions, up to 10 times. The main activities accounting for this increase are gold mining and processing, with artisanal and small-scale gold mining contributing with 32% of the emissions, and releases originating from fuel combustion for energy generation purposes, representing 28.3% (Gworek et al. 2017, Streets et al. 2017). Metallurgy emits 13.2% of Hg, while that of the cement industry is estimated at 10.8%. Other activities contributing to Hg pollution are the production of plastics, chlor-alkali products, batteries, pesticides, dentistry, control and measurement equipment's, lamps, electrical and electronic devices, preservatives in paints, cosmetics, and applications in traditional Chinese medicine and in religious ceremonies in Latin America and India (Gworek et al. 2017, Li et al. 2020).

Air

Gaseous Hg species represent about 98% of Hg in the air, whereas Hg in organic compounds, mainly as methylHg, represent only 0.3–1% of the total Hg content in this environmental part (Gworek et al. 2017). Volcanoes, geologic deposits of Hg and volatilization from the ocean, are all natural sources of atmospheric Hg (EPA 2021). Human activities have been estimated to cumulatively increase atmospheric Hg concentrations by 300–500% over the past century (Gworek et al. 2020).

Background Hg levels in the troposphere of the northern hemisphere are estimated at 2 ng/m^3(WHO 2000). Normal Hg concentration in Canada was determined as 1.84 ng/m^3, whereas in the spring, it fell to a level of much less than 1 ng/m^3; in cities, Hg levels in the atmosphere tends to be lower than 4 ng/m^3, but in industrial areas the Hg contamination in the air may exceed 5 ng/m^3 (Gworek et al. 2017). In heavily polluted areas values even higher can be found, with air levels of up to 10 µg/m^3 earlier detected near rice fields where Hg fungicides were used. Even higher total gaseous Hg levels were determined in areas close by Hg mines, it was noted that values of 8–243 µg/m^3 were found in Italy in 1982 and 0.1–50 µg/m^3 in Spain in 1994 (WHO 2000, Gworek et al. 2017).

Assuming an ambient air level of 0.010 µg/m^3, the average daily intake of inorganic Hg by inhalation would amount to about 0.2 µg; daily amounts absorbed into the bloodstream from the atmosphere were estimated to be about 32–64 ng in remote areas and about 160 ng in urban areas (WHO 1996, EEB 2019).

Coal burning-related Hg emissions are not directly regulated by the European Union (EU), but Germany has recently updated its long-standing Emission Limit Value (ELV) as 10–30 mg/Nm^3 for all coal-fired plants. In the USA all combustion plants with a size exceeding 25MWth are required to meet an approximate Hg limit of 1.5 µg/Nm^3 (converted) for hard coal, and 4.8 µg/Nm^3 (converted) for lignite (EEB 2019). The World Health Organisation (WHO) estimate a tolerable concentration of 0.2 µg/m^3 for long-term inhalation exposure to elemental Hg vapour (NSW Health 2013). According with this organization (2000) it does not seem appropriate to set air quality guidelines for methylHg compounds since, if present in the atmosphere, would make a negligible contribution to total human intake.

Workplace

Workers at risk of being exposed to Hg include the ones employed in areas where electrical equipment or automotive parts are manufactured, in fluorescent light bulb recycling services, in chemical processing plants that use Hg and workers in medical, dental (or other health services) who work with equipment containing the metal (Branco et al. 2017, NIOSH 2019). The National Institute for Occupational Safety and Health (NIOSH) recommend an airborne exposure limit (REL) of 0.05 mg/m^3 as Hg vapour averaged over a 10-hr work shift, and 0.1 mg/m^3 as Hg not to be exceeded at any time; the Threshold Limit Value (TLV) decided by American Conference of Governmental Industrial Hygienists (ACGIH) is 0.025 mg/m^3 averaged over an 8-hr work shift. The European Occupational Safety and Health Administration (OSHA) sets an airborne permissible exposure limit (PEL) of 0.1 mg/m^3 averaged over an 8-hr work shift (NJDH 2009). In countries of the European Member States (e.g., Spain) the regulated value for alkyl compounds of Hg is 0.01 mg/m^3(HBM4EU 2021).

Soil

Soil plays an important role in the biogeochemical Hg circulation as it accumulates the metal and is a source for other environmental sections. Hg occurs naturally in soils from geologic sources or as result of natural events such as forest fires and volcanic eruptions (Gworek et al. 2020). Ionized forms of Hg are largely adsorbed by soils and sediments and desorbed slowly, it is acknowledged that iron oxides adsorb Hg in neutral soils and that in acid soils, most Hg is adsorbed by organic matter. When organic matter is not present, Hg becomes more mobile, evaporating to the atmosphere or leaching to groundwater (EPA 2021). In reducing conditions of many permanently or periodically flooded soils, Hg may be biogeochemically transformed into organo-Hg forms, of which the toxicological relevant methylHg is the most common (as will be described next). However, soils with greater amounts of larger organic matter molecules, are less at risk of Hg methylation as they have lower bioavailability. Currently the global amount of Hg accumulated in soils is very large, posing risks to global public health (O'Connor et al. 2019). The average background concentration of Hg in soil ranges from 0.03 to 0.1 mg/kg whereas in contaminated sites concentrations 2- to 4-orders of magnitude higher can be found (Gworek et al. 2020). In several European agricultural soils, while their pollution by Hg was recently found to be relatively low, usually below 0.1 mg/kg, the ranges of the results were relatively wide; maximum values found to exceeded 1 mg/kg probably indicating a threat to food production (Gworek et al. 2020). Korea has alarming levels of soil contamination in farmlands and industrial areas or factories were also determined as 4 mg/kg and 16 mg/kg, respectively (Ye et al. 2016).

Water

Water pollution with Hg can occur from run-off water, contaminated by either natural or anthropogenic sources or from air deposition. The biggest risk to human health is Hg in aquatic environments, as it stays there for a very long time, 20–30 yr in the upper ocean and 100s of yr in the deep ocean; in water media Hg gets converted by microorganisms to the very toxic organic form methylHg (EPA 2021, HBM4EU 2021). The process tends to occur in environments with low oxygen levels, low pH, Hg bioavailability, high levels of dissolved organic compounds, which favours sulphate-reducing bacteria largely responsible for methylation. Those conditions are found primarily in deep sea environments, coastal marine sediments and some freshwater lakes (Gworek et al. 2020).

Naturally occurring levels of Hg in groundwater and surface water are less than 0.5 µg/L (WHO 1996). In drinking-water and rainwater the range is usually 5–100 ng/L, while the average value is about 25 ng/L. It is estimated that Hg intake from drinking-water is about 50 ng/d, mainly and only a small fraction of Hg^{2+} is absorbed; when a level in drinking-water of 0.5 µg/L is assumed, the average daily intake of inorganic Hg from this source would amount to about to 1 µg (WHO 1996, 2000). According to the Australian Drinking Water Guidelines (2011), the health guideline value for total Hg in drinking water is 1μg/L, while a maximum contaminant level of 2 μg/L is set by the US Environmental Protection Agency (EPA) (WHO 1996).

Food

Food is the main source of Hg for non-occupationally exposed populations, with fish and fish products accounting for most of the organic Hg in food (70–90%) (WHO 1996, 2000, Branco et al. 2017). This occurs as methylHg formed in aquatic environments bioaccumulates in fish and shellfish, implying that these organisms contain higher concentrations than the surroundings. Furthermore, methylHg biomagnifies, which means that large predatory fish are more likely to have high levels of Hg because of eating many smaller fish that have ingested plankton containing the metal (WHO 2017). Normal concentrations of Hg in edible tissues of various species of fish cover a wide range, from 50 to 1400 ng/g w.w. (WHO 2000). MethylHg deposited on land also enters the food-chain, an example being rice grown on contaminated soil, as it is grown in water and Hg is absorbed in the grain. (HBM4EU 2021). When methylHg is ingested by humans, it is absorbed more readily and excreted more slowly than other forms of Hg and moreover, the absorption of organic Hg from food is six times greater than of inorganic Hg (EPA 2021).

The average daily intake of Hg from food is in the range of 2–20 μg/d but may be much higher in regions where ambient waters have become contaminated with Hg and where fish constitute a high proportion of the diet (WHO 1996). The World Health Organization (WHO) considers an acceptable intake of total Hg as 2 μg/kg b.w./day (NSW Health 2013). European maximum levels (MLs) for total Hg content for fish and seafood are 0.5–1.0 mg/kg w.w. Additionally, the FAO/WHO Expert Committee on Food Additives (JECFA) established a provisional acceptable weekly intake (PTWI) for MeHg of 1.6 μg/kg b.w./wk and a PTWI for total Hg of 4 μg/kg b.w./wk; in the USA the National Research Council (NRC) established an intake limit of 0.7 μg/kg b.w./wk, while Japan suggests a higher methylHg exposure limit, which is 2.0 μg/kg body weight/wk (Ye et al 2016, Kuras et al. 2017). Despite the Hg content in fish, the European Food Safety Authority (EFSA) recommend weekly intakes of fish between 1–2 servings and 3-4 servings to gain health benefits, such as improved neurodevelopment in children and reduced risk of coronary heart disease in adults (HBM4EU 2021).

Consumer products

Fatalities and severe poisonings resulted from heating metallic Hg and Hg-containing objects at home. Incubators used to house premature infants were found to contain Hg vapour at levels approaching occupational threshold limit values; at the time the source was Hg droplets from broken Hg thermostats (WHO 2000). Amalgam surfaces release Hg vapour into the mouth, constituting the predominant source of human exposure to inorganic Hg in the general population. Depending upon the number of amalgam fillings, the estimated average daily absorption of Hg vapour from dental fillings varies between 3000 and 17 000 ng (HBM4EU 2021). Exposure to organomercurials might occur through the use of skin-lightening creams and other pharmaceuticals containing Hg; a known example is Thiomersal, which is used for the preservation of vaccines and immunoglobulins, at a level of 100 µg thiomersal per injection (WHO 2000, Branco et al. 2017).

Several regulations prohibit Hg in cosmetic products, in electrical and electronic equipment and also restricting its use in batteries and accumulators (HBM4EU 2021).

Toxicity

The risk of toxicity induced by Hg is determined by the route of exposure, and among the Hg chemical species some of them are more toxic or bioavailable than others (WHO 1996). In general terms Hg affects immune, genetic and enzyme systems and damages the nervous system, among other effects. While there are many similarities in the toxic effects of the various Hg species, there are also relevant differences (EPA 2021).

Acute toxicity

The ingestion of acute lethal toxic doses of any form of Hg result in similar terminal signs and symptoms, such as shock, cardiovascular collapse, acute renal failure and severe gastrointestinal damage. Clinical symptoms include pharyngitis, dysphagia, abdominal pain, nausea and vomiting, bloody diarrhoea and shock. Swelling of the salivary glands, stomatitis, loosening of teeth, nephritis, anuria and hepatitis may also occur (WHO 1996).

Mercury vapours

Accidental inhalation of Hg vapours is most likely to occur in the case of acute poisoning. Although the actual fatal level of Hg vapour is not known, exposure to more than $1-2$ mg/m^3 for a few hours causes acute chemical bronchiolitis and pneumonitis, which in extreme cases can be fatal due to respiratory failure. Acute effects occur from the inhalation of air containing Hg vapour at concentrations ranging from 0.05 to 0.35 mg/m^3, while exposure for a few hours to $1-3$ mg/m^3 may trigger pulmonary irritation and destruction of the lung tissue (WHO 1996, Park and Zheng 2012). Two hr after exposure, the lung injury may appear as hyaline membrane formation, and eventually extensive pulmonary fibrosis occurs (Asano et al. 2001, Bernhoft 2012). Nephrotic syndrome may also be elicited, characterized by excessive loss of protein (mainly albumin) in the urine and oedema. Depending on the dose, oral mucosa may also be affected (WHO 1996).

At relatively high doses classical signs and symptoms of elemental Hg vapour poisoning are primarily related to the central nervous system, such as tremors, mental disturbances syndrome of psychological abnormalities labelled erethism, which is manifested as excessive shyness, loss of confidence, vague fears, irritability, insecurity and suicidal melancholia (WHO 1996, Langford and Ferner 1999). Cognitive decrements and emotional alterations, which develop more insidiously, may be the most harmful effects of current exposures in the workplace. The peripheral nervous system may also be involved, as evidenced by decreased nerve conduction velocity (WHO 1996). Changes in renal and immune functions, endocrine and muscle function and several types of dermatitis have been described, while digestive, cardiovascular and reproductive systems may also be affected (Bernhoft 2012, WHO 1996, 2017, Teixeira et al. 2018).

Mercury salts

The toxicity of Hg salts varies with their solubility usually being Hg^{1+} compounds of low solubility and significantly less toxic than Hg^{2+} compounds; the lethal dose of Hg^{2+} chloride may be as small as 0.5 g (Langford and Ferner 1999). Inorganic salts of Hg are corrosive to the skin, eyes and gastrointestinal tract (WHO 2017). Acute single oral doses can induce severe gastrointestinal toxicity, with extensive precipitation of enterocyte proteins, abdominal pain, vomiting and bloody diarrhoea with potential necrosis of the gut mucosa; death may occur from peritonitis or from septic or hypovolemic shock and surviving patients usually develop renal tubular necrosis with anuria (WHO 2000, Bernhoft 2012).

A probable lethal dose for Hg chloride is reported from a range of 1 to 4 g but fatalities from estimated ingestions of 0.5 g are also documented (Beasley et al. 2014).

Organic mercury

Deaths have also resulted from 3 mon exposure to diethylHg at an estimated concentration of 1 mg/m^3; the lethal dose of methylHg is estimated to be 200 mg, even though paraesthesia of the hands, feet and mouth may occur at a total body burden of 40 mg (NIOSH 1994).

Chronic toxicity

At chronic low-level exposures to Hg the significant target organs for toxic effects are the central nervous system and the kidneys; nonspecific symptoms like weakness, fatigue, anorexia, weight loss and gastrointestinal disturbance have also been described (Bernhoft 2012, Park and Zheng 2012).

Inorganic mercury

However, if elemental Hg is ingested, it is absorbed relatively slowly and may pass through the digestive system without causing damage (Park and Zheng 2012, EPA 2021).

Chronic poisoning with Hg salts is rare and usually involves concomitant occupational exposure to Hg vapour (Langford and Ferner 1999, Bernhoft 2012). Hg salts can induce kidney toxicity, while immune dysfunctions include hypersensitivity reactions, such as asthma and dermatitis and disruption of lymphocyte subpopulations. Other effects such as thyroid dysfunction, inhibition of spermatogenesis, atrophy and capillary damage in the thigh muscle have also been witnessed; brain dysfunction is less evident than with other forms of Hg (Bernhoft 2012).

Organic mercury

Methyl- and ethylHg compounds have been the cause of the largest number of cases of Hg poisoning and long-term fatalities in the general population due to consumption of contaminated fish or consumption of bread prepared from cereals treated with alkylHg fungicides. The earliest effects are nonspecific, and include paraesthesia, malaise and Hunter-Russel syndrome concentric constriction of the visual field, deafness, dysarthria and ataxia. In the worst cases, the patient may go into a coma and ultimately die (WHO 1996).

Low levels of ethylHg containing thiomersal are used as a preservative in medical preparations, including vaccines, and are of particular concern since they have been linked to autism (Lohren et al. 2015).

Damage induced by methylHg is almost exclusively limited to the nervous system, which is selective to certain areas of the brain involved with sensory functions and coordination, such as the neurons in the visual cortex and the granule cells of the cerebellum. Constricted visual field and ataxia appear to have a latent period of weeks to months and are usually irreversible. The peripheral nervous system may also be affected, especially at high doses (WHO 2000, Lohren et al. 2015). People living in Minamata (Japan) who depended on seafood as a large part of their food supply, were exposed to MeHg by ingestion of polluted fish for almost 20 yr. This resulted in chronic MeHg poisoning with adverse conditions which were known as the "Minamata Disease". Children born from mothers exposed to MeHg showed extensive spongiosis of the cerebral cortex, a characteristic feature of foetal Minamata Disease; serious disturbances in mental and motor developments were observed in these children (WHO 1996, Ekino et al. 2007).

According to the International Agency for Research on Cancer (IARC) methylHg compounds are possibly carcinogenic to humans, while metallic Hg and inorganic Hg compounds could not be classified (IARC 1993).

Biomarkers

Hg toxicity continues to be a global health concern. Traditional biomarkers (BM) are practical and provide a reliable measure of exposure but a need for more sensitive and refined tools to assess the effects and/or susceptibility remains (Branco et al. 2017). There are three types of available BMs for Hg, and studies on novel candidates will be shown next .

Biomarkers of exposure

The most common BMs of exposure to Hg include measurements of total Hg levels in blood, urine and hair, in the absence of speciation analysis. Other exposure BMs such as the levels of Hg in faeces, nails, placental cord-blood or breast milk or nails, are not used often but have been proved quite useful in specific situations (Clarkson and Magos 2006).

Mercury levels in blood

Blood is responsible for Hg distribution to target organs, and currently the concentration of Hg in blood is accepted as a reliable BM of exposure to all forms of Hg. However, blood Hg reflects recent exposure to methylHg and/or Hg^0, only being useful for a short time after an acute exposure. Concentrations of Hg in this tissue (and in urine) may be low soon after the exposure has ceased, while concentrations in critical organs may still be high (Satoh 2000, WHO 2000, Boerleider et al 2017). Despite this, blood Hg levels may be useful in the case of chronic exposure, with human data demonstrating how dietary intakes may be transferred into blood Hg; blood Hg additionally reflects the systemic available Hg that may reach the brain after passing the blood-brain barrier (Grandjean et al. 2005, Boerleider et al 2017).

While assessing Hg vapours , it is recommended that after an acute exposure the first sample should be taken within 24 hr, and speciation should be carried out to eliminate the possible influence of dietary intake of Hg from fish. If the source of exposure is removed (e.g., a broken thermometer), follow-up blood samples can contribute to ensure that Hg was eliminated from the body (WHO 2000, Boerleider et al. 2017).

Erythrocyte Hg concentrations are more specific for methylHg assessments as approximately 90% of methylHg is found in red blood cells bound to the haemoglobin, whereas inorganic Hg (Hg^0 and Hg^{2+}) are evenly distributed between these cells and the plasma (Satoh 2000). Therefore, erythrocytic Hg concentration is more specific for methylHg exposure, and exposure to Hg vapour or other inorganic Hg compounds should be assessed in serum samples (WHO 2000).

The concentration of Hg in the whole blood is usually lower than 10 µg/L and in populations with limited fish consumption these values normally do not exceed 2µg/L. Blood Hg concentration of more than 200 µg/L may be associated with health effects in adults, and concentrations of 40 to 50 µg/L in pregnant woman could be associated with a toxic risk for the foetus (Satoh 2000, WHO 2000, Kim et al. 2012, Ye et al. 2016). An Occupational Biological Limit Values (BLV) has been established in Europe, which is 10 µg/L of blood for elemental Hg and inorganic divalent Hg compounds (SCOEL 2007).

Mercury levels in urine or faeces

The determination of urinary Hg concentrations is a quick way for identifying exposures and reflects average exposure over the previous few months in those chronically exposed, although only on a group basis; high individual (including diurnal) variability of urine levels exist (WHO 2000, Pesch et al. 2002, SCOEL 2007, Ye et al. 2016). In addition, since organic Hg represents only a very small portion of urinary Hg (most is excreted in the faeces), urinary Hg is more useful for the analysis of metallic or inorganic Hg compounds (Ye et al. 2016). For these reasons the concentration of Hg in urine is the most common BM of exposure to Hg^0, in occupational exposures and through dental amalgams (Branco et al. 2017).

Concentrations found in people without known exposure to Hg is less than 5 µg/L; sons of workers at a thermometer plant were reported to exhibit median urine Hg levels of 25 µg/L without signs of Hg intoxication in clinical examinations (WHO 1996, 2000). Mild proteinuria may occur in most sensitive adults at Hg urine values of 50–100 µg/L following chronic occupational exposures, and when urinary Hg levels exceeds 100 µg/L, neurological symptoms can develop; tremor and psychomotor disturbances usually appear at urine values of more than 300 µg/L. Levels of urinary Hg of 800 µg/L or above can be fatal (WHO 2000, Ye et al. 2016). After correcting the values for creatinine, neurotoxic effects attributed to Hg^0 are evident in subjects with urinary Hg levels exceeding 35 µg/g creatinine, even though neurobehavioral effects are described in the 20–30 µg/g creatinine range or lower (WHO 2000, Branco et al. 2017). Occupational Biological Limit Values (BLV) established in Europe are 30 µg/g creatinine in urine, for elemental Hg and inorganic divalent Hg compounds (SCOEL 2007).

MethylHg is excreted mainly through the faeces, making it easy to detect. It should be noted however, that since methylHg that remains in the lower gastrointestinal tract is subject to demethylation to inorganic Hg, all the Hg is in the inorganic form in people exposed only to methylHg (Rafati-Rahimzadeh et al. 2014).

Mercury levels in hair or nails

Keratin comprises 80–90% of hair composition and this compound is rich in sulphhydryl groups, which easily combines with Hg; methylHg bound to keratin constitutes more than 80% of the hair metal burden (Clarkson and Magos 2006, Ye et al. 2016). It has a long half-life in hair making this tissue useful for evaluating exposures that occurred months earlier, and when Hg is measured along the length of a hair strand, and can be used as an indicator of past blood levels (WHO 2000, Nuttall 2006). Since the migration of Hg to hair is irreversible and stays stable there for long periods it makes it easy to transport and store samples. MethylHg accumulate at higher concentrations in hair that ethylHg or inorganic Hg (which is incorporated poorly in hair) and thus, hair Hg concentrations are used as a BM of chronic exposure to methylHg reflected mainly in the uptake via fish consumption (WHO 2000, Pesch et al. 2002, Nuttall 2006, Ye et al. 2016, Branco et al. 2017, Brooks et al. 2018). Hair follicles also seem to accumulate the same transportable species that reach the brain, and accordingly, hair Hg levels correlate well with the levels in this organ, although the ratio may vary according to an individual's characteristics such as age, gender and genetics (Clarkson and Magos 2006). It has been considered that total Hg levels in hair and blood can be used in addition to BMs of Hg intoxication, the ratios considered by the Food and Drug Administration (FDA) and the WHO 250:1 and 250–300: 1, respectively (Ye et al. 2016).

Normal levels of Hg in hair range between 1 and 2 µg/g, but individuals who consume large amounts of fish may have hair Hg levels of more than 10 µg/g (Clarkson and Magos, 2006). A Lowest Observable Adverse Effect Level (LOAEL) for neurotoxic effects (paraesthesia) in adults set at 50 µg/g (Clarkson and Magos, 2006). In conditions of moderate Hg poisoning, the concentrations of Hg found in hair can range between 200 and 800 μg/g, and in severe intoxication go up to 2400 μg/g (Ye et al. 2016).

There is a great deal of apprehension from threats of Hg exposure to the developing foetus, as both elemental and organic forms of Hg can cross the placenta during gestation resulting in a far higher dose-to-weight ratio than in adults (Dack et al. 2021) The quantification of Hg in maternal hair during pregnancy is often used, providing information about Hg exposure during this period through segmental analysis (when considering a hair monthly growth rate of 1 cm); this measurement tends to correlate well with both maternal and foetal blood Hg (Pinheiro et al. 2020, Dack et al. 2021). The WHO recommends a limit of hair Hg of 1 mg/kg in pregnant woman and considers that a level of 10 mg/kg or above can increase the risk of foetal neurological defects (Ye et al. 2016, Pinheiro et al. 2020).

Samples of nails give an indication of long-term exposure to Hg, with dentists and smelter miners having higher levels of Hg in fingernails; a study revealed that dentists had values 5.86 times greater than controls (Clarkson 1993, Kwaansa-Ansah et al. 2019). Higher concentrations of toenail Hg were also associated with several metabolic risk factors, including higher systolic and diastolic blood pressure, fasting blood glucose and levels of triglycerides (Park and Seo 2016).

It should be stressed that despite the uses of the BMs mentioned so far and some associations with adverse effects that can be established, due to inter-individual human variability, they do not necessarily reflect the presence or absence of toxicity (Clarkson and Magos, 2006).

Mercury levels in placental cord blood or human milk

It has been debated that the analysis of placental cord blood is more likely to provide a better precision than maternal hair, to assess prenatal exposure to methylHg from maternal seafood consumption (Grandjean et al. 2005). Placental blood cord Hg is almost entirely of the methylated form (which crosses the placenta) and the determination of their levels reflect Hg exposure across the third trimester (Grandjean et al. 2005, Dack et al. 2021). There is a high affinity of methylHg to foetal haemoglobin, as well as larger haematocrit and greater haemoglobin concentration in new-borns, resulting in higher Hg concentrations in placental cord blood than in maternal blood; a ratio of 1.65 is described (Stern et al. 2003, Grandjean et al. 2005, Pinheiro et al. 2020, Iwai-Shimada et al. 2021). Regarding inorganic Hg, although this form is confined within the placenta, placental blood cord: maternal blood ratio of inorganic Hg was already found to be positively associated with that of methylHg (Iwai-Shimada et al. 2021). The safety limit in umbilical cord blood recommended by EPA is 5.8 μg/L, considered a potentially increased risk of adverse effects on pregnancy outcomes (Pinheiro et al. 2020). However, in a study performed in Poland 75% of all placental blood cord determinations exceeded this value, suggesting a general high exposure to Hg (Kozikowska et al. 2013).

Maternal Hg can be transmitted to children during the time of breastfeeding, since it goes through the mammary glands of lactating mothers. Two major forms of Hg can enter breast milk: the first is methyHg which does not enter breast milk at high rates as it is attached to red blood cells, but small amounts can be readily absorbed by the infant; the second form is inorganic Hg which enters breast milk easily but is not well absorbed in the infant's gastrointestinal system (Yurdakök 2015). Studies on Hg human breast milk levels reported concentrations between 0.2 and 5.5 ug/L, while others on mothers living in gold mining areas in Indonesia, Tanzania and Zimbabwe revealed values up to 4 µg/L in most of the samples (Bose-O'Reilly et al. 2008, Yurdakök 2015). Other studies found about 37 and 54% of the analyzed milk samples, had Hg levels higher than the normal mean concentration (1.7 µg/L) as mentioned by WHO (Yurdakök 2015, Vahidinia et al. 2019, Mahmoudi et al. 2020). Maternal seafood intake and amalgam fillings explained 46% of the observed variation in breast milk Hg concentrations in a Norwegian study (Vollset et al. 2019).

A relationship between different BMs used for characterizing methylHg exposures in adults and foetuses is shown in Fig. 5.

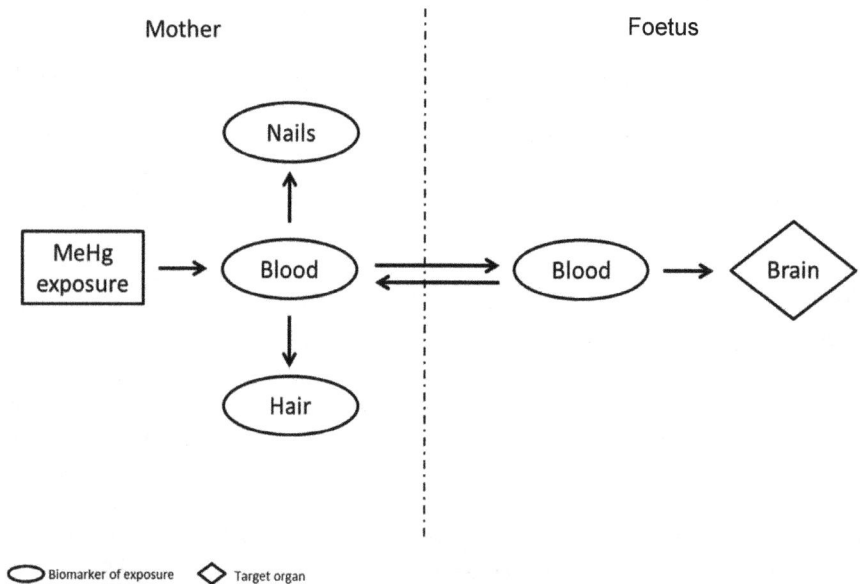

Figure 5. Relationship among BMs used to monitor methylHg exposures (Ruggieri et al. 2017).

Micro RNAs

As mentioned in other chapters, miRNAs are endogenous noncoding RNAs which may play crucial gene-regulatory roles by binding to the 3' UTR of protein-coding genes to mediate their posttranscriptional repression; several miRNAs exist. In an *in vitro* model of central nervous system differentiation exposed to methylHg, differential regulation of several miRNAs (miR-141, miR-196b, miR-302b, miR-367 and miR-372) were observed; these targets were mapped to pathways important for axonal guidance, learning and memory (Miguel et al. 2018). miRNA-92a and miRNA-486 are also referred as potential diagnostic BMs for Hg poisoning, as suggested by studies performed with Hg exposed workers who exhibited changes in the expression of these circulating molecules. Both induce apoptosis by jointly activating the nuclear factor kappa-light-chain-enhancer of activated B cells (NF-κB) and ciclo-oxigenase-2 (COX-2). In turn NF-κB signalling plays a critical role in mediating the expression of numerous proinflammatory gene products, including COX-2 and inducible Nitric Oxide Synthase (iNOS), which are involved in inflammatory responses. While exposure to Hg could induce the upregulation of miRNA-92a and miRNA-486, it was also demonstrated that Hg can induce NF-κB activation, resulting in increased expression of COX-2 (Ding et al. 2017).

Biomarkers of susceptibility

At present it is clear that the genetic background is one of several factors influencing human susceptibility to Hg-related outcomes (Andreoli and Sprovieri 2017). Genetic factors are known to affect the absorption, distribution, biotransformation, excretion and consequently, the toxicity of Hg species.

Glutathione

The response to Hg may be influenced by molecular variants in several regulatory glutathione (GSH) genes, involved in Hg toxicokinetics (Guan et al. 2015, Andreoli and Sprovieri 2017). Subjects with certain GSH variants may tolerate higher Hg exposures, due to faster elimination and/or better antioxidative GSH-associated capacity. Inversely, GSH deficiency is known to be associated with sensitivity to both Hg chloride and methylHg (Andreoli and Sprovieri 2017). The genetic deficiency in glucose-6-phosphate dehydrogenase (G-6-PDH) is another example; G-6-PDH is involved in the production of GSH, influencing the metabolism and the detoxification of Hg. Glutathione S-transferase (GST) (with the subclasses alpha, mu, pi, omega, theta and zeta) is a set of cytosolic enzymes that catalyze the conjugation of GSH to a wide variety of substrates, including Hg, and represents another example (Klautau-Guimarães et al. 2005, Andreoli and Sprovieri 2017). Several types of allelic variations have been identified within GSTs genes, so far the better studied being GSTM1 (GST-mu1), GSTT1 (GST-theta1) and GSTP1 (GST-p1) . The first two genes encode isoforms with impaired catalytic activity, resulting in phenotypes expressing less or no functional enzymes and this, leads to lowered levels of GSH-conjugates. There is reduced excretion and/or increased retention of Hg in hair, blood, erythrocytes and urine among exposed individuals exhibiting these genotypes. Regarding GSTP1, this gene encodes a GSH subunit able to alter the sensitivity to Hg; two important GSTP1 polymorphisms (rs1138272 and rs1695) have been proposed to influence methylHg BM levels in epidemiological studies (Andreoli and Sprovieri 2017). Notably, an epistatic effect of GSTT1 and GSTM1 deletion polymorphism is a risk factor for increased susceptibility to Hg exposure (Gundacker et al. 2007). In addition, Glutamate-Cysteine Ligase (GCL), which is a high quality limiting enzyme in the GSH synthesis, and also a determinant to Hg metabolism. The enzyme consists of two subunits, respectively coded by GCLC and GCLM genes, for which there are some functionally significant polymorphisms. Minor alleles of two Single-Nucleotide Polymorphisms (SNP) (rs17883901 in GCLC; rs41303970 in GCLM) decrease the promoter activity and have been related to increased methylHg and elemental Hg BM levels. A minor allele of another GCLC SNP (rs1555903) was also significantly related to Hg retention in the umbilical cord (Andreoli and Sprovieri 2017).

Metallothioneins

Metallothioneins (MT) are the body's natural chelating agents, involved in the transport, storage and detoxification of metals (Andreoli and Sprovieri 2017). Since MTs contain a high proportion of cysteine groups to which certain metals (including Hg) preferably bind, they play a role in preventing, at least partially, Hg accumulation in the liver and blood, and in decreasing renal toxicity (Aschner et al.2006, Gundacker et al. 2007, Peixoto et al. 2007). Genetic variants of MT are reported to affect Hg toxicokinetics (Woods et al. 2013). As mentioned in the earlier chapters, there are four major isoforms of MT (MT-1 through MT-4) identified in mammals. A relevant role is played by the MT1M (rs2270837) and MT2A (rs10636) variants on the relationship between lower urinary Hg level and personal exposure to elemental Hg. Similarly, MT1A (rs8052394) and MT1M (rs9936471) polymorphisms are involved in modifying hair Hg levels due to daily methylHg intake from the consumption of fish. The MT4 (rs11643815) polymorphism is another significant predictor of the Hg body burden. The molecular effects of these polymorphisms are still not fully understood, but a hypothesis is that changes in the molecular structures of the different MT isoforms could affect Hg retention in terms of metal–binding capabilities, and subsequently in BM levels (Andreoli and Sprovieri 2017). Very few associations of MTs polymorphisms and somatic traits have been reported, but it is documented that children with the variants MT1M (rs2270837) and MT2A (rs10636) exhibit significant modifications of Hg effects on multiple neurobehavioural functions (Woods et al. 2014). It was proposed that MT expression could depend on GST polymorphisms with important implications on the overall metal detoxification capacity of the human organism, namely Hg detoxification (Gundacker et al. 2007).

Selenoproteins

Selenium is an essential micronutrient which exerts its biological functions predominantly by means of selenium-dependent proteins, known as selenoproteins. These are important in protecting against Hg toxicity through binding Hg through their highly reactive selenol group, as well as by exerting an antioxidative role through eliminating reactive oxygen species (ROS) induced by Hg *in vivo*. Elevated concentrations of methylHg may directly downregulate the expression of many selenoprotein-coding genes, leading to the depletion of potentially important metabolites for the organism's response to metal toxicity. An example is the gene codifying selenoprotein pi (SEPP-1), this protein being the main one for selenium transport in the blood. Since the protein comprises up to 10 selenol groups, it is especially equipped for binding more Hg (Andreoli and Sprovieri 2017). Elevated elemental Hg exposure increases SEPP1 expression and the ability of SEPP1 to bind Hg, is evident in highly exposed miners. SEPP-1 3′UTR CT or TT genotypes may affect Hg binding and subsequent distribution to various tissues; the 3′UTR T allele is known to be linked to greater SEPP1 expression among people supplemented with selenium (Nordberg et al. 2007, Goodrich et al. 2011). Other studies have shown that concentrations of Hg in blood and urine among workers exposed to Hg^0 were positively correlated with the levels of mRNA for SEPP1 and also mRNA for glutathione peroxidase (GSH-Px), proposing that exposure to Hg^0 alters gene expression of antioxidant enzymes and the level of selenium-containing selenoproteins (Kuras et al. 2018).

ATP-binding cassette transporter

The ATP-Binding Cassette transporter superfamily (ABCs) can move several chemicals, including Hg, across cells of the surface membrane. The human genome carries 49 ABCs genes, which are arranged in seven subfamilies and are designated from A to G. Functional suppression of ABCs activity has proved to increase Hg content in cells and sensitivity to Hg toxicity *in vitro* (Andreoli and Sprovieri 2017). Research on associations between polymorphisms in the potential Hg-transporter ABCC2 gene and neurological effects in artisanal small-scale gold miners showed that certain ABCC2 polymorphisms may influence the neurotoxic effects in Hg-burdened subjects; ABCC2 alleles associated with worse neurological performance, were also correlated with higher levels of urinary Hg concentrations in the same population (Kolbinger et al. 2019).

Biomarkers of effect

There are no specific BMs of Hg effect, but several ones have been proposed for a long time. The general laboratory tests to evaluate Hg intoxication include complete blood cell count, electrolyte assays and renal and hepatic function tests. Electrocardiography (ECG), Pulmonary Function Test (PFT), cardiovascular monitoring, electroneuromyography and neuropsychological tests are also used in these evaluations (Ye et al. 2016).

Heme synthesis

Metal-specific changes in the urinary porphyrin excretion pattern associated with prolonged exposure to metals, such as Hg, have been described. These changes involves both metal-directed impairment of specific heme biosynthetic pathway enzymes in target tissues, as well as metal-facilitated oxidation of reduced porphyrins (porphyrinogens) (Pingree et al. 2001). In the case of exposures to Hg there is characteristically increased urinary concentrations of specific porphyrins: pentacarboxyporphyrin, coproporphyrin and the atypical keto-isocoproporphyrin; there is selective interference of Hg with the fifth (uroporphyrinogen decarboxylase) and sixth (coproporphyrinogen oxidase) enzymes of the heme biosynthetic pathway in kidney cells, which are an important principal target organ for Hg (Pingree et al. 2001, Heyer et al. 2006,). Such changes have been described as a BM of prolonged exposure to all forms of Hg, and a pattern of dose- and time-related porphyrinogenic response to Hg is described for animals and humans (Heyer et al. 2006). Thus, it is considered that the urinary porphyrin profile may serve as a useful BMs of Hg exposure in both clinical and epidemiologic studies (Woods et al 1993, Pingree et al. 2001, Heyer et al. 2006).

Oxidative stress

Hg induces oxidative stress, with data existing on children accidently exposed to elemental Hg with decreased catalase and glutathione peroxidase (GSH-Px) activities, which are enzymes involved in scavenging ROS (Day 2009, Dalkiran et al. 2021). Malondialdehyde (MDA) concentrations has long been used as a lipid peroxidation marker in oxidative stress assessment, and the levels of MDA were previously found to be augmented in the serum of children exposed to mehylHg through contaminated fish consumption (Carvalho et al. 2019, Morales and Munné-Bosch 2019, Dalkiran et al. 2021). Other research works showed that chronic low dose Hg-induced oxidative stress assists in damage to genetic material in exposed workers (Riverón et al. 2014). However, coproporphyrins or oxidative stress markers may not be sensitive enough and lack specificity (Branco et al. 2017).

Selenium and selenoproteins

Selenium and selenoproteins levels have proposed that BMs of effect, with *in vitro* studies suggesting that selenium and Hg could form Hg–Se complexes in a reducing environment with a molar ratio of 1, and then bound with plasma SEPP or GSH-Px (Chen et al. 2006, Yoneda and Suzuki 1997). Strong interactions of SEPP and Hg at high Hg exposure concentrations is documented. In mine workers exposed to Hg, augmented serum selenium concentrations were noted, and the finding that urinary selenium was only slightly higher, suggested that Hg could interfere with selenium metabolism by increasing their retention (Chen et al. 2006).

Selenoproteins, thioredoxin reductases (TrxR) are a family of naturally occurring selenoproteins which are an integrant part of the thioredoxin system (Branco and Carvalho 2019). They play important roles to several cellular functions such as protein repair and regulation of the cellular cycle and are very sensitive to Hg compounds; the inhibitory effects of mercurials on the thioredoxin system is demonstrated both *in vitro* and *in vivo* (Mustacich and Powis 2000, Chen et al. 2006, Branco et al. 2012). *In vivo* studies also showed that on exposure to Hg, histopathological alterations correlate with the level of TrxR activity, and point to the potential use of this enzyme as a BM of Hg toxicity (Branco et al. 2012).

Final Remarks

This chapter presented various forms of Hg chemical species, including inorganic and organic forms. Among inorganic forms, emphasis was given for acute toxicity of Hg vapour inhalation in occupational contexts, while among organic forms emphasis was given to chronic exposures to methylHg through the intake of contaminated fish.

The risk of toxicity induced by Hg is determined by the route of exposure, but also by the involved Hg chemical species, since some of them are more toxic or bioavailable than others. In addition, despite there are many similarities in the toxic effects of the various Hg species, there are also relevant differences. Relevance was given to Hg-induced toxicity in nervous and respiratory systems. Most common BMs of exposure to Hg were presented, such as measurements of Hg in blood, which reflect elemental and methylHg recent exposures, unlike Hg levels in hair or urine, which indicate long-term organic or inorganic Hg exposures, respectively. Less traditional BMs were described , with reference to the utility of blood cord or breast milk Hg levels to assess foetus and new-borns exposures. Despite the available traditional BM are practical and provide reliable measure of exposure, finding more sensitive ones to assess susceptibility and/or Hg specific effects is a current challenge. It is known that genetic factors affect Hg toxicokinetics and consequently Hg toxicity, efforts have been made to explore molecular variants in several genes which can serve as BMs of susceptibility to Hg toxic effects; these include several regulatory GSH, MT and selenoprotein-coding genes. There are no specific BMs of Hg effect, in spite of BMs for specific Hg effects have been proposed for a long time. Some authors consider that changes in oxidative stress markers or in urinary porphyrin profiles might serve as effect BMs for Hg. Other BMs promising higher specificity for Hg-induced effects are selenium and selenoproteins levels as well as changes in miRNA patterns. Continuous efforts on the adoption of measures to reduce Hg emissions are certainly necessary as well as in the development of reliable BMs of susceptibility and specific Hg effects.

References

Andreoli, V. and F. Sprovieri. 2017. Genetic aspects of susceptibility to mercury toxicity: An overview. Int. J. Environ. Res. Public Health. 14(1): 93.

Asano, S., K. Eto, E. Kurisaki, H. Gunji, K. Hiraiwa, M. Sato et al. 2001. Acute inorganic mercury vapor inhalation poisoning. Pathol. Int. 50(3): 169–174.

Aschner, M., T. Syversen, D.O. Souza and J.B. Rocha. 2006. Metallothioneins: Mercury species-specific induction and their potential role in attenuating neurotoxicity. Exp. Biol. Med. 231(9): 1468–73.

Beasley, D.M., L.J. Schep, R.J. Slaughter, W.A. Temple and J.M. Michel. 2014. Full recovery from a potentially lethal dose of mercuric chloride. J. Med. Toxicol. 10(1): 40–44.

Bernhoft, R.A. 2012. Mercury toxicity and treatment: A review of the literature. J. Environ. Pub. Health. 2012: 460508.

Boerleider, R.Z., N. Roeleveld and P.T.J. Scheepers. 2017. Human biological monitoring of mercury for exposure assessment. AIMS Environ. Sci. 4(2): 251–276.

Bose-O'Reilly, S., B. Lettmeier, G. Roider, U. Siebert and G. Drasch. 2008. Mercury in breast milk—A health hazard for infants in gold mining areas? Int. J. Hyg. Environ. Health. 211(5-6): 615–23.

Branco, V. and C. Carvalho. 2019. The thioredoxin system as a target for mercury compounds. Biochim. Biophys. Acta Gen. Subj. 1863(12): 129255.

Branco, V., P. Ramos, J. Canário, J. Lu, A. Holmgren and C. Carvalho. 2012. Biomarkers of adverse response to mercury: Histopathology versus thioredoxin reductase activity. BioMed. Res. Int. 2012: 359879.

Branco, V., S. Caito, M. Farina, J. Teixeira da Rocha, M. Aschner and C. Carvalho. 2017. Biomarkers of mercury toxicity: Past, present, and future trends. J. Toxicol. Environ. Health B. Crit. Ver. 20(3): 119–154.

Brooks, B., M. Mealey, G. Pitz, A. Asato, S. Bailey and S. Pollack. 2018. Biomonitoring pilot study hair mercury levels in clients attending the special supplemental nutrition program for women, infants and children (wic) program. Hawaii State Department of Health, Honolulu, USA.

Carvalho, L., S.S. Hacon, C.M. Vega, J.A. Vieira, A.L. Larentis, R. Mattos et al. 2019. Oxidative Stress levels induced by mercury exposure in amazon juvenile populations in Brazil. Int. J. Environ. Res. Public Health. 16(15): 2682.

Chen, C., H. Yu, J. Zhao, B. Li, L. Qu, S. Liu et al. Chai. 2006. The roles of serum selenium and selenoproteins on mercury toxicity in environmental and occupational exposure. Environ. Health Perspect. 114(2): 297–301.

Clarkson, T.W. 1993. Mercury: major issues in environmental health. Environ. Health Perspect. 100: 31–38.

Clarkson, T.W. and L. Magos. 2006. The toxicology of mercury and its chemical compounds. Crit. Rev. Toxicol. 36(8): 609–662.

Dack, K., M. Fell, C.M. Taylor, A. Havdahl and S.J. Lewis. 2021. Mercury and prenatal growth: A systematic review. Int. J. Environ. Res. Public Health. 18: 7140.

Dalkiran, T., K.B. Carman, V. Unsal, E.B. Kurutas, Y. Kandur and C. Dilber. 2021. Evaluation of oxidative stress biomarkers in acute mercury intoxication. Folia Medica 63(5): 704–709.

Day, B.J. 2009. Catalase and glutathione peroxidase mimics. Biochem. Pharmacol. 77(3): 285–296.

Ding, E., J. Guo, Y. Bai, H. Zhang, X. Liu, W. Cai et al. 2017. MiR-92a and miR-486 are potential diagnostic biomarkers for mercury poisoning and jointly sustain NF-κB activity in mercury toxicity. Sci. Rep. 7: 15980.

[EEB] European Environmental Bureau, Air & mercury. Cutting mercury emissions, improving people's health, 2019. EEB, Brussels, Belgium.

[EPA] Environmental Protection Agency, CLU-IN, Contaminants, Mercury, 2021. EPA, Washington, USA.

Ekino, S., M. Susa, T. Ninomiya, K. Imamura and T. Kitamura. 2007. Minamata disease revisited: an update on the acute and chronic manifestations of methyl mercury poisoning. J. Neurol. Sci. 262(1-2): 131–44.

Goodrich, J.M., Y. Wang, B. Gillespie, R. Werner, A. Franzblau and N. Basu. 2011. Glutathione enzyme and selenoprotein polymorphisms associate with mercury biomarker levels in Michigan dental professionals. Toxicol. Appl. Pharmacol. 257(2): 301–308.

Grandjean, P., E. Budtz-Jørgensen, P.J. Jørgensen and P. Weihe. 2005. Umbilical cord mercury concentration as biomarker of prenatal exposure to methylmercury. Environ. Health Perspect. 113(7): 905–908.

Guan, C., J. Jing, C. Jia, W. Guan, X. Li, C. Jin et al. 2015. A GSHS-like gene from Lycium chinense maybe regulated by cadmium-induced endogenous salicylic acid and overexpression of this gene enhances tolerance to cadmium stress in Arabidopsis. Plant Cell. Rep. 34: 871–884.

Gundacker, C., G. Komarnicki, P. Jagiello, A. Gencikova, N. Dahmen, K.J. Wittmann et al. 2007. Glutathione-S-transferase polymorphism. Total Environ. 385: 37–47.

Gworek, B., W. Dmuchowski, A.H. Baczewska, P. Brągoszewska, O. Bemowska-Kałabun and J. Wrzosek-Jakubowska. 2017. Air contamination by mercury, emissions and transformations—A review. Wat. Air and Soil Poll. 228(4): 123.

Gworek, B., W. Dmuchowski and A.H. Baczewska-Dąbrowska. 2020. Mercury in the terrestrial environment: A review. Environ. Sci. Eur. 32: 128.

[HBM4EU] European Human Biomonitoring Initiative, Mercury, 2021. German Environment Agency, Dessau-Roßlau, Germany.

Heyer, N.J., A.C. Bittner Jr., D. Echeverria and J.S. Woods. 2006. A cascade analysis of the interaction of mercury and coproporphyrinogen oxidase (CPOX) polymorphism on the heme biosynthetic pathway and porphyrin production. Toxicol. Lett. 61(2): 159–66.

[IARC] International Agency for Research on Cancer, Beryllium, Cadmium, Mercury, and Exposures in the Glass Manufacturing Industry, 1993. IARC, Lyon, France.

Iwai-Shimada, M., Y. Kobayashi, T. Isobe, S.F. Nakayama, M. Sekiyama, Y. Taniguchi et al. 2021. Comparison of simultaneous quantitative analysis of methylmercury and inorganic mercury in cord blood using LC-ICP-MS and LC-CVAFS: The pilot study of the Japan environment and children's study. Toxics 9(4): 82.

Kim, B.G., E.M. Jo, G.Y. Kim, D.S. Kim, Y.M. Kim, R.B. Kim, B.S. Suh and Y.S. Hong. 2012. Analysis of methylmercury concentration in the blood of Koreans by using cold vapor atomic fluorescence spectrophotometry. Ann. Lab. Med. 32(1): 31–37.

Klautau-Guimarães, M.N., R. D'Ascenção, F.A. Caldart, C.K. Grisolia, J.R. de Souza, A.C. Barbosa, C.M.T. Cordeiro and I. Ferrari. 2005. Analysis of genetic susceptibility to mercury contamination evaluated through molecular biomarkers in at-risk Amazon Amerindian populations. Genet. Mol. Biol. 28(4): 827–832.

Kolbinger, V., K. Engström, U. Berger and S. Bose-O'Reilly. 2019. Polymorphisms in potential mercury transporter ABCC2 and neurotoxic symptoms in populations exposed to mercury vapor from goldmining. Environ. Res. 176: 108512.

Kozikowska, I., L.J. Binkowski, K. Szczepańska, H. Sławska, K. Miszczuk, M. Śliwińska et al. 2013. Mercury concentrations in human placenta, umbilical cord, cord blood and amniotic fluid and their relations with body parameters of newborns. Environ. Pollut. 182: 256–62.

Kuras, R., B. Janasik, M. Stanislawska, L. Kozlowska and W. Wasowicz. 2017. Assessment of mercury intake from fish meals based on intervention research in the polish subpopulation. Biol. Trace Elem. Res. 179(1): 23–31.

Kuras, R., E. Reszka, E. Wieczorek, E. Jablonska, J. Gromadzinska, B. Malachowska et al. 2018. Biomarkers of selenium status and antioxidant effect in workers occupationally exposed to mercury. J. Trace. Elem. Med. Biol. 49: 43–50.

Kwaansa-Ansah, E.E., E.K. Armah and F. Opoku. 2019. Assessment of total mercury in hair, urine and fingernails of small-scale gold miners in the Amansie West District, Ghana. J. Health Pollut. 9(21): 190306.

Langford, N.J and R.E. Ferner. 1999. Toxicity of mercury. J. Hum. Hypertens. 13: 651–656.

Li, F., C. Ma and P. Zhang. 2020. Mercury deposition, climate change and anthropogenic activities: A review. Front. Earth Sci. 8: 316.

Lohren, H., L. Blagojevic, R. Fitkau, F. Ebert, S. Schildknecht, M. Leist and T. Schwerdtle. 2015. Toxicity of organic and inorganic mercury species in differentiated human neurons and human astrocytes. J. Trace Elem. Med. Biol. 32: 200–208.

Mahmoudi, N., A.J. Jafari, Y. Moradi and A. Esrafili. 2020. The mercury level in hair and breast milk of lactating mothers in Iran: a systematic review and meta-analysis. J. Environ. Health Sci. Eng. 18(1): 355–366.

Miguel, V., J.Y. Cui, L. Daimiel, C. Espinosa-Díez, C. Fernández-Hernando, T.J. Kavanagh et al. 2018. The role of MicroRNAs in environmental risk factors, noise-induced hearing loss, and mental stress. ARS. 28(9): 773–796.

Morales, M. and S. Munné-Bosch. 2019. Malondialdehyde: Facts and artifacts. Plant Physiol. 180(3): 1246–1250.

Mustacich, D. and G. Powis. 2000. Thioredoxin reductase. Biochem. J. 346(Pt 1): 1–8.

[NIOSH] National Institute for Occupational Safety and Health, Mercury compounds [except (organo) alkyls] (as Hg), 1994. U.S. Department of Health & Human Services, Washington, USA.

[NIOSH] National Institute for Occupational Safety and Health, Mercury, 2019. NIOSH, Washington, USA.

[NJDH] New Jersey Department of Health. 2009. Hazardous Substances Fact sheet. Mercury, Elemental and Inorganic Compounds. New Jersey Department of Health, New Jersey, USA.

Nordberg, G.F., B.A. Fowler, M. Nordberg and L.T. Friberg. 2007. Introduction—General considerations and international perspectives. pp. 1–9. *In*: G.F. Nordberg, B.A. Fowler, M. Ng and L.T. Friberg [eds.]. Handbook on the Toxicology of Metals. Academic Press, Cambridge, USA.

[NSW Health] New South Wales Health, Mercury and health, 2013. NSW Health, Canberra, Australia.

Nuttall, K.L. 2006. Interpreting hair mercury levels in individual patients. Ann. Clin. Lab. Sci. 36(3): 248–61.

O'Connor, D., D. Hou, Y.S. Ok, J. Mulder, L. Duan, Q. Wu, S. Wang, F.M.G. Tack and J. Rinklebe. 2019. Mercury speciation, transformation, and transportation in soils, atmosphericflux, and implications for risk management: A critical review. Environ. Int. 126: 747–761.

Park, J.D. and W. Zheng. 2012. Human exposure and health effects of inorganic and elemental mercury. J. Prev. Med. Public Health (6): 344–352.

Park, K. and E. Seo. 2016. Association between toenail mercury and metabolic syndrome is modified by selenium. Nutrients 8(7): 424.

Peixoto, N.C., M.A. Serafim, E.M. Flores, M.J. Bebianno and M.E. Pereira. 2007. Metallothionein, zinc, and mercury levels in tissues of young rats exposed to zinc and subsequently to mercury. Life Sci. 81(16): 1264–71.

Pesch, A., M. Wilhelm, U. Rostek, N. Schmitz, M. Weishoff-Houben, U. Ranft and H. Idel. Mercury concentrations in urine, scalp hair, and saliva in children from Germany. J. Expo. Sci. Environ. Epidemiol. 12: 252–258.

Pingree, S.D., P.L. Simmonds, K.T. Rummel and J.S. Woods. 2001. Quantitative evaluation of urinary porphyrins as a measure of kidney mercury content and mercury body burden during prolonged methylmercury exposure in rats. Toxicol. Sci. 61(2): 234–240.

Pinheiro, M.C.N., S.R. Carneiro and C.E.P. Corbett. 2020. Umbilical cord tissues as matrices to predict prenatal exposure to mercury—review. Ann. Pediatr. Child Health 8(7): 1197.

Rafati-Rahimzadeh, M., M. Rafati-Rahimzadeh, S. Kazemi and A.A. Moghadamnia. 2014. Current approaches of the management of mercury poisoning: Need of the hour. Daru. 22(1): 46.

Riverón, F.G., F.J. Arencibia, D.I.M. Fernández, M.N.P. del Castillo, G.R. Gutiérrez, B.A. Pandolfi et al. 2014. Oxidative stress and genotoxicity biomarkers in cuban workers with occupational chronic exposure to Mercury. RCST. 15(1): 35–41.

Ruggieri, F., C. Majorani, F. Domanico and A. Alimonti. 2017. Mercury in children: Current state on exposure through human biomonitoring studies. Int. J. Environ. Res. Public Health 14(5): 519.

Satoh, H. 2000. Occupational and environmental toxicology of mercury and its compounds. Ind. Health. 38: 153–164.

[SCOEL] Scientific Committee on Occupational Exposure Limits, Recommendation from the Scientific Committee on Occupational Exposure Limits for elemental mercury and inorganic divalent mercury compounds, 2007. SCOEL, Brussels, Belgium.

Sharma, B.M., O. Sáňka, J. Kalina and M. Scheringer. 2019. An overview of worldwide and regional time trends in total mercury levels in human blood and breast milk from 1966 to 2015 and their associations with health effects. Environ. Int. 125: 300–319.

Stern, A.H. and A.E. Smith. 2003. An assessment of the cord blood: Maternal blood methylmercury ratio: implications for risk assessment. Environ Health Perspect. 111(12): 1465–1470.

Streets, D.G., H.M. Horowitz, D.J. Jacob, Z. Lu, L. Levin, A.T. Schure and E.M. Sunderland. 2017. Total mercury released to the environment by human activities. Environ. Sci. Technol. 51(11): 5969–5977.

Teixeira, F.B., a. C.A. de Oliveira, L.K.R. Leão, N.C.F. Fagundes, R.M. Fernandes, L.M.P. Fernandes et al. 2018. Exposure to inorganic mercury causes oxidative stress, cell death, and functional deficits in the motor cortex. Front. Mol. Neurosci. 11: 125.

Vahidinia, A., F. Samiee, J. Faradmal, A. Rahmani, M.T. Javad and M. Leili. 2019. Mercury, Lead, Cadmium, and Barium levels in human breast milk and factors affecting their concentrations in hamadan, Iran. Biol. Trace Elem. Res. 187(1): 32–40.

Vollset, M., N. Iszatt, Ø. Enger, E. Lovise, F. Gjengedal and M. Eggesbø. 2019. Concentration of mercury, cadmium, and lead in breast milk from Norwegian mothers: Association with dietary habits, amalgam and other factors. Sci. Total Environ. 677: 466–473.

[WHO] World Health Organization, Background document for development of WHO Guidelines for Drinking-water Quality, 1996. WHO, Geneva, Switzerland.

[WHO] World Health Organization, Mercury, 2000. In: Air quality guidelines for Europe. WHO, Geneva, Switzerland.

[WHO] World Health Organization, Mercury and health 2017. WHO, Geneva, Switzerland.

Woods, J.S., M.D. Martin, C.A. Naleway and D. Echeverria. 1993. Urinary porphyrin profiles as a biomarker of mercury exposure: studies on dentists with occupational exposure to mercury vapor. J. Toxicol. Environ. Health. 40(2-3): 235–46.

Woods, J.S., N.J. Heyer, J.E. Russo, M.D. Martin, P.B. Pillai and F.M. Farin. 2013. Modification of neurobehavioral effects of mercury by genetic polymorphisms of metallothionein in children. Neurotoxicol. Teratol. 39: 36–44.

Woods, J.S., N.J. Heyer, J.E. Russo, M.D. Martin and F.M. Farin. 2014. Genetic polymorphisms affecting susceptibility to mercury neurotoxicity in children: summary findings from the Casa Pia Children's Amalgam clinical trial. Neurotoxicol. 44: 288–302.

Ye, B.J., B.G. Kim, M.J. Jeon, S.Y. Kim, H.C. Kim, T.W. Jang, H.J. Chae, W.J. Choi, M.N. Ha and Y.S. Hong. 2016. Evaluation of mercury exposure level, clinical diagnosis and treatment for mercury intoxication. Ann. Occup. Environ. Med. 28: 5.

Yoneda, S. and K.T. Suzuki. 1997. Detoxification of mercury by selenium by binding of equimolar Hg-Se complex to a specific plasma protein. Toxicol. Appl. Pharmacol. 143(2): 274–80.

Yurdaköok, K. 2015. Lead, mercury, and cadmium in breast milk. JPNIM. 4(2): e040223.

Part III
Biomarkers of Essential Metals with Potential for Toxicity

CHAPTER 10

Copper

Sources of Exposure

Copper (Cu) occurs on the Earth's crust at concentrations between 25–75 mg/kg in three oxidation states: Cu^0 (solid metal), but mostly Cu^{1+} (cuprous ion) or Cu^{2+} (cupric ion). Cu can be found as the free metal or associated with other elements in compounds that comprise different minerals (Flemming and Trevors 1989, ATSDR 2004b, Romić et al. 2014). Natural emissions include windblown dust, volcanic eruptions, native soils, decaying vegetation, forest fires and sea spray. Since Cu is an element, it does not break down in the environment and thus their levels rise as consequence of human activities. They occur primarily in Cu smelters and ore processing facilities, and due to metal plating, steelworks, refineries and domestic waste emissions, combustion of fossil fuels and wastes, wood preservatives, sewage sludge and pesticides such as algicides, fungicides and molluscicides (Flemming and Trevors 1989, ATSDR 2004b, Obrador et al. 2013). Cu compounds can also be added to fertilizers and animal feeds as a nutrient and used as food additives (e.g., nutrient and/or colouring agents) (WHO 2004). As it is able to conduct heat and electricity very well, most Cu is used in electrical equipment such as in wiring and motors; other uses comprise construction, such as roofing and plumbing and industrial machinery as heat exchangers (RSC 2022).

Air

The most relevant sources of Cu release to the atmosphere are Cu smelting, Cu and iron ore processing and combustion of fossil fuels and agricultural chemicals (EPA 1985, Hsu et al. 2018). In air Cu is related to Particulate Matter (PM), dusts and mists, being estimated that 13 times more Cu are emitted to air by human activities than by natural processes (Apori et al. 2018, Hsu et al. 2018). Cu concentrations in air ranges from a few to about 200 ng/m^3, with average values of 14 ng/m^3 reported in Canada and levels ranging 2 to 123 ng/m^3 described for the Czech Republic (WHO 2004, Hsu et al. 2018). Near smelters which process Cu ore into metal, the concentrations may reach 5,000 ng/m^3; populations living near sources of Cu are potentially exposed to high levels of Cu in dust by inhalation (ATSDR 2004b, Hsu et al. 2018).

Workplace

People working in mining Cu industries breath Cu-containing dust or are exposed by skin contact. Occupational exposure to soluble or not strongly attached to dust or dirt forms of Cu, occur more usually in agriculture, water treatment and industries such as electroplating (ATSDR 2004b). In Europe, the Occupational Safety and Health Administration (OSHA) sets a legal airborne permissible exposure limit (PEL) of 1 mg/m^3 for Cu dusts and mists, and a level of 0.1 mg/m^3 for Cu fumes, averaged over an 8-hr work shift. In the USA the National Institute of Occupational Safety and Health (NIOSH) agency establishes the same levels as airborne exposure limit (REL) but averaged over a 10-hr work shift, while the American Conference of Governmental Industrial Hygienists (ACGIH) threshold limit value (TLV) is 1 mg/m^3 for Cu dust and mists, and 0.2 mg/m^3 Cu fumes, averaged over a 10-hr work shift. Cu fumes are generally generated through high temperature operations such as welding brazing, soldering, plating and cutting; the reason why values established for Cu dust and mists and Cu fumes are different, is because the induced health effects are different (NJH 2008).

Soil

Cu is among the most frequently reported metal with potential hazards to soils, and unfortunately, soil Cu concentrations has increased dramatically since the Industrial Revolution (Apori et al. 2018). Once in this environmental media Cu becomes strongly attached to organic material and other components, such as clay, sand, in the top layers of the soil and may not move very far (ATSDR 2004a). Natural sources include weathering of natural high background rocks and metal deposit, but anthropogenic pollution occurs mainly due to pig and poultry manures, cow dung manure and the use of Cu-based fungicides to control crops diseases (Apori et al. 2018). High concentrations of Cu may also be found in soils from industrial atmospheric emissions which settles from air, or from wastes from mining and other Cu industries disposed of on soil (ATSDR 2004b). Elevated levels of Cu in agricultural soils may additionally result from spreading sludge from sewage treatment plants manure (Romic et al. 2014). Generally, soils contain between 2 and 250 mg/kg of Cu, the background values in China of 20.7 mg/kg have been described (ATSDR 2004b, Shi et al. 2008). Soils within an urban area in this country were shown to be significantly polluted with Cu (59.3 mg/kg), and this pollution was more severe in roadside dust (196.8 mg/kg) originating mainly from traffic contaminants (Shi et al. 2008). Concentrations close to 17,000 mg/kg were already determined near Cu and brass production facilities. Children constitute a concern due to hand to mouth contact and eating the contaminated dirt and dust; Cu concentrations of 2,480–6,912 mg/kg measured near Cu smelters may result in an intake of 0.74–2.1 mg/d for a child ingesting 300 mg of soil (ATSDR 2004b). It is proposed by some authors that 5–30 mg/kg is the optimal range of Cu levels for croplands and concerning soils, a general a threshold value of 100 mg/kg and a guideline value of 150 mg/kg have been proposed by Finnish and Swedish legislations, respectively; the threshold value is the level beyond which further assessment is needed in the area, while the guideline value is considered as denoting an ecological or health risk (Ballabio et al. 2018).

Water

Cu in surface water, groundwater, seawater and drinking-water is primarily present in complexes or in particulate matter (WHO 2004). Cu levels in surface and groundwater are generally very low, but high amounts may enter through mining, farming, manufacturing operations and municipal or industrial wastewater releases into rivers and lakes (CDC 2015). Cu concentrations in water may range from a few μg/L to 10 mg/L. In an unpolluted zone of the river Periyar (India) Cu levels were found ranging from 0.0008 to 0.010 mg/L, S and 0.0005 to 1 mg/L in several determinations in the USA and 0.003 to 0.019 mg/L in the river Stour, UK (ATSDR 2004b, WHO 2004). In groundwater the average Cu concentration is generally 0.005 mg/L although some groundwater contains levels of Cu up to 2.8 mg/L (ATSDR 2004b).

Drinking water can constitute a relevant source of excess Cu for the general public. The metal may enter by directly contaminating well water, but the primary cause is the corrosion of interior Cu plumbing, particularly if the water is acidic (CDC 2015, Bost et al. 2016). In such cases it is recommended that the water is allowed to run for 15–30 sec to reduce the concentrations below the acceptable drinking water standards (ATSDR 2004b). Measurements conducted in Europe, Canada and the USA indicate that Cu levels in drinking water can range from ≤ 0.005 to > 30 mg/L; there are large variation in residences, an example being values ranging from 0.02 to 3.5 mg/L in Berlin (WHO 2004, Bost et al. 2016). Regulations or guidelines for Cu in drinking water are settled by 104 countries, with a median value of 1.5 mg/L and a range of 0.05–3 mg/L (Taylor et al. 2020). The World Health Organization (WHO) recommended a limit for Cu in drinking water of 2 mg/L; the USA limits are 1–3 mg/L, while many European countries have a limit of 1 mg/L (Danzeisen et al. 2007).

Food

The major source of Cu intake is dietary, approximately 75% from solid food and 25% from drinking water (Hsu et al. 2018). The estimated daily intake of Cu from food is 1.0–1.3 mg/d for adults, while the density of Cu from a vegan diet is more than twice that of an omnivorous diet (0.7 and 2.0 mg/1,000 kcal, respectively); moreover, food products are seldom reinforced with Cu in the European Union (Bost et al. 2016). Careful control of Cu uptake is important for patients suffering from Cu metabolic disorders, such as Wilson's Disease (WD), an inherited autosomal recessive error in Cu metabolism characterized by excess Cu deposition in several organs, such as the liver and brain (Veríssimo et al. 2005). In most foods Cu is present bound to macromolecules rather than as a free ion. The liver and other organ meats, seafood, nuts and seeds (including whole grains) are good sources of dietary Cu, with about 40% from yeast breads, white potatoes, tomatoes, cereals, beef and dried beans and lentils; the use of vitamin/mineral supplements will increase exposure by about

2 mg/d (ATSDR 2004b). Nevertheless, Cu content in foodstuff varies according to local conditions, soil Cu concentration and anthropogenic activities, which may affect the Cu content in cereals, fruit and vegetables and, to a lesser extent, in meat or other animal products (Bost et al. 2016, Mwesigye et al. 2019). Soil pollution with Cu is quite appropriate for levels found in food, with studies indicating leafy vegetables exhibiting concentrations higher than the safe limit suggested by WHO/FAO (40 mg/kg d.w.) for plant foodstuffs (Filimon et al. 2021).

Cu toxicity can also be caused by consuming acidic foods cooked in uncoated Cu cookware (Royer and Sharman 2021). Childhood cirrhosis in children from India is a well-known case caused by toxic excesses of hepatic Cu derived from milk boiled in Cu vessels, although some researchers assume that this condition is rather an inherited disorder (Scheinberg and Sternlieb 1994). Additional sources of Cu in food are alcoholic beverages still distilled in Cu apparatus (Veríssimo et al. 2005).

In reality most concerns regarding exposure to Cu have been oriented in the aspect of Cu essentiality, and much less considering its toxicity. The amount of Cu present in the diet usually does not exceed by far the Recommended Dietary Allowances (RDAs) of the Food and Nutrition Board of the Institute of Medicine or the USA and Canadian agencies, which is 900 µg/d for adults with an upper limit of 10 mg (ATSDR 2004b, Danzeisen et al. 2007). In the view of Cu toxicity some authors have proposed a dose of 2.7 mg//d or 0.04 mg/kg b.w./day, which could be considered as a chronic reference dose (RfD); this dose also protects children and those with genetic susceptibility to increased Cu intake (Taylor et al. 2020).

Essentiality and Toxicity

The human body contains approximately 100 mg of Cu, which is an essential trace element, its deficiency leading to several adverse effects on health, but can also be toxic when present in excessive amounts (Bost et al. 2016) (Fig. 6). The metal is essential for a wide range of metabolic processes which include iron metabolism and immune function. The metal is also incorporated into several metalloenzymes involved in haemoglobin formation, drug/xenobiotic metabolism, carbohydrate metabolism, catecholamine biosynthesis, cross-linking of collagen, elastin, and hair keratin and in antioxidant defence (ATSDR 2004b, Bost et al. 2016). At least 30 Cu-containing enzymes function as redox catalysts or dioxygen carriers (Flemming and Trevors 1989). Cu-dependent enzymes, such as cytochrome c oxidase, superoxide dismutase, monoamine oxidase, dopamine β-monooxygenase and ferroxidases, all function to reduce activated oxygen species or molecular oxygen; ceruloplasmin is the most abundant Cu-dependent ferroxidase enzyme (ATSDR 2004b, Bost et al. 2016).

Deficiency

Cu deficiency is rare although might occur in infants, children and adults on synthetic diets and parenteral or enteral nutrition. Nevertheless, the most common cause of Cu deficiency is malabsorption, due to various acquired malabsorptive processes and

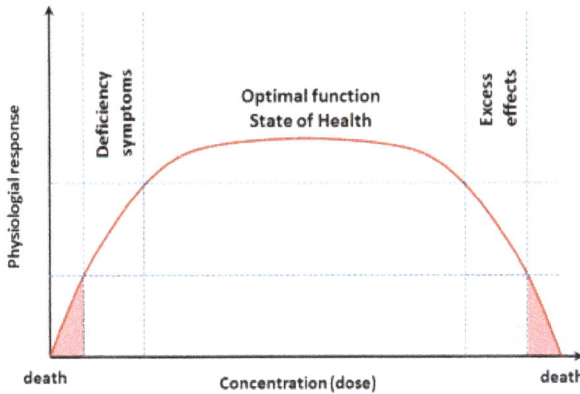

Figure 6. Dose response curve of copper, as an essential metal with potential for toxicity (Zoroddu et al. 2019).

losses from the body as a result of hemolytic anaemias, gut infections and parasitism, which cause protein-losing enteropathies. (Aggett 2013, Chaudhry and Ravich 2014). The best recognized disorder leading to Cu deficiency is Menkes disease, described as a genetically based inability to transport Cu across the intestinal barrier due to a mutation in the ATP7A gene (Chaudhry and Ravich 2014). Cu deficiency causes neurological manifestations that include myelopathy, peripheral and optic neuropathy (Hart et al. 2014). Other disorders include normocytic, hypochromic anaemia, leukopenia and osteoporosis (ATSDR 2004b).

Acute toxicity

Before discussing acute copper toxicity, it should be noted that Cu toxicosis can be classified according to another criteria: primary, when it results from an inherited metabolic defect and secondary, when it results from high intake or increased absorption or reduced excretion due to underlying pathologic processes (Royer and Sharman 2021). Cu overload is much less likely than Cu deficiency, at least in part since there are very efficient homeostatic mechanisms in the body (Danzeisen et al. 2007). However, several human data exist on high-dose acute poisoning in cases of suicidal intent with the ingestion of Cu compounds or accidental consumption of Cu-contaminated foods and beverages (NRC 2000). The acute lethal dose for adults is estimated to be between 4 and 400 mg/kg/b.w.; subjects ingesting large doses of Cu may present gastrointestinal ulcerations and bleeding, haematuria, intravascular haemolysis, methemoglobinemia, hepatocellular toxicity, acute renal failure and oliguria, and cardiotoxicity with hypotension, tachycardia and tachypnoea; central-nervous-system manifestations, include dizziness, headache, convulsions, lethargy, stupor and coma (NRC 2000, WHO 2004). At lower doses, Cu can cause symptoms typical of food poisoning with headache, nausea, vomiting and diarrhoea (WHO 2004).

Chronic toxicity

The incidence of Cu poisoning varies largely by the region, being unusual in western countries and more frequent in South Asian countries with a higher prevalence in rural populations. The risks are higher for neonates and infants as they have an immature biliary excretion system and enhanced intestinal absorption. Cu overload is a feature of Indian childhood cirrhosis (as already mentioned), endemic Tyrolean infantile cirrhosis and idiopathic Cu toxicosis (Royer and Sharman 2021). Cu overload can lead to cellular oxidative stress (notably, in the same way as Cu deficiency) with ensuing oxidative damage to membranes or macromolecules; the metal binds to the sulphydryl groups of several enzymes, such as glucose-6-phosphatase and glutathione reductase, interfering with their protection of cells from free radical damage (ATSDR 2004b, Danzeisen et al. 2007). The liver is the primary target of Cu because it is the first area of Cu deposition after it enters the blood (Gray et al. 2017). Inflammation, necrosis and altered serum markers of liver damage have been observed in rats fed on diets with Cu sulphate levels at least 100 times higher than the nutritional requirement (ATSDR 2004b). Patients with any liver disease, including all forms of hepatitis or alcohol abuse, are sensitive to Cu-related liver toxicity (Danzeisen et al. 2007). Another suitable target of Cu overload is the brain, with *in vivo* studies showing that chronic Cu toxicity causes impaired spatial memory and neuromuscular coordination, swelling of astrocytes and decreased serum acetylcholinesterase activity; Cu deposition in the choroid plexus, neuronal degeneration and augmented levels of Cu and zinc in the hippocampus are also seen (Danzeisen et al. 2007, Pal et al. 2013). Damage to proximal convoluted tubules in the kidneys has been observed in rats, with a latency period for the kidneys being effected longer than for the liver. The metal is also irritating to the respiratory tract, with signs like coughing, sneezing, runny nose, pulmonary fibrosis and increased vascularity of the nasal mucosa previously reported in workers exposed to Cu dust; anemia, immunotoxicity and developmental toxicity have also been reported (ATSDR 2004b). The Environmental Protection Agency (EPA) considers Cu not classifiable as to human carcinogenicity, and the metal is not listed as a human carcinogen by the Occupational Safety and Health Administration (OSHA), the National Toxicology Program (NTP), the International Agency for Research on Cancer (IARC), the American Conference of Governmental Industrial Hygienists (ACGIH) or the European Union (ATSDR 2004b, Teck 2018).

Biomarkers

Among the several biomarkers (BM) applied to access Cu status, the great majority are still controversial, present several limitations and seem to be appropriated to monitor Cu deficiencies rather than excessive levels (Danzeisen et al. 2007). It suggests that the BMs studied so far are not sensitive enough to detect an increase in body Cu before the appearance of functional or clinical effects or that the homeostatic mechanisms are so strong that no significant changes in body Cu occur with mild-to-moderate Cu exposure (Olivares et al. 2008). There is an urgent need to establish highly sensitive BMs for Cu overload since there is no current way to evaluate Cu excess (Pal et al. 2013). Concerning BMs of susceptibility of specific BMs of Cu effects, even less progress has been achieved.

Biomarkers of copper status

When a clinical suspicion of Cu toxicity is raised, the measurement of urine and blood Cu levels in addition to serum ceruloplasmin (Cp) concentrations, is the most common procedure; faecal evaluation of Cu levels may be obtained, for acute Cu poisoning. During hemolytic crisis, methemoglobinemia determinations to measure red blood cells lysis and blood glutathione concentrations are also used. Other laboratory tests include measures of the kidney function, haemolysis and liver damage (Royer and Sharman 2021).

Copper levels

Blood and urine Measurement of Cu levels in blood have been used to assess Cu status, with normal values for serum and whole blood described as averaging 151.6 and 217 µg/100 mL, respectively. In subjects who intentionally ingested a single dose of 1–30 g of Cu, values in serum and whole blood were found ranging from 2,390 to 3,460 and 3,830 to 6,840 µg/ mL, respectively (ATSDR 2004b, Jewell 2019). Serum free Cu (non-Cp bound Cu) levels > 1.6 mmol/L has been proposed as a diagnostic test for detection of WD (Pal 2014). However increased serum Cu levels may only be reflective of recent exposure since serum ionic Cu rapidly diminishes to normal levels following an acute dose. In addition, a relationship between blood Cu levels and the severity of toxic symptoms has not been established (ATSDR 2004b).

When compared to faecal excretion, urinary Cu excretion is low and reported as only 10–25 g/d (Brewer 2002, Bost et al. 2016). Elevated 24-hr urine Cu content can be used to assess Cu overload, particularly in patients with WD, but may be unsuitable to detect subtle increases in Cu load (Bertinato et al. 2010).

Hair and nails Cu content of hair on the scalp is 10–100 times higher than in serum, while showing much more variable values, which could range from 7 to 95 g/g. Despite any correlation being observed between Cu content in hair of the scalp and Cu intake, assessed with three 24-hr interviews, healthy young men with a controlled study diet exhibited an increase of Cu content of hair on the scalp of 9.2 to 21.1 g/g, when the intake of Cu shifted from 1.6 to 7.8 mg/day (Bost et al. 2016). Increased hair Cu levels have also been reported in workers exposed to 0.64–1.05 mg/m³ of an unspecified Cu compound, exhibiting concentration of Cu in hair of 705.7 µg/g, which was higher than the 8.9 µg/g concentration found in non-exposed workers (ATSDR 2004b). Sex, use of hormonal contraception, cancer and other pathological situations and environmental pollution, may affect the Cu content of hair on the scalp and thus, hair Cu levels are generally not regarded as a suitable marker of Cu status (Bost et al. 2016). While there is practically no data regarding the use of Cu levels in nails to assess Cu status, except for one study with preschool children which showed that the levels of Cu in toenail samples ranged from 3.0 to 18.6 µg/g. (ATSDR 2004b).

Cuproenzymes

Several other markers have been assayed in human studies in the view that many proteins bind and transport Cu, and that many enzymes need Cu for functioning. Expression of liver proteins in several organs, such as metallothionein (MT) and the Cu transporter-1 has been shown to be regulated by Cu status. However, their value as BMs is limited since they are located within organs that are not readily accessible. Hence, traditional approaches to the measurement of Cu status still include the determination of cuproenzymes levels in peripheral samples (Danzeisen et al. 2007).

Ceruloplasmin One of these cuproenzymes is ceruloplasmin (Cp), which is a Cu-binding protein containing most of the Cu present in the plasma (about 95%); the enzyme exhibits a ferroxidase activity important for iron metabolism (Bost et al. 2016). The amount of Cu associated with Cp is approximately 3.15 mg/mg Cp and Cp serum levels are 20 to 40 mg/dL in normal adults (McPherson 2021, Pal 2014). Additionally, the ferroxidase enzymatic activity of Cp may be clinically used to estimate plasma Cp levels; depressed activity is widely used for assessing Cu deficiency (Bertinato et al. 2010) Limitation of Cp measurements to assess Cu status include the fact that Cp is generally lower in men than in women; in women it increases with oestrogen, pregnancy and contraceptive pills; Cp is an acute-phase reactant, augmented during inflammation, infections, rheumatoid arthritis, myocardial complications and cancer; Cp is also age dependent, subject to seasonal changes and a minimum of 4 wk is needed to observe a variation in their levels (Bost et al. 2016). Furthermore, high levels of dietary Cu do not affect either mRNA transcription or protein translation of Cp, and therefore Cp does not respond to augmented Cu. A peculiar characteristic of Cp is that though it contains the predominance of serum Cu, absence or abnormality in its functioning does not lead to alterations in Cu homeostasis (rather, it exerts preferential effects on Fe homeostasis); and although Cp may be affected by Cu supplementation, this was only observed in unhealthy, highly weakened individuals (Pal 2014). All these limitations have led to believe that Cp determinations may not serve to assess Cu overload (Danzeisen et al. 2007, Pal 2014, Bost et al. 2016).

The non–Cp-bound Cu pool was proposed as a marker of Cu overload. However, calculations of this pool, by subtracting the value obtained from total Cu in the serum based on the assumption that each molecule of Cp binds 6 Cu atoms, led to inconsistent results (Olivares et al. 2008).

Superoxide dismutases Superoxide dismutases (SODs) are a group of metalloenzymes that form the front line of defence against reactive oxygen species (ROS)-mediated injury; SOD catalyzes the dismutation of the superoxide anion free radical into molecular oxygen and hydrogen peroxide. Based on the metal cofactors present in their active sites, SODs can be classified into four distinct groups, including Cu-Zinc-SOD (SOD1) (Younus 2018). As with Cp, SOD1 may indicate Cu deficiency (although negative or equivocal results have been reported) but is not responsive to excess Cu (Prohaska and Broderius, 2006, Danzeisen et al. 2007). Even though it was already suggested that erythrocytic SOD1 levels could reflect chronic Cu intakes, SOD1 proved not to be either a reliable or sensitive indicator. The fact that SOD1 is also an acute-phase reactant and responds to a variety of health and stress conditions, probably contributed to the variable results achieved so far (Danzeisen et al. 2007). Moreover, SOD activity is possibly less sensitive than Cp and this may be because its synthesis depends on Cu release from the liver and delivery to the bone marrow (Prohaska and Broderius 2006).

Diamine oxidase Diamine oxidase (DAO) is another Cu-dependent enzyme responsible for the oxidative deamination of diamines such as cadaverine, putrescine and some of their derivatives and histamine (Kehoe et al. 2000). Plasma DAO activity was reported to decrease on low-Cu diets, while serum DAO was found to increase after *in vivo* 3 mg Cu supplementation for 6 wk; however, since this supplementation study was considered at high risk of bias due to an incomplete report of outcomes, it was concluded by some researchers that DAO is not a suitable marker of Cu status (Danzeisen et al. 2007, Bost et al. 2016).

Peptidylglycine a-amidating monooxygenase Plasma and tissue Peptidylglycine A-amidating Monooxygenase activity (PAM) is a Cu dependent enzyme responsible for the post-translational modification of many important neuropeptides, including oxytocin, vasopressin, the adrenocorticotropic hormone ACTH, gastrin and many other molecules (Prohaska and Broderius 2006, Bousquet-Moore et al. 2010). Their levels in the plasma also have the potential to be used as a BMs of Cu status, with the advantage that a small amount of blood from a finger-prick sample may be sufficient to detect changes. In Menkes' diseased subjects (suffering from genetic Cu deficiency) plasma PAM is Cu depleted, which results in decreased enzymatic activity; notably, the *in vitro* addition of Cu to blood samples from these patients repletes the activity of the enzyme. Moreover, PAM was suggested to be less sensitive to endocrine changes than Cp and may even be a useful marker for Cu deficiency during development and childhood. However, PAM has only been tested in Cu deficiency and its activity is considered unlikely to increase in parallel with elevated Cu (Prohaska and Broderius 2006).

Lysyl oxidase Another explored cuproenzyme is lysyl oxidase, which is involved in the formation of cross-links in collagen and elastin, exhibiting the highest concentrations in connective tissues, such as the tendon and skin. The functional activity of the enzyme is decreased in dietary Cu deficiency. Although it is described that increased dietary Cu causes activation of lysyl oxidase in the tendons of chickens, these tissues are not suited for convenient accessibility, turning such determinations in men as limiting (Danzeisen et al. 2007). The effect of Cu intake on skin lysyl oxidase activity was addressed in humans, but in only one study (Bost et al. 2016).

Cytochrome c oxidase Cytochrome C Oxidase (CCO) is the terminal complex of the eukaryotic oxidative phosphorylation in mitochondria (Hordyjewska et al. 2014, Watson and McStay 2020). Some studies demonstrated that the activities of CCO were reduced in response to Cu depletion (Hordyjewska et al. 2014). Platelet CCO has been investigated *in vivo* as a BM for Cu status, it was noted that a significant reduction of platelet CCO activity was correlated with decreased Cu concentrations in the liver. In humans, it was observed a CCO activity dropped to 49% during a Cu depletion stage and increased back to 60% of the control level during Cu repletion.

Furthermore, CCO was found to be the most sensitive marker, when compared with other measured BMs which included the plasma Cu, Cp or erythrocyte SOD1 levels (Danzeisen et al. 2007).

General limitations pointed to the use of cuproenzymes in Cu status evaluations should be considered. Standardized assays are non-existent; there is high interindividual variability of enzymatic activities; some of these measurements are affected by other conditions; some enzymes are labile; and even for Cu deficiency, the performance (sensitivity, specificity and likelihood ratio) of these tests to detect mild changes has not been appropriately studied (Olivares et al. 2008).

Other biomarkers of copper status

Liver Cu concentration is the best indicator of Cu status and is even used to compare the performance of any test used to detect Cu overload; hepatic Cu concentration quantification by a liver biopsy is considered as the gold standard for confirming WD (Olivares et al. 2008, Ferenci et al. 2013, Pal 2014). On the other hand, a liver biopsy is an invasive procedure only justified when there is strong evidence of liver damage due to excessive Cu accumulation (Olivares et al. 2008, Bertinato and Zouzoulas 2009, Ferenci et al. 2013, Pal 2014).

As mentioned, all BMs represented so far seem to be more suitable to evaluate low Cu status. Efforts have been made to develop BMs suitable for detecting Cu overloads, and in this sense, the quantification of various Cu chaperones at protein and mRNA levels have also been tentatively used to evaluate Cu excess in humans (Pal 2014). Cu chaperones compose a specific class of proteins which assure safe handling and specific delivery of potentially harmful Cu ions to a variety of essential proteins (Paluma 2013). To ensure safe intracellular Cu delivery, chaperones need to respond to Cu exposure in a very sensitive way. They are not present in the plasma, but their expression in erythrocytes and white cells provides a possibility for easy-access measurements (Danzeisen et al. 2007). So far, three Cu chaperones are identified: Cox17, which brings Cu to the mitochondria, where it is ultimately incorporated into CCO; the Cu chaperone for SOD (CCS), which provides Cu to the metal-binding site of SOD1; and the third one, Atox1, which brings Cu to P-type Cu-transporting ATPases (Danzeisen et al. 2007, Suzuki et al. 2013). So far, the most promising candidate for accurate and sensitive Cu status detection is the CCS, since their levels are known to increase in tissues of overtly Cu-deficient rats. Moreover, it seems that increased erythrocyte CCS expression is more sensitive to mild reductions in Cu status, when compared to plasma Cu concentration or Cp activity (Iskandar et al. 2005, Danzeisen et al. 2007, Bertinato et al. 2010). As CCS protein elevation in Cu deficiency was found to be concomitant with a decrease in SOD1 levels, some authors proposed using erythrocyte CCS: SOD1 ratio for the determination of Cu deficiency (Harvey and McArdle 2008). Regarding Cu overload, it is acknowledged that Cu binding to CCS decreases CCS stability and promotes its degradation by the 26S proteasome (which also explains the upregulation of CCS under conditions of Cu deficiency); in accordance, *in vivo* Cu excess results in high Cu in the liver accompanied by lower CCS in liver and erythrocytes (Bertinato et al. 2010). CCS expression was already measured by real-time RT-PCR in peripheral mononuclear cells isolated from healthy men supplemented with Cu (8 mg/d) for 6 mon. Peripheral mononuclear cell CCS mRNA transcripts decreased significantly in these men after this time, firmly suggesting that this BM may serve as a useful indicator of Cu exposure. In another study which used a similar protocol, although no change in CCS expression was observed in low-Cp phenotype subjects, in high-Cp individuals CCS decreased significantly in response to the Cu supplementation (Danzeisen et al. 2007). An interesting rationale was considered. It was proposed that reduced CCS protein in erythrocytes was associated with increased body Cu load and not with high dietary Cu intake *per se*. Additionally, given that mature erythrocytes are anucleated, regulation of CCS content in response to Cu overload may occur during erythropoiesis in maturing erythrocytes that have a nucleus and can efficiently support protein turnover. Hence, if CCS regulation by Cu occurs in maturing erythrocytes, a measurable decrease in CCS content in mature circulating erythrocytes would not be detected immediately following Cu excess. According to this rationale, a detectable decrease in CCS would depend on the synthesis of significant amounts of new erythrocytes exposed to high Cu during maturation. Therefore, lower erythrocyte CCS content would be indicative of chronic Cu overload (Bertinato et al. 2010). In effect, the usefulness of CCS as a BM for Cu status should be validated, since it appears to be the most promising potential BM responding to both Cu deficiency and excess (Danzeisen et al. 2007, Harvey and McArdle 2008, Pal 2014).

Biomarkers of susceptibility

There are no validated BMs of susceptibility for both Cu deficiency and overload. As mentioned, subpopulations with a genetic predisposition to have an elevated risk of Cu accumulation are known to exist, and these include patients suffering from the autosomal recessive disorder WD. These subjects have mutations in the Cu ATPase ATP7B, which is a Cu pump responsible for Cu excretion; hence WD patients accumulate Cu to toxic levels (Danzeisen et al. 2007, Ryan et al. 2019). The prevalence of WD heterozygotes, individuals who carry one WD and one healthy allele of ATP7B, is not known. This is partially because heterozygotes appear perfectly healthy, but they may have mild abnormalities in Cu metabolism, which represents a potential public health problem (Danzeisen et al. 2007). An attempt to find unspecific BMs of susceptibility was described in a study involving Cu supplementation of a population carrying Cu-related polymorphisms leading to high or low serum Cp; liver aminotransferases glutamic-oxaloacetic transaminase, glutamic-pyruvic transaminase and γ-glutamyl transferase, were found to be increased and this remained below the clinical cut-off level to indicate liver pathology. The authors proposed that the measured BMs could be useful in detecting excess Cu intake within these phenotypic subgroups (Danzeisen et al. 2007).

Biomarkers of effect

No Cu-specific BMs of effects have yet been identified (Danzeisen et al. 2007). Since liver damage is a symptom of Cu toxicity, alanine aminotransferase (ALT) and aspartate aminotransferase (AST) are elevated in serum when liver damage occurs, and these BMs can be used to assess Cu toxicity. However, increased ALT and AST levels are not specific for Cu toxicity and are only increased once tissue damage has already occurred (Bertinato et al. 2010). Some constraints exist namely in WD assessments, such as the fact that these enzymes show rapid decline during Coombs-negative hemolytic anemia and because amino-transferase have an erythrocyte fraction in WD significantly affected by Cu-induced haemolysis (Hayashi et al. 2019).

A few other determinations have been proposed as functional indicators of Cu status. This is the case of urinary pyridinoline and deoxypyrodinoline, which are BMs of bone resorption and may be useful functional indicators of Cu status, although no data is available regarding excessive Cu levels. Increased bone resorption associated with Cu depletion in adult males and a reduced rate of bone loss at the lumbar spine in Cu-supplemented middle-aged women are documented; however, increases in bone turnover occurred already with no change in serum Cu or Cp (Bonham et al. 2002, Harvey and McArdle 2008). The complex nature of bone metabolism suggests that these BM are non-specific for Cu status and will be influenced by a number of nutritional and environmental factors, such as vitamin D levels and sunlight exposure (Danzeisen et al. 2007, Harvey and McArdle 2008).

Additionally, a physiological system with potentially large scope for establishing functional marker(s) of Cu status, is the immune system (Bonham et al. 2002). This system is compromised by Cu deficiency and many immune markers such as interleukin 2 (IL-2) production, neutrophil function and the phenotypic profiling of lymphocyte subsets are sensitive to mild Cu deficiency as well as to Cu repletion (Bonham et al. 2002, Danzeisen et al. 2007). IL-2 is the principal cytokine responsible for the progression of T lymphocytes and are known to decrease in both severe and marginal Cu deficiency without significant alterations of the conventional indicators of Cu status such as tissue Cu, tissue Cu–Zn-SOD or serum Cp activity. In infants' Cu deficiency or supplementation, neutrophil numbers can respectively decrease or increase, suggesting that this marker may constitute a sensitive indicator (Bonham et al. 2002). Lymphocyte subset phenotyping are other possible marker(s) of Cu status, since there is overall reduction in total T lymphocyte numbers, particularly within the CD4+ subset, by low Cu intake (Bonham et al. 2002, Danzeisen et al. 2007).

But in general terms while severe Cu toxicity is relatively easy to recognize due to the obvious clinical signs, so far it is virtually impossible to identify marginal Cu excess or the biological effects appearing from this condition (Pal 2014).

Final Remarks

In this chapter both sides of Cu were presented, depending on the dose: as an essential trace element and as a toxic chemical. With regards Cu essentiality, its role in a wide range of metabolic processes leading to a deficiency can result in severe health problems. Although Cu deficiency is rare it could occur in infants, children and in adults from malabsorption, such as in the genetic Menkes Disease. When in excessive levels Cu is toxic, mainly hepatotoxic; Cu toxicosis may be primary (inherited) such as in Wilson Disease, or secondary, namely from high intake through Cu contaminated food and water, which are the main sources for the general population. Cu overload is much less likely than Cu deficiency, partially due to efficient homeostatic mechanisms in the body. Currently BMs for assessing Cu status are mainly focused on Cu deficiency and so far, no laboratory indicators are universally accepted as an early marker of Cu excess; it is possible to detect Cu excess (and even deficiency) only in their extremes due to ensuing tissue damage, for example, an increase in liver enzymes. BMs studied so far seem to be not sensitive enough to detect an increase in body Cu before the appearance of functional or clinical effects. A good BM of Cu is needed to monitor and avoid chronic health problems and to give an 'early warning' before any tissue damage occurs. Moreover, in the absence of strong, sensitive and specific BMs, it is difficult to know the Cu status, either in relation to deficiency or excess, is a significant public health problem. Research to find a marker of Cu status may need to move away from traditional markers of Cu status and a long way still needs to be followed, as concerns to susceptibility and effect BMs for Cu.

References

Aggett, P.J. 2013. Copper. pp. 397–403. *In*: B. Caballero [ed.]. Encyclopedia of Human Nutrition. Academic Press, Cambridge, USA.

Apori, O.S., E. Hanyabui1 and Y.J. Asiamah. 2018. Remediation technology for copper contaminated soil: A review. Asian J. Soil Sci. 1(3): 1–7.

[ATSDR] Agency for Toxic Substances and Disease Registry, Copper CAS # 7440-50-8, 2004a. U. S. Department of Health and Human Services, Public Health Service, Atlanta, USA.

[ATSDR] Agency for Toxic Substances and Disease Registry, Toxicological profile for copper, 2004b. U. S. Department of Health and Human Services, Public Health Service, Atlanta, USA.

Ballabio, C., P. Panagos, E. Lugato, J.-H. Huang, A. Orgiazzi, A. Jones et al. 2018. Copper distribution in European topsoils: An assessment based on LUCAS soil survey. Sci. Total Environ. 636: 282–298.

Bertinato, J and A. Zouzoulas. 2009. Considerations in the development of biomarkers of copper. J. AOAC Int. 92(5): 1541–1550.

Bertinato, J., L. Sherrard and L.J. Plouffe. 2010. Decreased erythrocyte CCS content is a biomarker of copper overload in rats. Int. J. Mol. Sci. 11(7): 2624–2635.

Bonham, M., J.M. O'Connor, B.M. Hannigan and J.J. Strain. 2002. The immune system as a physiological indicator of marginal copper status? Br. J. Nutr. 87: 393–403.

Bost, M., S. Houdart, M. Oberli, E. Kalonji, J.F. Huneau and I. Margaritis. 2016. Dietary copper and human health: Current evidence and unresolved issues. J. Trace Elem. Med. Biol. 35: 107–115.

Bousquet-Moore, D., R.E. Mains and B.A. Eipper. 2010. Peptidylglycine α-amidating monooxygenase and copper: a gene-nutrient interaction critical to nervous system function. J. Neurosci. Res. 88(12): 2535–2545.

Brewer, G.J. 2002. Diagnosis of Wilson's disease: An experience over three decades. Gut. 50(1): 136.

[CDC] Centers for Disease Control and Prevention, Copper and Drinking Water from Private Wells, 2015. CDC, Atlanta, USA.

Chaudhry, V. and W.J. Ravich. 2014. Other neurologic disorders associated with gastrointestinal disease. pp. 281–292. *In*: M.J. Aminoff's [ed.]. Neurology and General Medicine. Churchill Livingstone, London, UK.

Danzeisen, R., M. Araya, B. Harrison, C. Keen, M. Solioz, D. Thiele et al. 2007. How reliable and robust are current biomarkers for copper status? Br. J. Nutr. 98(4): 676–683.

[EPA] Environmental Protection Agency, Sources of Copper Air Emissions, 1985. EPA, Washington, USA.

Ferenci, P., K. Caca, G. Loudianos, G. Mieli-Vergani, S. Tanner, I. Sternlieb et al. 2003. Diagnosis and phenotypic classification of Wilson disease. Liver Int. 23: 139–42.

Filimon, M.N., I.V. Caraba, R. Popescu, G. Dumitrescu, D. Verdes, D. Petculescu et al. 2021. Potential ecological and human health risks of heavy metals in soils in selected copper mining areas-a case study: The Bor area. Int. J. Environ. Res. Public Health. 18(4): 1516.

Flemming, C.A. and J.T. Trevors. 1998. Copper toxicity and chemistry in the environment: A review. Water Air Soil Pollut. 44: 143–158.

Gray, J.P., N. Suhali-Amacher and S.D. Ray. 2017. Metals and metal antagonists. pp. 197–208. *In*: S.D. Ray [ed.]. Side Effects of Drugs Annual, A Worldwide Yearly Survey of New Data in Adverse Drug Reactions. Elsevier, Amsterdam, Netherlands.

Hart, P., C.M. Galtrey, D.C. Paviour and M. HtutHart. 2014. Biochemical aspects of neurological disease. pp. 683–701. *In*: W.J. Marshall, M. Lapsley, A.P. Day and R.M. Ayling. [eds.]. Clinical Biochemistry: Metabolic and Clinical Aspects. Churchill Livingstone, London, UK.

Harvey, L.J. and H.J. McArdle. 2008. Biomarkers of copper status: A brief update. Br. J. Nutr. 99(3): S10–3.

Hayashi, H., K. Watanabe, A. Inui, A. Kato, Y. Tatsumi, A. Okumura et al. 2019. Alanine aminotransferase as the first test parameter for wilson's disease. J. Clin. Transl. Hepatol. 7(4): 293–96.

Hordyjewska, A., L. Popiołek and J. Kocot. 2014. The many "faces" of copper in medicine and treatment. Biometals. 27(4): 611–621.

Hsu, H.W., S.C. Bondy and M. Kitazawa, M. 2018. Environmental and dietary exposure to copper and its cellular mechanisms linking to alzheimer's disease. Toxicol. Sci. 163(2): 338–345.

Iskandar, M., E. Swist, K.D. Trick, B. Wang, M.R. L'Abbé and J. Bertinato. 2005. Copper chaperone for Cu/Zn superoxide dismutase is a sensitive biomarker of mild copper deficiency induced by moderately high intakes of zinc. Nutr. J. 4: 35.

Jewell, T 2019. What to Know About Copper Toxicity. https://www.healthline.com/health/copper-toxicity#foods-list.

Kehoe, C.A., M.S. Faughnan, W.S. Gilmore, J.S. Coulter, A.N. Howard and J.J. Strain. 2000. Plasma diamine oxidase activity is greater in copper-adequate than copper-marginal or copper-deficient rats. J. Nutr. 130(1): 30–33.

McPherson, R.A. 2021. Specific proteins. *In*: R.A. McPherson and M. Pincus [eds.]. Henry's Clinical Diagnosis and Management by Laboratory Methods. Elsevier, Amsterdam, Netherlands.

Mwesigye, A.R., S.D. Scott, E.H. Bailey and S.B. Tumwebaze. 2019. Uptake of trace elements by food crops grown within the Kilembe copper mine catchment, Western Uganda. J. Geochem. Explor. 207: 106377.

[NJH] New Jersey Department of Health, Copper. 2008. NJH, Trenton, USA.

[NRC] National Research Council, Committee on Copper in Drinking Water, Copper in Drinking Water, 2000. Academic Press, Cambridge, USA.

Obrador, A., D. Gonzalez and J.M. Alvarez. 2013. Effect of inorganic and organic copper fertilizers on copper nutrition in spinacia oleracea and on labile Copper in Soil. Agric. Food Chem. 61(20): 4692–4701.

Olivares, M., M.A. Méndez, P.A. Astudillo and F. Pizarro. 2008. Present situation of biomarkers for copper status. Am. J. Clin. Nutr. 88(3): 859S–862S.

Pal, A. 2014. Copper toxicity induced hepatocerebral and neurodegenerative diseases: An urgent need for prognostic biomarkers. Neurotoxicology. 40: 97–101.

Pal, A., R.K. Badyal, S.V. Vasishta, Attri, B.R. Thapa and R. Prasad. 2013. Biochemical, histological, and memory impairment effects of chronic copper toxicity: A model for non-wilsonian brain copper toxicosis in wistar rat. Biol. Trace. Elem. Res. 153: 257–268.

Paluma, P. 2013. Copper chaperones. The concept of conformational control in the metabolism of copper. FEBS Lett. 587(13): 1902–1910.

Prohaska, J.R. and M. Broderius. 2006. Plasma peptidylglycine alpha-amidating monooxygenase (PAM) and ceruloplasmin are affected by age and copper status in rats and mice. Comp. Biochem. Physiol. B Biochem. Mol. Biol. 143(3): 360–366.

Romić, M., L. Matijević, H. Bakić and D. Romić. 2014. Copper accumulation in vineyard soils: distribution, fractionation and bioavailability assessment. pp. 799–825. *In*: M. Hernández-Soriano [ed.]. IntechOpen, London, UK.

Royer, A. and T. Sharman. 2021. Copper Toxicity. StatPearls Publishing [Internet].

[RSC] Royal Society of Chemistry, Copper, 2022. RSC, London, U.K.

Ryan, A., S.J. Nevitt, O. Tuohy and P. Cook. 2019. Biomarkers for diagnosis of Wilson's.

Scheinberg, I.H. and I. Sternlieb. 1994. Is non-Indian childhood cirrhosis caused by excess dietary copper? Lancet. 344(8928): 1002–4.

Shi, G., Z. Chen, S. Xu, J. Zhang, L. Wang, C. Bi et al. 2008. Potentially toxic metal contamination of urban soils and roadside dust in Shanghai China. Environ. Pollut. 156(2): 251–60.

Suzuki, Y., M. Ali, M. Fischer and J. Riemer. 2013. Human copper chaperone for superoxide dismutase 1 mediates its own oxidation-dependent import into mitochondria. Nat. Commun. 4: 2430.

Taylor, A.A., J.S. Tsuji, M.R. Garry, M.E. McArdle, W.L. Goodfellow Jr, W.J. Adams et al. 2020. Critical review of exposure and effects: Implications for setting regulatory health criteria for ingested copper. Environ. Manage. 65(1): 131–159.

Teck 2018. Copper metal safety data sheet. https://www.teck.com/media/Copper-Metal-TAMI-2018-SDS-.pdf.

Veríssimo, M.I.S., J.A.B.P. Oliveira and M.T.S.R. Gomes. 2005. The evaluation of copper contamination of food cooked in copper pans using a piezoelectric quartz crystal resonator. Sens. Actuators B Chem. 111-112: 587–591.

Watson, S.A. and G.P. McStay. 2020. Functions of cytochrome c oxidase assembly factors. Int. J. Mol. Sci. 21(19): 7254.

[WHO] World Health Organization, Copper in Drinking-water. Background document for development of WHO Guidelines for Drinking-water Quality, 2004. WHO, Geneva, Switzerland.

Younus, H. 2018. Therapeutic potentials of superoxide dismutase. Int. J. Health Sci. 12(3): 88–93.

Zoroddu, M.A., J. Aaseth, G. Crisponi, S. Medici, M. Peana and V.M. Nurchi. 2019. The essential metals for humans: A brief overview. J. Inorg. Biochem. 195: 120–129.

CHAPTER 11

Iron

Sources of Exposure

Iron (Fe) is one of the Earth's most rich resources, since it is the fourth most abundant element, by mass, on the Earth's crust. The most common Fe-containing ore is haematite, although Fe is also found widely distributed in other minerals such as magnetite and taconite; Fe is found most in nature in the form of its oxides (WHO 2003, IDPH 2010, RSC 2022). Being a transition metal Fe can readily contribute and accept electrons to participate in oxidation–reduction reactions, which are essential for several fundamental biologic processes (Dev and Babitt 2017).

It is estimated that 98% of mined Fe ore goes to manufacturing steel, which is one of the biggest industries in the world and has occupied a prominent place since the beginning of the industrial revolution (Mele and Magazzino 2020, Sreenivasan 2021). Steel is used in construction, automobiles, machinery, electrical equipment and a number of products ranging from cookware to furniture; in 2050 the demand for steel could increase by five times (Sreenivasan 2021). Fe as Fe oxide nanoparticles, which includes Fe_2O_3 and Fe_3O_4, are particularly relevant in the environment since they can be emitted during volcanic eruptions, but can be also found from vehicle and industry emissions as air pollution particulates. They are chemically synthesized for a wide variety of applications, such as paint, ink, polish, rubbers, plastics, component of cement, cosmetics, inter alia for drinking-water pipes, electronics, as a catalyst and in medical devices (WHO 2003, Pelclova et al. 2015, SWA 2019). These nanoparticles have been widely researched for applications in Magnetic Resonance Imaging (MRI) as they are mainly superparamagnetic. Fe_3O_4 is viewed as very promising in medical applications because of its proven biocompatibility (Lodhia et al. 2010). On the other hand, the abundant presence of Fe-rich air pollution nanoparticles, emitted from the mentioned industries and traffic-related sources, are a major potential concern. Fe oxide nanoparticles belong to 13 priority nanomaterial groups identified by the Organization for Economic Co-operation and Development (OECD) that need safety evaluations. To illustrate, the *in situ* identification of the composition of exogenous nanoparticles in human myocardial mitochondria led to discovering that it is dominated by the presence of Fe (Pelclova et al. 2015, Maher et al. 2020).

Air and workplace

In remote areas Fe levels in air are about 50–90 ng/m^3 and at urban sites 1.3 µg/m^3, while concentrations up to 12 µg/m^3 have been reported in the vicinity of steel producing plants. The intake of Fe from air is estimated to be about 25 µg/d in urban areas (WHO 2003).

Fe oxide may be present in welding fumes in some workplaces, which were recently reclassified as carcinogenic to humans (Group 1) by the International Agency for Research on Cancer (IARC) (Falcone et al. 2018, SWA 2019). Fe ore miners were seen to have a 70% greater mortality from lung cancer than the general population; yet other reports suggest that Fe oxide is not a human carcinogen (Falcone et al. 2018). The fact is that there is very little information about the potential risks of occupational exposure to these chemicals during production and processing, and studies on the effect of occupational exposure have yet to be published (Pelclova et al. 2015). Some exposure limits for Fe oxide (measured as Fe) exist: the Occupational Safety and Health Administration (OSHA) airborne Permissible Exposure Limit (PEL) is 10 mg/m^3 averaged over an 8-hour work shift, the National Institute for Occupational Safety and Health (NIOSH) Airborne Exposure Limit and the European Health and Safety Authority (HSA) Occupational Exposure Limit Value are 5 mg/m^3 averaged over a 10-hr work shift and an 8-hr reference period, respectively. The Australian interim Total Weight Average (TWA) is also 5 mg/m^3 for respirable particulate matter, recommended to protect for lung inflammation and pulmonary siderosis in exposed workers (NJDH 2007, SWA 2019, HSA 2021).

Soil and water

Fe oxide minerals are usually found in soils varying their concentrations from 0.1 to 50% and in many cultivated soils Fe is relatively abundant having, on average, a total concentration of 20 to 40 g/kg (Colombo et al. 2014, Oppong-Anane et al. 2018). Contamination with Fe is found in landfill leachates in different locations of the world (Oppong-Anane et al. 2018).

In water Fe is mainly present in two forms: soluble ferrous Fe or insoluble ferric Fe. Water containing ferrous Fe is clear and colourless as the Fe is completely dissolved, but when exposed to air turns cloudy, a reddish-brown substance begins to form, and this sediment is oxidized (IDPH 2010). Fe contamination of water can either be geogenic or via industrial effluents and domestic waste, may represent several problems for water filtration systems used in agricultural, industrial and municipal applications (Ityel 2011, Kumar et al. 2017). The presence of high Fe concentrations in water changes it colour, taste and odour, leaves stains on clothes and corrodes water pipelines, augmenting even more Fe concentrations in drinking water (IDPH 2010, Kumar et al. 2017, HSE 2019). Excess Fe also promotes undesirable bacterial growth within distribution systems, resulting in the deposition of a slimy coating on piping. Fe (as Fe^{2+}) levels of 40 µg/L can be detected in distilled water, while in mineralized spring water the taste threshold value was already determined as 0.12 mg/L. In anaerobic groundwater Fe is usually present in concentrations of

0.5–10 mg/L, although concentrations up to 50 mg/L were already measured (WHO 2003). In drinking-water Fe is seldom found at concentrations greater than 10 mg/L, with levels as little as 0.3 mg/L could cause water to turn a reddish-brown in colour (IDPH 2010). Since in general the levels are normally less than 0.3 mg/L, contributing to about 0.6 mg of the daily intake; most tap water in United States supplies contains approximately 5% of the dietary requirement for Fe (WHO 2003, IDPH 2010). However Fe concentrations can be higher in countries where Fe salts are used as coagulating agents in water-treatment plants and where cast Fe, steel and galvanized Fe pipes are used for water distribution (WHO 2003). The recommended limit for Fe in water in the USA is 0.3 mg/L, and since the metal is considered a secondary or "aesthetic" contaminant, this value is based on taste and appearance rather than on any detrimental health effect. Private water supplies are not subject to these rules, but guidelines can be used to evaluate water quality (WDH 2017). Europe resolved on a limit of 0.2 mg/L, again considering that above this value the colour, taste and smell of the water may be affected. The international expert committee of the Food and Agriculture Organization (FAO) and the World Health Organization (WHO) consider that drinking water with Fe levels of up to 2 mg /L does not pose an appreciable risk to the general population (HSE 2019).

Food

The major source of exposure to Fe is food, with liver, red meat, beans, nuts, green leafy vegetables and fortified breakfast cereals presenting a relatively high Fe content (WHO 2003, Lynch et al. 2018). Liver, kidney, fish and green vegetables contain 20–150 mg/kg of Fe, whereas red meats and egg yolks contain 10–20 mg/kg. Rice and many fruits and vegetables have lower Fe contents, 1–10 mg/kg (WHO 2003). There are two main forms of food Fe: heme Fe in meat products, and non-heme Fe in both plant and animal foods (including meat) and in the different forms of Fe used for food fortification. Heme Fe is always well absorbed, while the absorption of non-heme Fe depends on the Fe status of the individual consuming the meal and on the meals' composition (Lynch et al. 2018). Nevertheless, Fe absorption is very variable. While homeostatic mechanisms increase intestinal Fe absorption in Fe deficiency, downregulation at high intake levels seems insufficient to prevent accumulation of high Fe stores; unlike other minerals there is no regulated Fe excretion, including in Fe overload (Schümann 2001, Lynch et al. 2018). Estimates of the minimum daily requirement for Fe depending on age, sex, physiological status and Fe bioavailability range from about 10 to 50 mg/d , with daily intakes of food reported as ranging from 10 to 14 mg (WHO 2003, Lynch et al. 2018). However, during the last decades efforts regarding dietary Fe focused mainly on prevention of deficiencies, especially during growth and pregnancy (Schümann 2001). But while Fe is an essential element in human nutrition, it can be toxic if present in excessive levels; thus, risk values should be protective against deficiency but also against toxicity (WHO 2003, EPA 2006). Namely in Ethiopia, which is described as the highest per capita Fe intake in the world, the average daily intake of Fe is 471 mg/d; increased stored Fe in the liver and adverse health effects have not only been observed, because of the low bioavailability of the Fe in Ethiopian food (EPA 2006). A case of exposure to excessive levels of Fe leading to adverse effects, with chronic hemochromatosis, is described among the South African Bantu population from an excessive intake of absorbable Fe in an alcoholic beverage prepared in Fe pots or drums; the beer's high acidity (pH 3.0–3.5) enhances Fe leaching from the vessels. Such African Fe overload, also called Bantu siderosis, affects up to 15% of adult males in rural societies (EPA 2006, Pietrangelo 2018). Nigeria's canned and non-canned beverages, also have Fe levels above the Maximum Contaminant Level (MCL) set by agencies like the US Environmental Protection Agency (EPA) (Odukudu et al. 2014). Taking the highest intake level of 18.7 mg/d and a body weight of 70 kg, a No Observed Adverse Effects Level (NOAEL) of 0.27 mg is established for chronic Fe toxicity (EPA 2006).

Consumer products

Fe oxides are common colourants in eye shadows, blushes and concealers, with average concentrations of 1401 ppm detected earlier in some cosmetics; in some products such as hair relaxers, disinfectant, tissue paper and toothpaste, levels of 0.927, 0.700, 0.623 and 0.608 ppm, were respectively determined. The significant amount of Fe present in these consumer products and their regular use may result in considerable dermal and/or inhalation exposures. It is also mentioned that the exposure to small doses of Fe from consumer products may result in cellular death or colourectal cancer due to cumulative effects. Considering the Joint FAO/WHO Expert Committee on Food Additives' Provisional Tolerable Daily Intake (PTDI) for Fe of 0.8 μg/g b.w./d , the estimated doses of exposure through the use of some the mentioned consumer products are calculated as far above this limit (Odukudu et al. 2014).

Health conditions

Some health conditions can result, more or less directly, in chronic excessive exposures to Fe (Ogun and Adeyinka 2022). Secondary Fe overload may occur in people with renal failure, cancer, chronic anaemias with ineffective erythropoiesis, such as thalassaemia intermedia, sideroblastic anaemias and chronic haemolytic anaemias. These chronic conditions could lead to the need of repeated treatments with parenteral Fe, blood transfusions, multiple infusions of intravenous Fe or prolonged ingestion of Fe supplements, which can result in Fe toxicity (Ogun and Adeyinka 2022). Transfusion Fe overload is directly associated with the number of blood transfusions; one unit of transfused blood contains about 200–250 mg of Fe, and patients who receive more than 10 to 20 units of blood are at a significant risk of Fe excess (Rasel and Mahboobi 2021, Yuen and Becker 2021).

Since Fe is found in many over-the-counter multivitamins, excess of pharmaceutical Fe may cause toxicity; even therapeutic doses to treat Fe deficiency may cause gastrointestinal side effects including nausea, vomiting, diarrhoea and gastrointestinal pain (EPA 2006, Yuen and Becker 2021). However, it is known that adults have often taken Fe supplements for extended periods without deleterious effects, and an intake of 0.4–1 mg/kg b.w./d is unlikely to cause adverse effects in healthy persons (WHO 2003). Moreover, for most patients, Fe deficiency is reversed within 6 mon of treatment, thus limiting the duration of the exposure (Schümann 2001, EPA 2006). Higher Fe levels are also present in patients with alcoholic liver disease, non-alcoholic fatty liver disease and hepatitis C viral infection (Milic et al. 2016).

Essentiality and Toxicity

Fe is an essential element in human nutrition, which is incorporated into proteins required for vital cellular and organismal functions including oxygen transport, mitochondrial respiration, intermediary and xenobiotic metabolism, nucleic acid replication and repair, host defence and cell signalling (WHO 2003, Dev and Babitt 2017).

Deficiency

Inadequate intake of bioavailable Fe (nutritional Fe deficiency) is the most common cause of deficiency and the most prevalent disorder of Fe balance worldwide; in fact, Fe deficiency is the only disorder of Fe balance in which nutrition has a primary role (Lynch et al. 2018). Its prevalence is highest among young children and women of childbearing age, particularly pregnant women (EPA 2006). A normal singleton pregnancy carried to term requires a transfer of 500–800 mg of maternal Fe, and it is estimated that the demand for absorbed Fe increases from 0.8 mg/d in early pregnancy to 7.5 mg/d in late pregnancy (Means 2020). In pregnant women Fe insufficiency augments the risks for a preterm delivery and delivering a low-birthweight baby. Young children are at great risk of Fe deficiency because of their rapid growth and increased Fe requirements; the deficiency causes developmental delays and behavioural disturbances (EPA 2006). Elderly people can also be affected due to decreasing Fe intake and absorption or increasing demand and loss, with multiple aetiologies often coexisting in an individual patient (Cappellini et al. 2020). The most widely recognized clinical manifestation of Fe deficiency is anaemia, a condition affecting nearly one quarter of the world's population, with 50% of the cases attributed to Fe deficiency (Dev and Babitt 2017). Major health consequences of anaemia include an increased risk of impaired cognitive and physical development in children, reduced physical performance and work productivity in adults, and cognitive decline in the elderly (EPA 2006, Dev and Babitt 2017).

Acute toxicity

Fe toxicity from intentional or accidental ingestion is a common form of poisoning, with accidental ingestions usual in children less than 6 yr; acute ingestion of Fe is especially hazardous to them (Yuen and Becker 2021). The average lethal dose is 200–250 mg/kg b.w. but deaths have been described following ingestion of doses as low as 40 mg/kg b.w. Haemorrhagic necrosis and sloughing of areas of mucosa in the stomach with extension into the submucosa were observed in autopsies (WHO 2003).

Chronic toxicity

Excess Fe is toxic since the redox cycling of Fe in the presence of oxygen and hydrogen peroxide catalyzes the production of free radicals in the Fenton reaction, which will damage DNA, protein and lipids (Dev and Babitt 2017). Although Fe absorption is regulated, excessive accumulation of Fe in the body resulting from chronic ingestion of high levels of Fe cannot be prevented by intestinal regulation; moreover, humans do not have a mechanism to increase excretion of absorbed Fe (EPA 2006).

However, there is a long-standing controversy as to whether a chronic overload due to oral intake is possible in individuals with a normal ability to control Fe absorption (EPA 2006). But it is known that excess tissue Fe levels are a risk factor for several diseases, such as diabetes, despite the mechanisms underlying this association not being completely understood; but the fact that even in an apparently healthy population, increased dietary heme Fe and high body Fe stores are associated with an increased risk of type 2 diabetes and other insulin resistance states (Huang et al. 2011, Dev and Babitt 2017).

Fe overload cardiomyopathy is a major cause of death in patients with diseases associated with chronic anaemia such as thalassemia or sickle cell disease after chronic blood transfusions. Excess free Fe that enters cardiomyocytes augments Reactive Oxygen Species (ROS) generated via Haber-Weiss and Fenton reactions, contributing to high morbidity and mortality rates (Gordan et al. 2018).

With Fe being the most abundant transition metal in the brain, Fe overload or pathological deposition was earlier associated with many neurodegenerative disorders, including Parkinson's disease and Alzheimer's disease. Whether Fe accumulation contributes to the pathogenesis of these neurological disorders or whether Fe accumulation occurs because of the pathogenic condition, is still a matter of debate (Dev and Babitt 2017, Yan and Zhang 2020).

Fe overload is also associated with an increased risk of hepatocellular carcinoma and possibly other cancers in patients with hereditary hemochromatosis and β-thalassemia. Fe excess is thought to contribute to cancer development via two main mechanisms: the pro-oxidant effects of Fe can damage DNA and then, promote oncogenesis; cancer cells have an enhanced dependence on Fe to maintain their rapid growth rate (Dev and Babitt 2017).

Biomarkers

As mentioned earlier, Fe is essential for many biological processes, but Fe levels must be firmly regulated to avoid harmful effects of both Fe deficiency and overload. For this reason, biomarkers (BM) which could allow accessing individual Fe status, susceptible subjects and monitor Fe effects in a realistic manner are obviously important.

Biomarkers of iron status

There is no single measure of Fe status applicable in every case and all measures are subject to potential confounding factors. Thus, combinations of measures of Fe stores, Fe supply to tissues and functional haemoglobin Fe are often needed to reach an accurate assessment (Pippard 2011).

Iron levels in blood and urine

Serum Fe levels can be determined in cases of Fe overdose and are useful in both confirming an ingestion and predicting serious toxicity, but only if blood samples are taken at the appropriate time; this is 4 to 6 hr after ingestion, which is the peak absorption for most preparations, with the exception of slow-release Fe ones (Liebelt 2007, Reynolds and Ventre 2007). Normal serum Fe concentrations are 50 to 150 µg/dL, with serum Fe levels more than 500 µg/dL mostly always associated with clinical toxicity, and 300 µg/dL often associated with clinical symptoms. In such acute cases, measurements should be repeated at 8 to 12 hr intervals in symptomatic patients or if there is a concern for ingestion of sustained release products. Other parameters such as arterial blood gas, complete blood count, serum electrolytes, glucose, blood urea nitrogen, creatinine and liver function tests should be assayed (Reynolds and Ventre 2007).

For cases other than acute poisoning, serum Fe level measurements have limited utility as they cannot always be correlated with the severity or the clinical stage of Fe intoxication; in fact, it is intracellular Fe and not the free circulating Fe in the blood, that is responsible for systemic toxicity (Liebelt 2007, Doig 2020). Moreover, this BM exhibits a high within-day and between-day variability and increases after recent ingestion of Fe-containing foods or supplements (Doig 2020).

When serum Fe levels are determined to access Fe deficiency, there are confounders such as the presence of inflammatory processes and certain other forms of chronic disease (Firkin and Rush 1997).

The assessment of Fe chelation efficacy in patients with chronic transfusion-dependent Fe overload, can be performed through measuring urinary Fe; this measurement provides indications as to whether therapy with chelators is able to promote significant Fe excretion through the kidneys (Wahidiyat et al. 2019).

Iron metabolism

Ferritin or transferrin (Tf) measurements are often used to assess Fe status in the body as depending on it, Fe is stored constrained to ferritin within mucosal cells and macrophages in the liver, spleen and bone or is transported in the plasma bound to Tf (a plasma glycoprotein responsible for ferric-Fe delivery through blood to various tissues) (Koperdanova and Cullis 2015, Ogun and Adeyinka 2022). Both ferritin and Tf levels are regulated post transcriptionally according to Fe intracellular concentrations, as mentioned that serum levels of ferritin and Tf, together with several red blood cell parameters, can be used clinically to evaluate Fe balance (EPA 2006, Sermini et al. 2017).

Ferritin Regarding ferritin, since it is an intracellular Fe storage protein it is accepted as marker of Fe stores; its level in serum has been mentioned by several authors as the most useful BM to measure Fe status. While low ferritin values provide absolute evidence of Fe deficiency, subnormal levels of ferritin can be detected when Fe stores are exhausted and even before the serum Fe level has become affected; and in this way represent the most sensitive index of early Fe deficiency (Firkin and Rush 1997).

The normal range of ferritin in serum depends on several variables including age and sex. The normal range is 25–155 µg/L in menstruating adult females, and 40–260 µg/L in adult males (Firkin and Rush 1997). Generally, concentrations > 300 µg/L in men and postmenopausal women, and > 200 µg/L in premenopausal women, are regarded as elevated (Ogun and Adeyinka 2022). Levels < 12 µg/L in children younger than 5 yr and < 15 µg/L for those at 5 yr and older, indicate low Fe stores and a high risk of Fe deficiency (Northrop-Clewes and Thurnham 2013, Ogun and Adeyinka 2022). However, normal serum ferritin can be seen in patients who are deficient in Fe and have coexisting diseases, such as hepatitis or anaemia. Other authors have claimed that low serum ferritin and Fe levels, with elevated Total Iron Binding Capacity (TIBC) (which will be described later), can diagnose Fe deficiency (Harper and Conrad 2020). Fe overload can be often indicated by raised ferritin levels, but this change is not specific since ferritin is an acute phase protein, is also released in inflammatory disorders, liver disease, alcohol excess or malignancy. Raised ferritin levels require further investigation in primary care to determine if they truly represent Fe overload; in such cases, it is recommended to conduct measurements of the soluble Tf receptor (also to be described later) or the ratio soluble Tf receptor: ferritin, as these BMs are less affected by inflammatory processes (Sermini et al. 2017, Ogun and Adeyinka 2022).

Transferrin, transferrin receptors and saturation With regards to Tf, when serum or plasma Fe concentrations falls progressively below the normal range due to exhaustion of Fe reserves, Tf levels rises towards, or above the upper limit of the normal range; in this way Tf increase is an indicator of Fe deficiency severity (Firkin and Rush 1997, Sermini et al. 2017, Faruqi and Mukkamalla 2022). Serum Tf concentrations augment in infants with increased severity of Fe deficiency. Nonetheless, serum Tf levels have not proved to be superior to serum ferritin for detecting Fe deficiency (Worwood et al. 2017).

The levels of receptors for Tf can also be used as BMs, as the major application of the serum Tf receptor assay for detecting patients with absence of stored Fe (ferritin and haemosiderin in cells) (Worwood et al. 2017). The presence of Tf receptors in enterocytes allow the entrance of Fe transported by Tf. It is through this mechanism that cells "detect" systemic Fe status, inducing the regulation of their levels via a negative loop. As Tf receptors mediate the capture of cellular Fe, they play a key role in Fe homeostasis. This BM provides the advantage of being less affected by inflammatory states than other markers of Fe status, such as serum ferritin (Sermini et al. 2017). However, circulating Tf receptor levels increase, not only in patients with simple Fe deficiency but also in patients with anaemia or chronic disease who lack stainable Fe in the bone marrow (Worwood et al. 2017). Furthermore, while this BM decreases when Fe status improves this could also occur under conditions of protein-energy malnutrition; therefore, if a patient has concurrent Fe deficiency, it is difficult to determine whether a low Tf receptor level reflects Fe status or protein status (Litchford 2008).

Tf saturation is another BM, considered by some authors as the best one to evaluate Fe status in the body and is often used in subjects with the genetic hereditary haemochromatosis type 1 (HFE-related hemochromatosis) (Pietrangelo 2012, Sermini et al. 2017). Tf saturation can be calculated as the ratio of serum Fe concentration and Total Iron Biding Capacity (TIBC) expressed as a percentage; according to other authors it can also be calculated using the serum Fe concentration and one of the following: TIBC, unsaturated Fe binding capacity or Tf levels (Pietrangelo 2012, Sermini et al. 2017, Worwood 2017, Faruqi and Mukkamalla 2022). A Tf saturation < 16% is usually considered to indicate an inadequate Fe supply for erythropoiesis; the European Association for the Study of the Liver (EASL) recommend thresholds of > 50% for men and > 45% for women (Sermini et al. 2017, Worwood 2017). In the view of Fe toxicity, Tf saturation is often elevated in young adults with hemochromatosis before the development of Fe overload or a rise in ferritin concentration. However, a wide biologic variability in Tf saturation may limit its use as a screening test (Pietrangelo 2012).

Serum iron binding capacity An alternative way of knowing the amount of Tf in serum is that it has a serum Fe binding-capacity , which is able to bind Tf to with Fe (Firkin and Rush 1997). The Fe binding-capacity is of two kinds , the aforementioned TIBC and the unsaturated Fe-binding capacity (UIBC). TIBC is the total serum Fe and correlates with Tf concentrations serving as a good indirect measurement of Tf; when Fe in serum decreases, TIBC values increase (Sherwood et al. 1998, Sermini et al. 2017, Faruqi and Mukkamalla 2022). Serum TIBC is an essential test used for the diagnosis of Fe deficiency anaemias and other disorders of Fe metabolism but is not useful in the management of acute Fe toxicity (Reynolds and Ventre 2007, Kundrapu and Noguez 2018, Faruqi and Mukkamalla 2022).

Since only one-third of Tf is saturated with Fe, the transferrin present in serum has an extra binding capacity (67%), and this is the UIBC (Faruqi and Mukkamalla 2022). The UIBC has been seen to be similar in screening performance to the Tf saturation and can be conducted at a lower cost, in populational studies. Both TIBC and UIBC are however subject to analytic and biological variability, which limit their usefulness as screening tests namely for hemochromatosis (Adams et al. 2007). Most laboratories define normal ranges for TIBC as 240 to 450 µg/dL, and for UIBC 111 to 343 µg/dL (Faruqi and Mukkamalla 2022).

Hormones Hepcidin, an acute phase reactant liver-derived peptide hormone, is another key regulator of systemic Fe homeostasis. Its unbalanced production contributes to the pathogenesis of a spectrum of Fe disorders including hemochromatosis, β-thalassemia and ferropenic anaemia (Zaritsky et al. 2009, Girelli et al. 2016, Sermini et al. 2017). This regulation is exerted by counteracting the function of ferroportin, which acts as the major cellular Fe exporter in the hepatocyte cell membrane. When hepcidin induces the internalization and degradation of ferroportin, such events result in increased intracellular Fe stores, decreased dietary Fe absorption and decreased circulating Fe levels (Tsuchiya and Nitta 2013). Hepcidin production is firmly regulated: it is increased by the plasma and liver Fe to prevent Fe absorption through a feedback mechanism to maintain stable body Fe levels, and by inflammation as a host defence mechanism to limit extracellular Fe availability to microbes; inversely, hepcidin is decreased by erythroid activity to ensure Fe supply for erythropoiesis (D'Angelo 2013, Girelli et al. 2016, Sermini et al. 2017). As hepcidin levels reflect the integration of multiple key signals involved in Fe regulation, which include both Fe deficiency and overload, its measurement is a useful clinical tool for the management of Fe disorders (Zaritsky et al. 2009, Girelli et al. 2016). Even in the absence of anaemia, hepcidin appears to be a sensitive indicator of Fe deficiency. Moreover, compared to haematocrit or haemoglobin, a decrease in hepcidin is an early marker of Fe deficiency when detected together with Tf saturation and decrease in ferritin . In pure Fe deficiency anaemia, serum and urinary hepcidin concentrations are significantly suppressed (D'Angelo et al. 2013). Although some patients, particularly the elderly, may have detectable hepcidin levels attributed to comorbidities such as renal, inflammatory or neoplastic diseases, the use of measuring basal hepcidin to personalize the optimal route of Fe administration has been seen in patients with iron deficiency anaemia. Hepcidin diagnoses have also proved useful in the evaluation of the appropriateness of oral Fe supplementation in children from regions with high infection burden engaged in host–pathogen battle for essential Fe. In Fe-loading anaemias, hepcidin measurement may also be valuable for identifying the most severely affected patients, helping to predict the development of Fe overload and guiding the therapy (Girelli et al. 2016).

Thus erythroferrone (ERFE) is the key hormone in Fe homeostasis which control the release of stored Fe. ERFE is produced from erythroblasts during erythropoiesis, and this production is stimulated by erythropoietin which acts directly on the liver to inhibit the production of hepcidin, leading to increased Fe delivery for intensified erythropoietic activity (Sermini et al. 2017, Almousawi and Sharba 2019). High levels of serum ERFE have been proposed as a new BM in patients with beta thalassemia, which is an inherited disorder that leads to Fe overload and anaemia with the need of frequent blood transfusions; changes in this BM proved to reflect mild or severe anaemia, and Fe overload in these patients (Almousawi and Sharba 2019). Notably ERFE levels were found to be correlated inversely with those of hepcidin and ferritin, and positively with those of soluble Tf receptor, among patients on haemodialysis (Honda et al. 2016).

Other biomarkers of iron status

When the Fe supply is insufficient or when Fe utilization is impaired, zinc is used in the biosynthetic route of heme instead of Fe, occurring an Fe-zinc substratecompetition for ferrochelatase (Mwangi et al. 2014, Sermini et al. 2017). It was suggested that this substitution is one of the first biochemical responses to Fe depletion, occurring mainly in the bone narrow (Sermini et al. 2017). Consequently, there is an increase in the levels of zinc protoporphyrin (ZPP), which becomes the main form of nonheme protoporphyrin present in Red Blood Cells (RBC) under conditions of Fe depletion

(Magge et al. 2013, Gammon et al. 2018). Thus, ZPP indicates the systemic supply of Fe to erythrocytes in the bone marrow, with raised levels of ZPP in RBCs indicating diminished Fe stores and a decline in the available Fe in the bone marrow (Mwangi et al. 2014, Gammon et al. 2018). Augmented ZPP levels can measure a Fe deficiency severity and has been used as a screening marker to manage Fe deficiency in children and pregnant women; it has also been recommended to be used in combination with haemoglobin concentrations in surveys to assess population Fe status. It is additionally reported that the erythrocytic ratio ZPP: heme reflects Fe status in the bone narrow (Mwangi et al. 2014, Sermini et al. 2017). Suppressed ZPP values on Fe overload are also described well, with a study revealing that intra-erythrocytic concentration ZPP measurements could even be used to distinguish hemochromatosis from other hyperferritinemic conditions; ZPP clearly discriminates hemochromatosis patients from individuals showing increased ferritin due to other disorders, since ZPP is significantly decreased in patients with hemochromatosis (Metzgeroth et al. 2007).

Liver Fe susceptometry using Superconducting Quantum Interference Devices (SQUID) provides an alternative non-invasive method for measuring liver Fe content through a biopsy. Although SQUID has been calibrated, validated and used for clinical studies, its complexity, high cost and limited availability have impeded its widespread use. MRI is widely available and accessible and has been shown to be very sensitive to the presence of Fe, constituting a gold standard for the assessment of tissue Fe concentrations. However, MRI-based susceptibility mapping in the abdomen is a challenging problem, due to the lack of focal sources of susceptibility, the presence of fat and air, as well as motion related production (Hernando et al. 2013, Wahidiyat et al. 2019).

Biomarkers of susceptibility

Despite there are no references on susceptibility BMs for both Fe deficiency or overload, some of the BMs of Fe status, including serum Fe, ferritin, Tf, Tf saturation and others, can be used as phenotypic proxies representing potential genetic instruments for Fe status (Gill et al. 2019). Mutations in different Fe metabolism genes can cause Fe deficiency or overload (Moksnes et al. 2021).

In primary Fe overload mostly known as hereditary haemochromatosis, Fe accumulation occurs due to excessive absorption of dietary Fe through inheritance of an autosomal recessive disorder affecting the hemochromatosis gene HFE on chromosome 6 (WHO 2003, EPA 2006, Ogun and Adeyinka 2022). In such conditions, Fe exceeds the buffering capacity of Tf, which due to its role carries Fe in extracellular fluids and blood, is crucial for preventing the presence of free Fe and ensuing cellular toxicity, owing to Fe-induced production of harmful ROS (Dev and Babitt 2017, Murakami et al. 2019). Thus, non-Tf-bound Fe may be absorbed into the liver, heart and endocrine glands, where excess Fe triggers oxidative damage and organ dysfunction, leading to cirrhosis, cardiomyopathy, diabetes mellitus and other endocrinopathies (Pinto et al. 2014, Dev and Babitt 2017). It has also been shown that heterozygosity or homozygosity for the C282Y variant of the HFE gene protects against the development of Fe deficiency, with rs1800562 Single Nucleotide

Polymorphism (SNP) in HFE being associated with four Fe measures: serum Fe, Tf saturation, TIBC and UIBC; in individuals who are homozygous for SNP rs1800562 on chromosome 6p22.2 (the C282Y mutation in the HFE gene) increase in Fe stores is likely to occur (McLaren et al. 2011, Benyamin et al. 2014).

Population-based studies have identified genetic variants in several other genes such as the ones encoding the transmembrane protease serine 6 (TMPRSS6), Tf, transferrin receptor 2 (TFR2), ERFE and IL6R67–70 330 for different Fe status BMs (Benyamin et al. 2014, Moksnes et al. 2021). Larger studies yielded additional loci: ABO, ARNTL, FADS2, NAT2, SLC40A1, TEX14, DUOX2, F5, SLC11A2 and TFRC associated with one or more Fe homeostasis BMs (ferritin, Fe or TIBC). While variants at DUOX2, F5, SLC11A2 and TMPRSS6 were associated with Fe deficiency anaemia, other variants at TF, HFE, TFR2 and TMPRSS6 were associated with Fe overload. These associations implicated proteins contributing to the main physiological processes involved in Fe homeostasis: Fe sensing and storage, absorption of Fe from the gut, Fe recycling, erythropoiesis and bleeding/ menstruation. Notably, the DUOX2 missense variant, which is present in 14% of the population, is associated with all Fe homeostasis BMs, and increases the risk of Fe deficiency anaemia by 29% (Bell et al. 2021). Genetic variants in Tf and HFE, have also been estimated to account for about 40% of genetic variation in Tf levels with dominant associations found between serum Tf and several Tf SNP (rs3811647, rs1358024, rs452586) (McLaren et al. 2011, Moksnes et al. 2021). Mutations in the hepcidin regulatory gene, TMPRSS6, were also found to lead to refractory Fe deficiency anaemia due to the inability to suppress hepcidin production in the liver; such patients do not respond to oral Fe supplementation (Dev and Babitt 2017, Moksnes et al. 2021). In this gene, the SNPs rs855791 on exon 17 and the rs4820268 on exon 13 were largely associated with lower serum Fe concentrations, mean corpuscular volume and haemoglobin levels. Associations between SNPs in TMPRSS6 and Tf saturation were also found in adolescent and adults (McLaren et al. 2011).

The inherited disorder porphyria cutanea tarda, a hepatic porphyria with cutaneous photosensitivity and liver dysfunction due to hepatic Fe deposition, may also result in Fe overload (Ogun and Adeyinka 2022).

Biomarkers of effect

Some authors have defended the particular biology of stored body Fe markers could make them inappropriate to indicate the health effects of Fe, as only free Fe can precipitate oxidative stress. Therefore, any measure of bound Fe, such as tissue ferritin or serum ferritin, will fail to identify any harmful effect of Fe (unless the measure of bound Fe is also a marker of free Fe) (Lee and Jacobs 2004). The fact that there are no effects BMs specific for Fe, although it can be stipulated that since the biological active Fe is able to cause harm, their assessment may indicate Fe effects.

During Fe overload, the bloodstream may represent the first part where defensive systems against Fe toxicity could dominate. But when large amounts of excess Fe are released into the circulation, they are likely to exceed the serum transferrin

Fe-binding capacity, leading to the appearance of various forms of Fe not bound to transferrin; it is under these circumstances that non-Tf bound Fe (NTBI) appears in the serum (Lee and Jacobs 2004). Therefore, NTBI is the fraction comprising Fe bound to proteins other than Tf; it represents Fe that is bound to ligands with substantially less affinity than for Tf (Melicine et al. 2021). When NTBI is present in the plasma, it is rapidly taken up by the liver leading to toxic cellular effects and worsen hepatic Fe overload (Lee and Jacobs 2004, Melicine et al. 2021). *In vivo*, NTBI is able to stimulate both the peroxidation of membrane lipids and the formation of the highly reactive and damaging hydroxyl radicals. Moreover, the uptake of NTBI into cells is not a feedback-regulated process, as this Fe form is able of freely enter the cell membrane. It is demonstrated that antioxidant depletion is inversely associated with NTBI, but not with serum ferritin, among beta-thalassaemic patients (Melicine et al. 2021).

Additionally, the labile Fe pool (LIP) comprises Fe bound mainly to low-molecular weight compounds and is in principle only found in intracellular sections (Lee and Jacobs 2004, Melicine et al. 2021). The cytosolic LIP, "strategically" located at the cross-roads of cell Fe metabolism, serves both as metabolic source of metal but also as indicator of cell Fe levels; cells sense and regulate LIP by balancing the uptake of circulating total body Fe stores, with the storage of unutilized cell Fe in shells of ferritin molecules (Cabantchik 2014). Since LIP are catalytically active and capable of initiating free radical reactions, it is generally accepted that a major and persistent rise in Labile Cells Iron (LCI) levels can compromise cell integrity through overriding cell antioxidant defences. The determination of LCI in living cells has been based on spectroscopic probes but no standard method exists at this time. These investigations must be coupled with a complete blood count to check for an abnormality of erythrocytes and haemoglobin and a measurement of C reactive protein to assess a possible inflammatory syndrome (Cabantchik 2014, Melicine et al. 2021).

As mentioned, when present in excess within cells and tissues, Fe disrupts redox homeostasis and catalyzes the propagation of ROS, leading to oxidative stress (Galaris et al. 2019). Free radicals generate lipid peroxidation, malondialdehyde (MDA) is one of the final products of polyunsaturated fatty acids peroxidation in the cells; MDA is usually known as a marker of oxidative stress (Gaweł et al. 2004). MDA is significantly increased in both transfusion-dependent and non-transfusion-dependent beta-thalassemia patients, concomitantly with ferritin. Hence, MDA may be used as a good surrogate marker of Fe overload in all types of beta-thalassemia, irrespective of severity and blood transfusion dependency and used as a prognostic BM of management, namely involving chelation therapy (Basu et al. 2021).

Final Remarks

This chapter covered the many uses of Fe in the community and environmental and public health concerns appearing from increased emissions of Fe due to anthropogenic activities, with relevance for the abundant presence of Fe-rich air pollution nanoparticles. On the other hand, nutritional Fe deficiency is the most common cause of deficiency and the most prevalent disorder of Fe balance worldwide and the only disorder of Fe balance in which nutrition has the primary role. The greatest concerns are young children and pregnant women. However, during the last decades efforts regarding dietary Fe focused mainly on the prevention of deficiencies, but when in excess Fe can be toxic being the major source of food, supplements or some health disorders and their treatments. In addition, acute Fe toxicity from intentional or accidental ingestion is still a common form of poisoning, particularly in children less than 6 yr. While homeostatic mechanisms increase intestinal Fe absorption in Fe deficiency, unlike other minerals there is no regulated Fe excretion, including in Fe overload; it is known that excess tissue Fe levels are a risk factor for several diseases. There is no single measure of Fe status that can be applied to every case, as combinations of Fe stores, Fe supply to tissues, and functional haemoglobin Fe measurements, often need to reach at a clear assessment. Additionally, serum Fe levels can be determined in cases of Fe overdose; ferritin or Tf measurements are often used to assess Fe status in the body, with ferritin generally accepted as a marker of Fe stores; Tf saturation is considered by some as the best BM to evaluate Fe status and is often used in subjects with the genetic HFE-related hemochromatosis; serum Fe binding capacity is an alternative means of expressing the amount of Tf in serum, with serum TIBC considered an essential test to diagnose Fe deficiency anaemias and other disorders of Fe metabolism; hepcidin, reflecting the integration of multiple key signals involved in Fe regulation, being a useful clinical tool for the management of Fe disorders. ZPP levels has been used as a screening marker to manage Fe deficiency in children and pregnant women. Mutations in various Fe metabolism genes can cause Fe deficiency or overload, as hereditary haemochromatosis, a well-known condition leading to Fe overload. Population-based studies have also identified genetic variants in several other genes, such as TMPRSS6, TFR2, ERFE and IL6R67–70 330, for different Fe status. Larger studies yielded additional loci, with associations of variants at DUOX2, F5, SLC11A2 and TMPRSS6 for Fe deficiency anaemia and variants at TF, HFE, TFR2 and TMPRSS6 associated with Fe overload. There are no effects BMs specific for Fe, although it can be implied that since the biological active Fe form is able to cause harm, their assessment may indicate Fe effects.

It is estimated that worldwide, more than 2 billion people are Fe-deficient; in developing countries, 40–45% of school-age children are anaemic, approximately 50% because of Fe deficiency (Zimmermann and Köhrle 2002). On the other hand, Fe overload has been more recently recognized as a relevant topic for public health, most particularly regarding overload on health disorders. However, concerns about the validity of current screening recommendations for these conditions have been referred to (Hulihan et al. 2011).

References

Adams, P.C., D.M. Reboussin, R.D. Press, J.C. Barton, R.T. Acton, G.C. Moses et al. 2007. Biological variability of transferrin saturation and unsaturated iron-binding capacity. Am. J. Med. 120(11): 999.e1–7.

Almousawi, A.S. and I.R. Sharba. 2019 Erythroferrone hormone a novel biomarker is associated with anemia and iron overload in beta thalassemia patients. J. Physics: Conf. Series. USA. 1294: 062045.

Basu, D., D.G. Adhya, R. Sinha and N. Chakravorty. 2021. Role of malonaldehyde as a surrogate biomarker for iron overload in the β-thalassemia patient: A systematic meta-analysis. Adv. Redox Res. 3: 100017.

Bell, S., A.S. Rigas, M.K. Magnusson, E. Ferkingstad, E. Allara, G. Bjornsdottir et al. 2021. A genome-wide meta-analysis yields 46 new loci associating with biomarkers of iron homeostasis. Commun. Biol. 4: 156.

Benyamin, B., T. Esko, J.S. Ried, A. Radhakrishnan, S.H. Vermeulen, M. Traglia et al. 2014. Novel loci affecting iron homeostasis and their effects in individuals at risk for hemochromatosis. Nat. Commun. 5: 4926.

Cabantchik, Z.I. 2014. Labile iron in cells and body fluids: Physiology, pathology, and pharmacology. Front. Pharmacol. 5: 45.

Cappellini, M.D., K.M. Musallam and A.T. Taher. 2020. Iron deficiency anaemia revisited. J. Intern. Med. 287(2): 153–170.

Colombo, C., G. Palumbo, J.Z. He, R. Pinton and S. Cesco. 2014. Review on iron availability in soil: Interaction of Fe minerals, plants, and microbes. J. Soils Sediments 14: 538–548.

D'Angelo, G. 2013. Role of hepcidin in the pathophysiology and diagnosis of anemia. Blood Res. 48(1): 10–15.

Dev, S. and J.L. Babitt. 2017. Overview of iron metabolism in health and disease. Hemodial. Int. 21(1): S6–S20.

Doig, K. 2020. Iron kinetics and laboratory assessment. pp. 264–281. *In*: E. Keohane, C. Otto and J. Walenga [eds.]. Rodak's Hematology Clinical Principles and Applications. Saunders, Philadelphia, USA.

[EPA] Environmental Protection Agency, Provisional Peer Reviewed Toxicity Values for Iron and Compounds (CASRN 7439-89-6), 2006. EPA, Washington, USA.

Falcone, L.M., A. Erdely, V. Kodali, R. Salmen, L.A. Battelli, T. Dodd et al. 2018. Inhalation of iron-abundant gas metal arc welding-mild steel fume promotes lung tumors in mice. Toxicology 409: 24–32.

Faruqi, A. and S.K.R. Mukkamalla. 2022. Iron Binding Capacity. In: StatPearls [Internet].

Firkin, F. and B. Rush. 1997. Interpretation of biochemical tests for iron deficiency: Diagnostic difficulties related to limitations of individual tests. Aust. Prescr. 20: 74–76.

Galaris, D., A. Barbouti and K. Pantopoulos. 2019. Iron homeostasis and oxidative stress: An intimate relationship. Biochim. Biophys. Acta - Mol. Cell. Res. 1866(12): 118535.

Gammon, R.R., T. Kozel, P. Morel and C. Kendrick. 2018. Evaluation of iron stores by zinc protoporphyrin analysis in blood donors. Lab. Med. 49(4): 311–315.

Gaweł, S., M. Wardas, E. Niedworok and P. Wardas. 2004. Malondialdehyde (MDA) as a lipid peroxidation marker. Wiad. Lek. 57(9-10): 453–455.

Gill, D., C.F. Brewer, G. Monori, D.A. Trégouët, N. Franceschini, C. Giambartolomei et al. 2019. Effects of genetically determined iron status on risk of venous thromboembolism and carotid atherosclerotic disease: A mendelian randomization study. J. Am. Heart Assoc. 8(15): e012994.

Girelli, D., E. Nemeth and D.W. Swinkels. 2016. Hepcidin in the diagnosis of iron disorders. Blood. 127(23): 2809–2813.

Gordan, R., S. Wongjaikam, J.K. Gwathmey, N. Chattipakorn, S.C. Chattipakorn and L.H. Xie. 2018. Involvement of cytosolic and mitochondrial iron in iron overload cardiomyopathy: An update. Heart. Fail. Rev. 23(5): 801–816.

Harper, J.L. and M.E. Conrad. 2020. What is the role of serum iron and ferritin testing in the diagnosis of iron deficiency anemia? *In*: E.C. Besa. Drugs & Diseases, Hematology, Iron Deficiency Anemia Q&A. Medscape, New York, USA.

Hernando, D., R.J. Cook, C. Diamond and S.B. Reeder. 2013. Magnetic susceptibility as a B0 field strength independent MRI biomarker of liver iron overload. Magn. Reson. Med. 70(3): 648–656.

Honda, H., Y. Kobayashi, S. Onuma, K. Shibagaki, T. Yuza, K. Hirao et al. 2016. Associations among erythroferrone and biomarkers of erythropoiesis and iron metabolism, and treatment with long-term erythropoiesis-stimulating agents in patients on hemodialysis. PLoS One. 11(3): e0151601.

[HSA] Health and Safety Authority, Code of Practice for the Safety, Health and Welfare at Work (Chemical Agents) Regulations (2001–2021) and the Safety, Health and Welfare at Work (Carcinogens) Regulations (2001–2019), 2021. HSA, Dublin, Ireland.

[HSE] Health Service Executive, Iron in Drinking Water, Frequently Asked Questions, 2019. Health Service Executive National Drinking Water Group, Dublin, Ireland.

Huang, J.D., B. Luo, M. Sanderson, J. Soto, E.D. Abel, R.C. Cooksey et al. 2011. Iron overload and diabetes risk: A shift from glucose to Fatty Acid oxidation and increased hepatic glucose production in a mouse model of hereditary hemochromatosis. Diabetes. 60(1): 80–87.

Hulihan, M.M., C.A. Sayers, S.D. Grosse, C. Garrison and A.M. Grant. 2011. Iron overload: What is the role of public health? Am. J. Prev. Med. 41(6:4): S422–S427.

[IDPH] Illinois Department of Public Health, Iron In Drinking Water Fact Sheet, 2010. IDPH, Springfield, USA.

Ityel, D. 2011. Ground water: Dealing with iron contamination. Filtr. Separat. 48(1): 26–28.

Koperdanova, M. and J.O. Cullis. 2015. Interpreting raised serum ferritin levels. BMJ. 351: h3692.

Kumar, V., P.K. Bharti, M. Talwar, A.K. Tyagi and P. Kumar. 2017. Studies on high iron content in water resources of Moradabad district (UP). India. Water Sci. 31(1): 44–51.

Kundrapu, S. and J. Noguez. 2018. Laboratory assessment of anemia. Adv. Clin. Chem. 83: 197–225.

Lee, D.H. and D.R. Jacobs Jr. 2004. Serum markers of stored body iron are not appropriate markers of health effects of iron: A focus on serum ferritin. Med. Hypotheses. 62(3): 442–445.

Liebelt, E.L. 2007. Iron. pp. 1119–1128. In: M.W. Shannon, S.W. Borron and M.J. Burns [eds.]. Haddad and Winchester's Clinical Management of Poisoning and Drug Overdose. Saunders, Philadelphia, USA.

Litchford, D.M. 2008. Nutritional issues in the patient with diabetes and foot ulcers. pp. 199–217. In: J.H. Bowker and M.A. Pfeifer [eds.]. Levin and O'Neal's The Diabetic Foot. Mosby, Maryland Heights, USA.

Lodhia, J., G. Mandarano1, N.J. Ferris and S.F. Cowell. 2010. Development and use of iron oxide nanoparticles (Part 1): Synthesis of iron oxide nanoparticles for MRI. Biomed. Imaging. Interv. J. 6(2): e12.

Lynch, S., C.M. Pfeiffer, M.K. Georgieff, G. Brittenham, S. Fairweather-Tait, R.F. Hurrell et al. 2018. Biomarkers of Nutrition for Development (BOND)-Iron review. J Nutr. 148(1): 1001S–1067S.

Magge, H., P. Sprinz, W.G. Adams, M. Drainoni and A. Meyers. 2013. Zinc protoporphyrin and iron deficiency screening: Trends and therapeutic response in an urban pediatric center. JAMA Pediatr. 167(4): 361–367.

Maher, B.A., A. González-Maciel, R. Reynoso-Robles, R. Torres-Jardón, and L. Calderón-Garcidueñas. 2020. Iron-rich air pollution nanoparticles: An unrecognised environmental risk factor for myocardial mitochondrial dysfunction and cardiac oxidative stress. Environ. Res. 188: 109816.

McLaren, C.E., C.P. Garner, C.C. Constantine, S. McLachlan, C.D. Vulpe, B.M. Snively et al. 2011. Genome-wide association study identifies genetic loci associated with iron deficiency. PLoS One. 31; 6(3): e17390.

Means, R.T. 2020. Iron deficiency and iron deficiency anemia: Implications and impact in pregnancy, fetal development, and early childhood parameters. Nutrients. 12(2): 447.

Mele, M. and C. Magazzino. 2020. A Machine Learning analysis of the relationship among iron and steel industries, air pollution, and economic growth in China. J. Clean. Prod. 277: 123293.

Melicine, S., K. Peoc'h and M. Ducastel. 2021. Exploration of iron metabolism: What is new? J. Lab. Precis. Med. 6.

Metzgeroth, G., B. Schultheis, A. Dorn-Beineke, R. Hehlmann and J. Hastka. 2007. Zinc protoporphyrin, a useful parameter to address hyperferritinemia. Ann. Hematol. 86(5): 363–368.

Milic, S., I. Mikolasevic, L. Orlic, E. Devcic, N. Starcevic-Cizmarevic, D. Stimac et al. 2016. The role of iron and iron overload in chronic liver disease. Med. Sci. Monit. 22: 2144–2151.

Moksnes, M.R., A.F. Hansen, S.E. Graham, S.A.G. Taliun, K.H. Wu, W. Zhou et al. 2021. Genome-wide meta-analysis of iron status biomarkers and the 2 effect of iron on all-cause mortality in HUNT. medRxiv. 237604296.

Murakami, Y., K. Saito, H. Ito and Y. Hashimoto. 2019. Transferrin isoforms in cerebrospinal fluid and their relation to neurological diseases. Proc. Jpn. Acad. B: Phys. Biol. Sci. 95(5): 198–210.

Mwangi, M.N., S. Maskey, P.E.A. Andang, N.K. Shinali, J.M Roth, L. Trijsburg et al. 2014. Diagnostic utility of zinc protoporphyrin to detect iron deficiency in Kenyan pregnant women. BMC Med. 12: 229.

[NJDH] New Jersey Department of Health, Iron Oxide. 2007. NJDH, Trenton, USA.

Northrop-Clewes, C.A. and D.I. Thurnham. 2013. Biomarkers for the differentiation of anemia and their clinical usefulness. J. Blood Med. 4: 11–22.

Odukudu, F.B., J.G. Ayenimo, A.S. Adekunle, A.M. Yusuff and B.B. Mamba. 2014. Safety evaluation of heavy metals exposure from consumer products. Int. J. Consum. Sud. 38(1): 25–34.

Ogun, A.S. and A. Adeyinka. 2022. Biochemistry, Transferrin. StatPearls, Treasure Island, Finland.

Oppong-Anane, A.B., K.Y. Deliz Quiñones, W. Harris, T. Townsend and J.-C.J. Bonzongo. 2018. Iron reductive dissolution in vadose zone soils: Implication for groundwater pollution in landfill impacted sites. Appl. Geochem. 94: 21–27.

[OSHA] Occupational Safety & Health Administration, Occupational Chemical Database iron salts, soluble (as Fe), 2021. OSHA, Washington, USA.

Pelclova, D., V. Zdimal, P. Kacer, Z. Fenclova, S. Vlckova, K. Syslova et al. 2015. Oxidative stress markers are elevated in exhaled breath condensate of workers exposed to nanoparticles during iron oxide pigment production. J. Breath Res. 10(2016): 016004.

Pietrangelo, A. 2012. Hemochromatosis. pp. 941–959. e6 *In*: A.J. Sanyal, T.D. Boyer, K.D. Lindor and N.A. Terrault [eds.]. Zakim and Boyer's Hepatology. Elsevier, Amsterdam, Netherlands.

Pietrangelo, A. 2018. Hemochromatosis. pp. 941–959. e6. *In*: A.J. Sanyal, T.D. Boyer, K.D. Lindor and N.A. Terrault [eds.]. Zakim and Boyer's Hepatology. Elsevier, Amsterdam, Netherlands.

Pinto, J.P., J. Arezes, V. Dias, S. Oliveira, I. Vieira, M. Costa et al. 2014. Physiological implications of NTBI uptake by T lymphocytes. Front. Pharmacol. 5: 24.

Pippard, M.J. 2011. Iron deficiency anemia, anemia of chronic disorders and iron overload. pp. 173–195. *In*: A. Porwit, J. McCullough and W.N. Erber [eds.]. Blood and Bone Marrow Pathology. Churchill Livingstone, London, UK.

Rasel, M. and S.K. Mahboobi. 2021. Transfusion Iron Overload. In: StatPearls [Internet].

Reynolds, M.E., M. Kathleen and M.D. Ventre. 2007. Iron overdose. pp. 318–319. *In*: L.C. Garfunkel, J.M. Kaczorowski and C. Christy [eds.]. Pediatric Clinical Advisor. Mosby, Maryland Heights, USA.

[RSC] Royal Society of Chemistry, Periodic Table – Iron, 2022. RSC, London, UK.

Schümann, K. 2011. Safety aspects of iron in food. Ann. Nutr. Metab. 45(3): 91–101.

Sermini, C.G., M.J. Acevedo and M. Arredondo. 2017. Biomarcadores del metabolismo y nutrición de hierro [Biomarkers of Metabolism and Iron Nutrition]. Rev. Peru Med. Exp. Salud. Publica. 34(4): 690–698.

Sherwood, R.A., M.J. Pippard and T.J. Peters. 1998. Iron homeostasis and the assessment of iron status. Ann. Clin. Biochem. 35 (Pt 6): 693–708.

Sreenivasan, V. 2021. Air Pollution in Steel Industry: Environmental, Health & Social Impact. https://www.devic-earth.com/blog/the-steel-industry-in-india-a-sustainability-review-and-best-practices.

[SWA] Safe Work Australia, Ion oxide fume and dust (Fe2O3) (as Fe). 2019. SWA, Canberra, Australia.

Tsuchiya, K. and K. Nitta. 2013. Hepcidin is a potential regulator of iron status in chronic kidney disease. Ther. Apher. Dial. 17(1): 1–8.

Wahidiyat, P.A., E. Wijaya, S. Soedjatmiko, I.S. Timan, V. Berdoukas and M. Yosia. 2019. Urinary iron excretion for evaluating iron chelation efficacy in children with thalassemia major. Blood Cells Mol. Dis. 77: 67–71.

[WDH] Wisconsin Department of Health, Iron In Drinking Water, 2017. WDH, Madison, USA.

[WHO] World Health Orgaization, Iron in Drinking-water Background document for development of WHO Guidelines for Drinking-water Quality. 2003. WHO, Geneva, Switzerland.

Worwood, M., A.M. May and B.J. Bain. 2017. Iron deficiency anaemia and iron overload. pp. 165–186. *In*: B.J. Bain, I. Bates and M.A. Laffan [eds.]. Dacie and Lewis Practical Haematology. Elsevier, Amsterdam, Netherlands.

Yan, N. and J. Zhang. 2020. Iron metabolism, ferroptosis, and the links with alzheimer's disease. Front. Neurosci. 29(13): 1443.

Yuen, H.W. and W. Becker. 2021. Iron Toxicity. *In*: StatPearls Publishers [Internet].

Zaritsky, J., B. Young, H.J. Wang, M. Westerman, G. Olbina, E. Nemeth, T. et al. 2009. Hepcidin—a potential novel biomarker for iron status in chronic kidney disease. Clin. J. Am. Soc. Nephrol. 4(6): 1051–1056.

Zimmermann, M.B. and J. Köhrle. 2002. The impact of iron and selenium deficiencies on iodine and thyroid metabolism: Biochemistry and relevance to public health. Thyroid. 12(10): 867–878.

CHAPTER 12

Magnesium

Sources of Exposure

Magnesium (Mg) is the eighth most abundant element on the Earth's crust, found combined in large deposits in minerals such as magnesite and dolomite. The sea contains trillions of tons of Mg, comprising the source of much of the 850,000 ton produced each year (RSC 2022). The metal exists in nature as Mg^{+2}, whereas the free elemental state is more reactive and can be produced artificially. Mining and production of Mg alloys are relevant sources of human exposure to the metal (Jaiswal et al. 2020). Mg was already weaponized to construct incendiary bombs, flares and ammunitions that were subsequently deployed in World War II, causing massive conflagrations and widespread devastations (Tan and Ramakrishna 2021). Nowadays Mg applications are large: various parts of automobiles and aircraft, as a deoxidizing and desulphurizing agent in the ferrous metal industry, and as a reducing agent in the production of titanium, zirconium and other reactive metals; pure Mg metal is used to protect steel structures from corrosion and has many applications in the chemical industry, including the use as Grignard reagents (HC 2009, Jaiswal et al. 2020, Jayasathyakawin et al. 2020, Prasad et al. 2022); Mg is also used in blasting compositions, incendiaries, pyrotechnics and signal flares, and as an additive to plastics and other materials as a fire protection measure or as a filler (Jaiswal et al. 2020, Lenntech 2022); along with with aluminum, Mg is additionally used in beverage cans, pressure die-cast products, electrical equipment, portable tools, sports equipment, office equipment and many other products (HC 2009, Tan and Ramakrishna 2021); it also reaches the environment from the application of fertilizers and from cattle feed (Lenntech 2022). With its superior ability to biodegrade *in vivo*, Mg compounds have received accelerated interest as a promising biomaterial by both research and industry communities (Tan and Ramakrishna 2021). Mg hydroxide, sulphate and chloride are also used in several medicines, such as laxatives and antacids and in supplements (Jaiswal et al. 2020).

In certain regions such as the magnesite mining areas of Liaoning Province in China, the dust emissions from mining enterprises are enormous, with some of the mining entities completely ignoring environmental protection; bringing significant destruction to the vegetation, landscape and water bodies (Wang et al. 2015). Nevertheless, according to some Mg-related industries Mg is considered the most eco-friendly and sustainable metal in the world. This is attributed to its widespread natural occurrence and the way it is harvested and processed; 100% can be recycled, and it dissolves naturally leaving no trace (Allite 2022). Additionally, Mg emissions have fallen by 82% since 1990, which was mainly due to a reduction of solid fuel use (NAEI 2014).

Air

In 2021 the Environmental Protection Agency (EPA) identified primary Mg refining facilities as a major source of hazardous air pollutant emissions, which notably were not Mg itself. Rather they included other dangerous chemicals such as chlorine, hydrochloric acid, dioxin/furan and trace amounts of several metals. The release of Mg to the atmosphere also comes from fertilizers, construction material, paper industry, blasting compositions, blasting flares and pyrotechnics; even so, Mg in air is not considered to contribute significantly to exposures of this element (HC 2009, Jaiswal et al. 2020). Oxides of Mg are released into the atmosphere naturally as mineral periclase, and as anthropogenic emissions, Mg oxide fume is diluted in air to levels which do not affect the environment (NPI 2018). Mg levels in air are closely related to the extent of particulate pollution: data regarding a semirural area showed a concentration of 0.9 $\mu g/m^3$; in Ontario, the average Mg concentration on a day with heavy particulate pollution was determined as 9 $\mu g/m^3$, and when there was little particulate pollution, 0.46 $\mu g/m^3$; several cities in the USA showed Mg concentrations in particulates ranging from 0.36 to 7.21 $\mu g/m^3$, while an industrialized area in northwest Indiana with cement and steel industries and combustion of coal was found to contain levels of Mg in particulates of 1.35 $\mu g/m^3$ (HC 2009).

Workplace

Occupational exposure to Mg and its compounds is very common in industries related to metallurgy, electronics, manufacturing or processing of Mg and its compounds (Jaiswal et al. 2020). In industrial workplaces where Mg, Mg compounds or alloys are used in "hot" operations, such as welding, brazing, soldering, plating, cutting and metallizing, Mg oxide fumes may also be formed; inhalation can be reduced by adequate ventilation at the site of formation and a respirator may be required in some conditions (NPI 2018). Worksafe Australia and the American Conference of Governmental Industrial Hygienists (ACGIH) has set an 8-hr Total Weight Average (TWA) Threshold Limit Values (TLV) for Mg oxide fume of 10 mg/m^3, while the Occupational Safety and Health Administration (OSHA) Permissible Exposure Limits (PEL) 8-hour TWA is 15 mg/m^3 (as total particulate) and the German MAK value is 4 mg/m^3 (inhalable fraction) (NPI 2018, WHO 2021).

Soil

High soil Mg may occur naturally due to rock parent material, but the use of wastewater containing high levels of Mg for irrigation or its disposal over decades have increased the Mg concentrations in soils (PDA 2017, Qadir et al. 2018). Other areas due to the regular use of magnesian limestone over many years has also raised Mg levels in this environmental section (PDA 2017). Very high levels of Mg in soils of some areas cause apprehension as emerging examples of deterioration and degradation have led to environmental and food security constraints; they are namely sites close-by magnesite mining industry, a rapidly developing industry particularly in the last two decades (Qadir et al. 2018). Soils polluted with the metal may be covered with Mg dust in a range of forms, some of which can form crusts through a series of sedimentary and chemical processes. The main component of the dust in mining areas is Mg oxide which may change into Mg carbonate and Mg hydroxide causing soil pH to increase, and serious damages to soil colloids resulting in soil compaction (Wang et al. 2015, Pan et al. 2016). High Mg soils cover extensive areas on the globe, occurring namely in Kazakhstan, Colombia, Iran, Pakistan, India, Australia and the USA (Qadir et al. 2018). In China, which in 2017 accounted for about 87% of the world's crude Mg production, most of smelting plants dumped smelting slags (including Mg slag) directly into wastelands or landfills due to the lack of mature and effective recycling methods (Wang et al. 2020). Concerning Mg amounts in soils, there are generally between 0.05 and 0.5% as total Mg, but only a small proportion is in forms available for plant uptake (PDA 2017). In Indian soils available Mg content can vary from 73 to 352 mg/kg, while in the Slovak Republic values ranging from 200–400 mg/kg are described; soils of a dumping ground of a magnesite mining factory already exhibited values of 8,400 to 83,100 mg/kg (Venkatesh et al. 2018, Štofejová et al. 2021).

Water

Mg and other alkali earth metals are responsible for hardness in water , as they are washed from rocks and subsequently end up in seawater, rivers and in rainwater. After sodium, Mg is the most found cation in oceans, where amounts of about 1,300 mg/L are present; rivers contain approximately 4 mg/L of Mg (Lenntech 2022). High concentrations of Mg relative to calcium are typically found in municipal and industrial wastewaters due to Mg-containing chemicals, such as the Mg oxide used in the tannery industry (Qadir et al. 2018). Some authors consider that Mg is particularly important in the context of water pollution, as it leads to the distribution of certain organisms in streams; others, contended that $MgSO_4$ exhibit low toxicity being of less significance as an environmental contaminant (van Dam et al. 2010, KWW 2022). Under natural conditions water from areas rich in Mg-containing rocks may contain Mg in the concentration range of 10 to 50 mg/L, but in surface water Mg levels may vary greatly with location and often with the season (HC 2009). Most Asian drinking-water supplies have Mg concentrations below 20 mg/L, and Dutch drinking water contains Mg concentrations between 1 and 5 mg/L (WHO 2009, Lenntech 2022). The intake of Mg from drinking-water depends on the hardness of the water and varies from 1.5 mg (soft water, 1 mg/L) to 37.5 mg (hard water, 25 mg/L) (HC 2009). Bottled water available in North America and Europe range from 1 to 120 and 1 to 126 mg/L, respectively (WHO 2009). Many publications document increased incidence of cardiovascular disease associated with deficient levels of Mg in drinking water. Inversely there is no direct evidence that increased hardness in water can cause adverse health effects, except for extremely high Mg water content (hundreds of mg/L) which can cause diarrhoea (Sengupta 2013, Rapant et al. 2017). Due to insufficient data to suggest either minimum or maximum concentrations, no guideline values are currently proposed by agencies such as the World Health Organization (WHO) or the Australian Government (HC 2009, Sengupta 2013, Rapant et al. 2017, NPI 2018). Based on the observation of decreased mortality from cardiovascular and oncological diseases when concentrations in groundwater were 28–54 mg/L, some authors proposed guidelines values for Mg > 25 mg/L (Rapant et al. 2017).

Food

Dietary sources of Mg are quite varied, with dairy products, vegetables, grain, fruits and nuts being the important contributors (WHO 2009, Jaiswal et al. 2020). Average Mg concentrations in some foods are as follows: seafood, 0.35 mg/g; meat, 0.27 mg/g; cereals and grains, 0.8 mg/g; dairy products, 0.15 mg/g; vegetables, 0.22 mg/g; fruits, 0.08 mg/g; nuts, 1.97 mg/g; and oils and fats, 0.007 mg/g (HC 2009). As Mg is present in such a variety of foods severe Mg deficiency is rarely seen in healthy people (Jaiswal et al. 2020). In spite of that, 45% of Americans are Mg deficient with 60% of adults not reaching the Average Dietary Intake (ADI) (Workinger et al. 2018). Adult women consume only 68% of the US Recommended Dietary Allowance (RDA) of Mg, and adult men 80%; a pregnant woman consumes only 34–58% of her RDA for the metal. This condition is similar in many countries of the world (WHO 2009). Processing techniques, such as grain bleaching and vegetable cooking, can cause a loss of up to 80% of Mg content, with modern dietary practices now estimated to consist of up to 60% processed foods (Workinger et al. 2018). The estimated daily intake of Mg from an average diet is 205 mg for children and between 200 and 300 mg for adults (HC 2009, Workinger et al. 2018). The Scientific Committee for Food of the European Commission set an acceptable range of 150–500 mg/day of Mg intake for adults, while the US RDA for adults is 400-420 mg daily for men and 310–320 mg for women; during pregnancy an intake of 350-360 mg daily and lactation, 310–320 mg (WHO 2009, HC 2022).

Excessive Mg levels due to a Mg rich diet is very uncommon. This condition is generally observed in people having health issues such as kidney failure or with intake of high Mg supplements (Jaiswal et al. 2020). There is a current concern regarding the increased intake of foods containing Mg in combination with unjustified dietary supplementation, since these supplements are widely available as solid oral forms containing Mg salts or multi-component preparations, could lead to hypermagnesemia (Niedworoki et al. 2011, Ajib and Childress 2021). A survey revealed uncontrolled excessive dietary supplementations or medical treatment with Mg, which could constitute a health threat; in this survey the average daily requirement for Mg was exceeded by 292 of the 949 respondents (Niedworoki et al. 2011). A Tolerable Upper Intake Level (UL), which is the maximum daily intake unlikely to cause harmful effects on health, is settled but only for supplements ; this value is 350 mg and was established choosing osmotic diarrhoea as the most sensitive toxic manifestation of excess Mg intake from supplements (WHO 2009, Niedworoki et al. 2011, HC 2022). Children who took Mg from supplements have a median daily intake of 23 mg, while the ULs for children from 1–3 and 4–8 yr old, 65 and 110 mg/day, respectively (IM 1997, WHO 2009).

Medicines

Mg salts, known as milk or cream of magnesia, are broadly used clinically for relief of gastrointestinal symptoms of dyspepsia, heartburn, gastroesophageal reflux disease and constipation, by acting as antacids and laxatives (Hillier 2008, Jaiswal et al. 2020). In the treatment of constipation , Mg is converted into forms able to promote the transfer of water to the intestinal lumen and increase the water content and volume of the stool; while Mg oxide use can lead to hypermagnesemia, but this is rare (Mori et al. 2021). In the case of antiacids, like several other Mg salts, the hydroxide form is rarely given alone, often being combined with aluminum hydroxide (Hillier 2008).

Essentiality and Toxicity

Mg is an essential metal required for fundamental processes such as nucleic acid synthesis and energy production, playing a pivotal role in the synthesis of ATP from ADP and inorganic phosphate (Schwalfenberg and Genuis 2017). Several biological processes including glycolysis, oxidative phosphorylation or phosphate transfer reactions, requires Mg as a direct or indirect factor (Hartwig 2001, Jaiswal et al. 2020).

Deficiency

Since under normal conditions Mg levels are homeostatically regulated, short-term dietary deficiencies can be overcome by the large available pool of Mg in the bones. Depletion can occur as a result of vomiting, diarrhoea, intestinal malabsorption, protein malnutrition and the action of certain drugs which increase the loss of Mg through the kidneys; this can be the case of diuretics and some antibiotics and chemotherapy treatments (HC 2009, WHO 2009). Hypomagnesaemia also occurs in diabetics, alcoholics and in those infected with human immunodeficiency virus-1 (HIV-1) (WHO 2009). Low Mg levels are associated with endothelial dysfunction, increased vascular reactions, elevated circulating levels of C reactive protein and decreased insulin sensitivity. Hypomagnesaemia has also been implicated in hypertension, coronary heart disease, type 2 diabetes mellitus and metabolic syndrome. It has been suggested that long-term Mg deficiency may be a factor in cardiovascular disease (Martin et al. 2009, WHO 2009). Symptoms of hypomagnesaemia are diverse and include weakness, lessening of muscle control, muscle cramps, vertigo, nystagmus, gastrointestinal disorders and psychiatric manifestations; in most severe cases seizures could occur (HC 2009, Martin et al. 2009).

Toxicity

Poisonous effects of Mg are very rare as the kidneys are efficient in eliminating excess Mg from the body (McDonnell et al. 2010, Jaiswal et al. 2020). However oral fatal dosing has been described, varying for a healthy adult from 0.5 to 5 g/kg b.w. The symptoms of hypermagnesemia may vary from mild to severe depending upon the age, sex or health conditions, among other factors (Jaiswal et al. 2020). Early-onset symptoms of acute toxicity are nausea and vomiting, flushing, weakness, impaired breathing, hypotension, hypocalcaemia, arrhythmia, asystole and urinary retention (Niedworoki et al. 2011, Azmeen et al. 2021). Prolonged effects include sleepiness and loss of tendon reflexes; other effects of acutely increased Mg concentration include dilation of pupils, facial flushing, paralysis and coma; one case report also described shock and acidosis, which occurred in spite of dialysis (Jaiswal et al. 2020, Azmeen et al. 2021). Even with early detection and treatment, mortality can be high (Azmeen et al. 2021).

In occupational frameworks intoxication is reported due to exposures of less than 10 min to freshly generated Mg oxide fumes at concentrations from 400 to 600 mg/m^3 .The symptoms of exposure include those of metal fume fever: fever, chills, muscular pain, nausea and vomiting, as well as leucocytosis; exposures to Mg oxide may also cause chronic respiratory disease (NIOSH 2011).

More moderate hypermagnesemia is often iatrogenic and is usually associated with excess Mg intake by patients with renal insufficiency or renal failure, due to significantly decreased ability to excrete Mg. The rectal administration of Mg medication is also a common cause for Mg poisoning (WHO 2009, Niedworoki et al. 2011, Jaiswal et al. 2020). Milder and readily observable adverse effects of Mg occur through intake of contaminated drinking water, which have a laxative effect particularly with Mg sulphate at concentrations above 700 mg/L; the human body can adapt to this effect with time (HC 2009). Laxative effects have also been associated with excess intake of Mg taken in the form of supplements, but not from Mg in diet (WHO 2009).

Biomarkers

At present measurements of total or ionized Mg in serum, plasma, cellular components, urine or Mg retention from a load test are conducted, but they may not always reflect Mg nutritional status. Additionally, while several BMs of susceptibility have been proposed, BMs for Mg are also needed which would reflect changes in biochemical processes where Mg is involved (Franz 2004).

Biomarkers of magnesium status

Mg have a unique mechanism of absorption and a sensitive compartmental handling in the body, which makes the diagnosis of their status quite difficult (Workinger et al. 2018). There is a lack of available clinical tests for the assessment of Mg that can identify which tissues are deficient and the status of Mg in these tissues (WHO 2009). Another difficulty is related to Mg metabolism since the equilibrium and exchange of Mg between body parts and tissue pools occurs slowly and due to this, determining Mg concentration in one tissue may not provide information about Mg status in another (Niedworoki et al. 2011). Furthermore, about 60% of the Mg is present in the bone, of which 30% is exchangeable and functions as a reservoir to stabilize the serum concentration; about 20% is in skeletal muscle, 19% in other soft tissues and less than 1% is in the extracellular fluid. Currently, there is no simple, rapid and accurate laboratory test to indicate the body Mg status (Swaminathan 2003).

Magnesium levels in blood

Whole blood It is difficult to evaluate Mg status from blood tests, although some authors claim that blood Mg concentration reflects the equilibrium between intestinal Mg reabsorption and renal Mg excretion (Swaminathan 2003, Wolf 2017). Blood Mg levels with values of 1.7–2.2 mg/dL are considered within the normal range and it is suggested that a blood level of Mg exceeding 2.6 mg/dL leads to hypermagnesemia, with toxic effects on the body and leading to severe health problems (Jaiswal et al. 2020). This BM have been shown to increase in response to Mg supplementation but does not signal that a complete equilibrium has been established (Workinger et al. 2018).

Plasma The measurement of Mg concentrations in plasma is used widely in nutritional research (Mataix et al. 2006). Normal plasma Mg levels are reported as approximately 2.1 mg/dL (1.7 mEq/L); hypomagnesaemia for plasma Mg levels is less than 1.7 mg/dL (0.70 mmol/L) and hypermagnesemia, for plasma Mg levels greater than 2.4 mg/dL (1.00 mmol/L) (Croker and Walmsley 1986, Niedworoki et al. 2011). However, this BM is considered as having limited sensitivity and specificity; in a western diet study enlisting people with a Mg intake similar to the mean values found for other western countries (20–30% below the RDA) any significant correlation was found between Mg intake and plasma levels; this was possibly due to the marked homeostatic control of the metal (Mataix et al. 2006)

Serum Total serum Mg concentration is the predominant clinical test used to assess Mg status (Costello and Nielsen, 2017). Levels between 2.2 and 2.7 mg/dL (0,75 and 0,96 mmol/L) are reported as the range observed in healthy adult subjects (Arnaud 2008, Costello and Nielsen 2017). Hypomagnesemia is explained for serum Mg level less than 1.7 mg/dL (1.4 mEq/L or 0.7 mmol/L), with serum Mg < 1.8 mg/dL (< 0·75 mmol/L) considered a useful measurement for severe deficiency; for values between 1.8 and 2.1 mg/dL (0,75 and 0,85 mmol/L) a loading test can identify subjects with Mg deficiency (Arnaud 2008, El Kateb et al. 2019). In fact Mg deficiency has been defined by clinical medicine as a serum Mg concentration below a laboratorial reference interval, but this view is controversial as this interval is biased by the large number of "normal" individuals who have a subtle chronic negative Mg balance due to significant decrease in Mg intake (Elin 2010); in addition, there may be subjects with serum Mg concentration within the reference interval that have a total body deficit for the metal, which is the case of chronic marginally negative Mg balance, a condition where serum concentrations may be supported by Mg from other tissue pools, particularly the bone (WHO 2009). Some of these "normal" serum Mg patients may even have signs or symptoms of low Mg and these include alcoholics, patients with poorly controlled diabetes or diarrhoea and refractory hypokalaemia or unexplained hypocalcaemia (Costello and Nielsen 2017, El Kateb et al. 2019). Such individuals need adjustments of their diet (or Mg supplementation) to achieve a normal Mg status for health, which is recommended by some authors as 2.1 mg/dL (0.85 mmol/L) (Elin 2010).

Inversely higher concentrations of Mg in serum are observed when subjects take Mg medication such as Mg-rich antacids or cathartics or in the case of renal failure (Arnaud 2008, Jaiswal et al. 2020). Hypermagnesemia is defined as a serum Mg concentration > 2.6 mg/dL (> 2.1 mEq/L) and clinical consequences may vary by serum levels. Levels > 5 mg/dL (> 4.0 mEq/L) can result in hyporeflexia, > 6 mg/dL (> 5.0 mEq/L) in prolonged atrioventricular conduction, > 12 mg/dL (>10.0 mEq/L) in complete blockage of the heart and concentrations > 16 mg/dL (> 13.0 mEq/L) may lead to cardiac arrest (HC 2009, Niedworoki et al. 2011). Skeletal muscle paralysis, respiratory depression, coma and death occur at plasma concentrations of 18 mg/dL (15 mEq/L) (18 mg/dL) (HC 2009).

Some researchers consider serum Mg as an unreliable test because serum Mg concentrations are not only dependent on dietary intake but also on intestinal absorption and kidney function; while some studies found correlations between serum and tissues values, others did not (Arnaud 2008, Costello and Nielsen (2017). Serum concentrations of Mg can also be affected by short term changes, like day to day and hour to hour variability of the amount of the metal absorbed and excreted through the kidneys (Workinger et al. 2018). It should also be considered that serum Mg represents only 0.3% of total body Mg (WHO 2009).

Blood cells Blood cells Mg levels are often mentioned as preferable to serum or plasma levels due to their higher Mg content (0.5 vs. 0.3%, respectively) (Workinger et al. 2018). Red Blood Cells (RBC) Mg concentrations can be revealed by the cell or haemoglobin; RBC membrane Mg levels have also been measured (Witkowski et al. 2011). RBC Mg levels correlate with Mg status particularly when subjects are placed on long-term (~3 mon) Mg repleted or depleted diets; after Mg supplementation with a median dose of 320 mg/d for a median duration of 2 mon (range: 3 wk–5 yr), the RBC Mg was already found to be significantly higher in treatment groups than in placebo groups (Zhang et al. 2016, Workinger et al. 2018). However, in some other studies decreased levels of Mg in RBC were observed only after several weeks of low dietary intake, this delay was attributed to the RBC pool; thus, this BM reflects long-term rather than current nutrient status (Niedworoki et al. 2011). In addition, most RBC studies do not validate the method through inter-compartmental sampling (e.g., urine and muscle) (Workinger et al. 2018). Notably the Mg content of white blood cells such as lymphocytes was shown *in vivo* to be a better index than RBC Mg of intracellular Mg in skeletal and cardiac muscle (Niedworoki et al. 2011).

Ionized Magnesium Ionized Mg (iMg) is an interesting BM as approximately 20 to 30% of the total Mg (tMg) circulating in serum is bound to proteins and thought to be physiologically inactive, while ionized Mg (iMg) comprises approximately 60–70% of circulating tMg. Since iMg is the biologically active fraction of circulating tMg, it is plausible that iMg may be a more physiologically relevant marker than tMg (Altura and Altura 1992, Rooney et al. 2020). iMg can be measured in whole blood, plasma or serum, and the values on these three matrices for a given person are virtually identical and held within a narrow range of 1.3-1.6 mg/dL (0.53–0.67 mM) in normal healthy controls (Altura and Altura 1992). There are some advantages in measuring iMg when compared with the determination of tMg. The response of the whole blood iMg to a single oral dose of Mg was earlier observed to be more pronounced, than the responses of serum Mg concentration and total urine Mg; this suggests that iMg could be a more sensitive nutrition BM for Mg intake (Zhan et al. 2020). Earlier observed cases of low tMg but normal iMg, indicate false positive hypomagnesemia, and it was also advised that tMg measurements could overestimate the incidence of hypomagnesemia when significant hypoalbuminemia was present (Saha et al. 1998). On the other hand, subjects suffering from cardiac disease, cardiopulmonary bypass, abnormal pregnancy, renal transplant, diabetics and asthmatics, often exhibit significant alterations from their regular iMg levels, without changes in tMg (Altura and Altura 1992). Limited consistency of iMg diagnosis as BM of Mg status is pointed out. Johansson and Whiss (2007) reported a weak correlation between iMg and tMg, which as observed in other works, was not only in critically ill patients but also in patients in whom Mg status was examined as a whole. Currently, whether iMg is a more appropriate measure of Mg status than tMg and its public health or clinical utility, remains to be determined (Rooney et al. 2020).

Magnesium levels in urine and faeces

Urinary Mg have been used frequently in many published clinical studies, many of them using 24 hr urine analyses (Workinger et al. 2018). Increased kidney reabsorption conserves Mg in the case of Mg deprivation leading to decreased urinary excretion of the metal, whereas urinary excretion increases when Mg intake is in excess; therefore, urine Mg reflects Mg intake and can provide information on an individual's Mg status (Witkowski et al. 2011, Malinowska et al. 2020). Epidemiological studies report that after Mg supplementation with a median dose of 480 mg/d for 3 mon, 24 hr urine Mg excretion was significantly elevated compared to the placebos, and increased by 32% after treatment when compared to baseline levels (Zhang et al. 2016). Other controlled metabolic experiments indicate that 40 to 80 mg (1.65 to 3.29 mmol) is the range of daily Mg excretion when Mg intakes are < 250 mg/d , and that 80 to 160 mg is the daily range when intakes are > 250 mg/d , independent of gender (Costello and Nielsen 2017). No standards have been set for urinary Mg excretion indicating deficiency, as urinary levels of 3.0–4.3 mg/dL are considered as within the normal range, while for values higher than 5.0 mg/dL toxic effects can be observed (Costello and Nielsen 2017, Jaiswal et al. 2020). Nevertheless, there is criticism regarding this BM due to wide fluctuations of renal Mg reabsorption and excretion and descriptions that Mg levels in urine that do not correlate with either the amount of ingested Mg or Mg blood levels (Workinger et al. 2018). Another restriction is the within-subject variation of 36% and a between-subject variation of 26% demonstrated after measuring 24 hr urinary Mg excretion in several subjects (Djurhuus et al. 1995). Several factors affecting kidney filtration, such as diabetes, diuretics and renal dialysis, largely influence urinary Mg which limit its utility as a BM in several pathophysiological conditions (Witkowski et al. 2011, Malinowska et al. 2020).

At present the combined determination of serum Mg concentration, 24 hr urinary Mg excretion and dietary Mg intake is considered the most practical method to obtain a sound assessment of Mg status (Costello and Nielsen 2017).

It could be possible to evaluate Mg status by measuring faecal Mg since concentrations of faecal Mg concentrations were shown to augment significantly compared with a placebo group after supplementation with Mg (Zhang et al. 2016). However, these methods require 3–7 d collection and are avoided by researchers and subjects (Workinger et al. 2018).

Magnesium levels in hair and nails

In vivo studies which implicated the administration of a diet supplemented with Mg revealed significant changes in Mg contents of various hair fractions after 2 or 3 mon. In accordance a 3 mon supplementation of children aged 2–6 yr with a Mg dose of 7 mg/kg b.w./d resulted in significantly augmenting Mg levels in hair (7.74 to 11·03 μg/g d.w.) (Arnaud 2008). In other studies, the sensitivity of hair Mg levels to metal intake modifications was smaller than the ones observed for other metals (Woźniak et al. 2019). Hair Mg concentrations were measured in a boy with congenital hypomagnesaemia at 8 and 12 yr of age and compared with control healthy children with the same hair colour; surprisingly, his Mg content of hair was higher than controls despite the lower Mg values in serum, erythrocyte, lymphocyte and teeth. The authors concluded that hair Mg was not suitable to reflect this deficiency (Arnaud 2008).

The response of the nail as an elemental BM is not as well characterized for some chemicals, and these chemicals includes Mg (He 2010). Even though the reliability over a 6 yr period of measurements in toenails of 16 trace elements (including Mg) has been reported, there is a need to validate the measurement of nail Mg and to demonstrate that the sample represents a period of either deficient or excessive Mg intake (Arnaud 2008).

Magnesium levels in muscle and bones

Muscle Mg concentration appears to be a good marker, but biopsies limit its usefulness, in the same way as as the case of bone Mg which is the most important but heterogeneous Mg part . In the near future, the development of new and non-invasive techniques such as Nuclear Magnetic Resonance (NMR) may provide a valuable tool for routinely analyzing Mg in such tissues (Arnaud 2008).

Biomarkers of susceptibility

Hypomagnesemia is by far the most researched condition as compared with hypermagnesemia, and the genetic causes of hypomagnesemia are broadly described. They comprise recessive or dominant disorders, are highly heterogeneous and known to involve genes encoding proteins manifested in the kidneys (Viering et al. 2017). Since primary hypomagnesemia can occur by renal and also intestinal Mg wasting, it is important to determine it origin (Meij et al. 2000). The most important clinical diagnostic tool for this differentiation, is the determination of the fractional excretion of Mg (FEMg), indicated by the formula: $[(Mg^{2+}_{urine} \times creatinine_{plasma}) / (0.7\ Mg^{2+}_{plasma} \times creatinine_{urine})] \times 100\%$; A FEMg > 4 % in a hypomagnesemia patient is consistent with renal Mg wasting, while a patient with a FEMg of < 2 % will likely have an extra-renal origin of their hypomagnesemia despite a FEMg < 4 % does not rule out renal Mg wasting (Viering et al. 2017).

Wolf (2017) listed 16 inherited forms of renal hypomagnesemia, with specific mutated gene loci and affected gene proteins. In broad terms, inherited forms of hypomagnesemia can be divided into: (i) hypercalciuric, mostly affecting the kidney Thick Ascending Limb (TAL) which absorbs the majority of filtered Mg; (ii) Gitelman-like, mostly affecting the kidney Distal Convoluted Tubule (DCT) in the distal tubule which plays an important role in determining the final urinary excretion of Mg; (3) and other forms of hypomagnesemia, mostly affecting the DCT (Dai et al. 2001, Wolf 2017).

Hypercalciuric hypomagnesemia may be attributed to mutations in the following genes: the CLDN16 which encode claudin-16, a tight-junction protein involved in paracellular reabsorption of Mg (HGNC 2022); the CLDN19 encoding the tight-junction gene claudin 19, whose mutations are associated with renal Mg wasting (UniProtKB 2022); the calcium sensing receptor gene (CASR) for which the functional activity of mutant CASR determines the serum Mg level and renal Mg handling (Kinoshita et al. 2014); and the chloride voltage-gated channel Kb (CLCNKB) whose mutations disrupt reabsorption of sodium chloride in distal nephron segments impairing the renal conservation of Mg (Quamme et al. 2008). The recessive mutations in CLDN16 and CLDN19 are the most frequent causes of hypercalciuric hypomagnesemia. These mutations disrupt the pore selectivity of the tight junctions which impairs paracellular Ca^{2+} and Mg^{2+} reabsorption in the TAL (Viering et al. 2017, Tinawi 2020). Gitelman-like hypomagnesemia result from mutations in the following genes: the CLCNKB; the solute carrier family 12 (sodium/chloride transporter), member 3 (SLC12A3); the potassium inwardly rectifying channel subfamily J member 10 (KCNJ10); the FXYD domain containing ion transport regulator 2 (FYXD2); the Hepatocyte Nuclear Factor 1-Beta (HNF1B); the pterin-4 alpha-carbinolamine dehydratase 1 (PCBD1); or the Transient Receptor Potential Melastatin 6 (TRPM6). The barttin CLCNK type accessory beta subunit

(BSND) encodes barttin which binds to chloride channels that transport chloride ions out of kidney cells, leading to Mg^{2+} wasting by indirect means (Giménez-Mascarell et al. 2018, HGNC 2022). SLC12A3 encodes the Na^+-Cl^--Cotransporter (NCC) exhibited on the apical membrane of the DCT may promote renal Mg^{2+} wasting, since NCC activity determines Mg^{2+} entry into DCT cells through the transient receptor potential channel subfamily M member 6 (TRPM6) (Grillone et al. 2016, Viering et al. 2017, Dong et al. 2020, Maeoka and McCormick 2020). KCNJ10 encodes Kir4.1, which is one of the components of the inwardly rectifying K^+ channel Kir4.1/Kir5.1, may alter mutations in this gene membrane voltage, thereby affecting transport processes such as those for Mg^{2+}, defining Mg^{2+} loss (Claverie-Martin et al. 2021, HGNC 2022). FYXD2 encodes the Na^+, K^+-ATPase gamma-subunit which is a member of the FXYD family of transmembrane proteins that participates in the tubular handling of Mg (Adalat et al. 2009, Meij et al. 2000, HGNC 2022). HNF1B regulates the transcription of FXYD2 as their mutation are the most common cause of genetic hypomagnesemia for pediatric nephrologists; heterozygous genotypes are also associated with a multi-system disorder and are the most usual genetic cause of congenital anomalies of the kidney and urinary tract (Adalat et al. 2009, Viering et al. 2017). Wild-type PCBD1 binds HNF1B to co-stimulate the FXYD2 promoter, the activity of which is instrumental in Mg^{2+} reabsorption in the DCT (Ferrè et al. 2014). TRPM6 is a member of the melastatin-related subfamily of Transient Receptor Potential (TRP) ion channels and a close homologue of TRPM7, which been characterized as a Mg and calcium permeable ion channel vital for cellular Mg homeostasis (Schlingmann et al. 2007); while TRPM7 is ubiquitously exhibited in human tissues and has been proposed as an indispensable cellular Mg entry pathway, the TRPM6 channel has a more restricted expression. Their highest levels are found in the distal small intestine and colon, but also in the distal convoluted tubule of the kidney where it is involved in the reabsorption of Mg. TRPM6 expression in the kidneys and intestines is regulated by dietary Mg availability (Ikari et al. 2008, Luongo et al. 2018). TRPM6 mutations cause a defect in the active transcellular pathway of intestinal Mg absorption (Naderi and Reilly 2008).

Other forms of genetic hypomagnesemia include mitochondrial hypomagnesemia, for which pathophysiological mechanisms are still not explained. At least three distinct mitochondrial syndromes accompanied by hypomagnesemia are identified: deletions in the mitochondrial genome; recessive mutations in the gene seryl-TRNA Synthetase 2 (SARS2) which takes part in the biosynthesis of selenocysteinyl-tRNA(sec) in mitochondria, and mutations in the mitochondrial tRNAIle gene MT-TI. Other hypomagnesemia is described involving mutation in other genes such as CNMM2 and Epidermal Growth Factor gene (EGF), both involved in Mg transport (Viering et al. 2017).

Biomarkers of effect

No validated BM of effects specific for Mg are described, although functionals BMs which could reflect changes in biochemical processes where Mg is involved are certainly needed. Some possible BMs include the following: Na/K ATPase, thromboxane B2 or the C-reactive protein, but again focused only on hypomagnesemia (Franz 2004).

Regarding Na/K-ATPase, hypomagnesemia is attributed to their inhibition or destabilization in DCT cells. Changes Na/K-ATPase activity affect intracellular K concentration and consequently the driving force for transepithelial Mg transport (Mayan et al. 2018). Platelet activity and thrombi formation, which can be measured by the levels of thromboxane B2, is known to be increased in Mg deficiency and decreased after Mg supplementation (Franz 2004). C-Reactive Protein (CRP) is a marker of inflammation and individuals with intakes below the RDA were found to have elevated CRP (King et al. 2005). A study registered on type 2 diabetes hypomagnesemic patients who had Mg supplementation for 3 mon demonstrated the efficacy on restoring circulating levels of CRP (as well as TNF-α, 8-isoprostane) and other glucose intolerance BMs (Zghoul et al. 2018).

Final Remarks

This chapter described the large applications of Mg, which according to some Mg-related industries is considered the most eco-friendly metal in the world. As Mg is present in a number of foods, severe Mg deficiency is rarely seen in healthy people, while excessive Mg levels due to Mg rich diet are also very uncommon. Both hypo- and hypermagnesemia are generally observed in people having health issues, or in the case of hypermagnesemia, due to the intake of high Mg supplements. A current concern is generated by the increased intake of foods containing Mg in combination with unjustified dietary supplementation. In other contexts, Mg oral fatal dosing is also described (even with early detection and treatment, mortality can be high) and in some workplaces exposures to Mg oxide fumes can lead to symptoms of metal fume fever. Currently the combined determination of serum Mg concentration, 24 hr urinary Mg excretion and dietary Mg intake is considered the most practical method to obtain a sound assessment of Mg status. Measurements of Mg concentrations in the plasma are used widely in nutritional research, whereas total serum Mg concentrations are the predominant clinical tests. Ionized Mg (iMg) as the biologically active fraction of circulating total Mg, is suggested as sensitive nutrition BM for Mg intake. In turn urinary Mg have been used frequently in many published clinical studies, many of them using 24 hr urine analyses, which may reflect Mg intake and provide information on an individual's Mg status. Hypomagnesemia is by far the most researched condition as compared with hypermagnesemia, and the genetic causes of hypomagnesemia are broadly described. They comprise recessive or dominant disorders, are highly heterogeneous, known to involve genes encoding proteins expressed in the kidneys, and can be classified as hypercalciuric, mostly affecting the kidney Thick Ascending Limb (TAL) which absorbs the majority of filtered Mg, Gitelman-like, mostly affecting the kidney Distal Convoluted Tubule (DCT) in the distal tubule which play an important role in determining the final urinary excretion of Mg and other forms of hypomagnesemia, which include mitochondrial hypomagnesemia. No validated BM of effects specific for Mg are described, although functionals BMs which could reflect changes in biochemical processes where Mg is involved are certainly needed. Some possible BMs include the following: Na/K ATPase, thromboxane B2 or the C-reactive protein, but again focused only on hypomagnesemia. Suitable BMs for hypermagnesemia of specific Mg effects are urgently required.

References

Adalat, S., A.S. Woolf, K.A. Johnstone, A. Wirsing, L.W. Harries, D.A. Long et al. 2009. HNF1B mutations associate with hypomagnesemia and renal magnesium wasting. J. Am. Soc. Nephrol. 20(5): 1123–1131.

Ajib, F.A. and J.M. Childress. Magnesium Toxicity. [Updated 2021 Nov 14]. *In*: StatPearls [Internet]. Treasure Island (FL): StatPearls Publishing; 2022 Jan-. Available from: https://www.ncbi.nlm.nih.gov/books/NBK554593/.

[Allite] Allite inc., Reintroducing Magnesium. 2022. Allite inc., Miamisburg, USA.

Altura, B.T. and B.M. Altura. 1992. Measurement of ionized magnesium in whole blood, plasma and serum with a new ion-selective electrode in healthy and diseased human subjects. Magnes. Trace Elem. 10(2-4): 90–98.

Arnaud, M.J. 2008. Update on the assessment of magnesium status. Br. J. Nutr. 99(3): S24–36.

Azmeen, A., D. Condit, E. Rosenberg, S. Mcpeck, A. Deengar, S. Sathyan et al. 2021. A case of catastrophic magnesium overdose. J. Crit. Care. 160(4): a954–a955.

Claverie-Martin, F., A. Perdomo-Ramirez and V. Garcia-Nieto. 2021. Hereditary kidney diseases associated with hypomagnesemia. Kidney Res. Clin. Pract. 40(4): 512–526.

Costello, R.B. and F. Nielsen. 2017. Interpreting magnesium status to enhance clinical care: Key indicators. Curr. Opin. Clin. Nutr. Metab. Care 20(6): 504–511.

Croker, J.W. and R.N. Walmsley. 1986. Routine plasma magnesium estimation: a useful test? Med. J. Aust. 145(2): 71, 74–6.

Dai, L.-J., G. Ritchie, D. Kerstan, H.S. Kang, D.E.C. Cole and G.A. Quamme. 2001. Magnesium transport in the renal distal convoluted tubule. Physiol. Rev. 81(1): 51–83.

Djurhuus, M.S., J. Gram, P.H. Petersen, N.A. Klitgaard, J. Bollerslev and H. Beck-Nielsen. 1995. Biological variation of serum and urinary magnesium in apparently healthy males. Scand. J. Clin. Lab. Invest. 55(6): 549–558.

Dong, B., Y. Chen, X. Liu, Y. Wang, F. Wang, Y. Zhao et al. 2020. Identification of compound mutations of SLC12A3 gene in a Chinese pedigree with Gitelman syndrome exhibiting Bartter syndrome-liked phenotypes. BMC Nephrol. 21: 328.

El Kateb, M., Joel M. Topf, in Nephrology Secrets (Fourth Edition). 2019. Disorders of magnesium metabolism. https://www.sciencedirect.com/topics/biochemistry-genetics-and-molecular-biology/magnesium-blood-level.

Elin, R.J. 2010. Assessment of magnesium status for diagnosis and therapy. Magnes. Res. 23(4): S194–198.

[EPA] Environmental Protection Agency, Primary Magnesium Refining: National Emissions Standards for Hazardous Air Pollutants (NESHAP), 2021. EPA, Washington, USA.

Ferrè, S., J.H. de Baaij, P. Ferreira, R. Germann, J.B. de Klerk, M. Lavrijsen et al. 2014. Mutations in PCBD1 cause hypomagnesemia and renal magnesium wasting. J. Am. Soc. Nephrol. 25(3): 574–586.

Franz, K.B. 2004. A functional biological marker is needed for diagnosing magnesium deficiency. J. Am. Coll. Nutr. 23(6): 738S–7341S.

Giménez-Mascarell, P., C.E. Schirrmacher, L.A. Martínez-Cruz and D. Müller. 2018. Novel aspects of renal magnesium homeostasis. Front. Pediatr. 6: 77.

Grillone, T., M. Menniti, F. Bombardiere, M.F. Vismara, S. Belviso, F. Fabiani et al. 2016. New SLC12A3 disease causative mutation of Gitelman's syndrome. World J. Nephrol. 5(6): 551–555.

Hartwig, A. 2001. Role of magnesium in genomic stability. Mutat. Res. 475(1–2): 113–121.

[HC] Harvard TH Chan, The Nutrition Source, Magnesium, 2022. HC, Cambridge, USA.

[HC] Health Canada, Guidelines for Canadian Drinking Water Quality: Supporting Documents—Magnesium, 2009. HC, Ottawa, Canada.

He, K. 2010. Trace elements in nails as biomarkers in clinical research. Eur. J. Clin. Invest. 41: 98–102.

[HGNC] HUGO Gene Nomenclature Committee, CLDN16 claudin 16 [Homo sapiens (human)], 2022. HGNC, European Molecular Biology Laboratory, European Bioinformatics Institute (EMBL-EBI), Cambridge, UK.

Hillier, K. 2008. Magnesium hydroxide. pp. 1–3. *In*: S.J. Enna and D.B. Bylund [eds.]. The Comprehensive Pharmacology Reference. Elsevier, Amsterdam, Netherlands.

Ikari, A., C. Okude, H. Sawada, T. Takahashi, J. Sugatani and M. Miwa. 2008. Down-regulation of TRPM6-mediated magnesium influx by cyclosporin A. Naunyn Schmiedebergs Arch. Pharmacol. 377(4-6): 333–343.

[IM] Institute of Medicine (US) Standing Committee on the Scientific Evaluation of Dietary Reference Intakes. 1997. Magnesium pp. 190–250. *In*: Standing Committee on the Scientific Evaluation of Dietary Reference Intakes [eds.]. Dietary Reference Intakes for Calcium, Phosphorus, Magnesium, Vitamin D, and Fluoride. National Academies Press, Washington, USA.

Jaiswal, A.K., R. Kumar, K. Bisht, D.K. Sharma, M. Gupta and A. Basnal. 2020. Magnesium poisoning with analytical aspects and its management. Indian J. Community Med. 7(2): 51–55.

Jayasathyakawin, S., M. Ravichandran, N. Baskar, C.A. Chairman and R. Balasundaram. 2020. Mechanical properties and applications of Magnesium alloy—Review. Mater. Today: Proc. 27(2): 909–913.

Johansson, M. and P.A. Whiss. 2007. Weak relationship between ionized and total magnesium in serum of patients requiring magnesium status. Biol. Trace Elem. Res. 115(1): 13–21.

King, D.E., A.G. Mainous, M.E. Geesey and R.F. Woolson. 2005. Dietary magnesium and C-reactive protein levels. J. Am. Coll. Nutr. 24(3): 166–171.

Kinoshita, Y., M. Hori, M. Taguchi, S. Watanabe and S. Fukumoto. 2014. Functional activities of mutant calcium-sensing receptors determine clinical presentations in patients with autosomal dominant hypocalcaemia. J. Clin. Endocrinol. Metab. 99(2): E363–68.

[KWW] Kentucky Water Watch, Magnesium and water quality, 2022. KWW, Frankfort, USA.

[Lenntech] Magnesium (Mg) and water. Magnesium and water: reaction mechanisms, environmental impact and health effects, 2022. Lenntech, Delfgauw, Netherlands.

Luongo, F., G. Pietropaolo, I. Dhennin-Duthille, H. Ouadid-Ahidouch, F.I. Wolf et al. 2018. TRPM6 is essential for magnesium uptake and epithelial cell function in the colon. Nutrients. 10(6): 784.

Maeoka, Y. and J.A. McCormick. 2020. NaCl cotransporter activity and Mg 2+ handling by the distal convoluted tubule. Am. J. Physiol. Renal Physiol. 319(6): F1043–F1053.

Malinowska, J., M. Małecka and O. Ciepiela. 2020. Variations in magnesium concentration are associated with increased mortality: Study in an unselected population of hospitalized patients. Nutrients. 12(6): 1836.

Martin, K.J., E.A. González and E. Slatopolsky. 2009. Clinical consequences and management of hypomagnesemia. J. Am. Soc. Nephrol. 20(11): 2291–2295.

Mataix, J., P. Aranda, M. López-Jurado, C. Sánchez, E. Planells and J. Llopis. 2006. Factors influencing the intake and plasma levels of calcium, phosphorus and magnesium in southern Spain. Eur. J. Nutr. 45(6): 349–354.

Mayan, H., Z. Farfel and S.J.D. Karlish. 2018. Renal Mg handling, FXYD2 and the central role of the Na, K-ATPase. Physiol. Rep. 6(17): e13843.

McDonnell, N.J., N.A. Muchatuta and M.J. Paech. 2010. Acute magnesium toxicity in an obstetric patient undergoing general anaesthesia for caesarean delivery. Int. J. Obstet. Anesth. 19(2): 226–231.

Meij, I.C., J.B. Koenderink, H. van Bokhoven, K.F. Assink, W.T. Groenestege, J.J. de Pont et al. 2000. Dominant isolated renal magnesium loss is caused by misrouting of the Na+, K+-ATPase gamma subunit. Nat. Genet. 26(3): 265–266.

Mori, H., J. Tack and H. Suzuki. 2021. Magnesium oxide in constipation. Nutrients 13(2): 421.

Naderi, A.S.A. and R.F. Reilly Jr. 2008. Hereditary etiologies of hypomagnesemia nature clinical practice. Nephrology 4(2): 80–89.

[NAEI] National Atmospheric Emissions Inventory, Pollutant Information: Magnesium, 2014. NAEI, UK.

Niedworoki, E., M. Muc-wierzg, E. Nowakowska-zajdeu, L. Duu and K. Klakla. 2011. Magnesium content in daily food portions and the influence of supplementation. Int. J. Immunopathol. Pharmacol. 24(4): 975–981.

[NIOSH] National Institute for Occupational Safety and Health, Magnesium oxide, 2011. NIOSH, Washington, USA.

[NPI] National Pollutant Inventory, Magnesium oxide fume, 2018. Department of Agriculture Water and the Environment, Canberra, Australia.

[OSHA] Occupational Safety and Health Administration, Magnesium oxide (Fume), 2021. OSHA, Washington, USA.

Pan, D., Y. Chu, X. Zhao and M. Shan. 2016. Research on Magnesium Pollution Status and Remediation of Tilth Soils in Haicheng Magnesium Mine. International Conference on Civil, Transportation and Environment (ICCTE 2016).

[PDA] Potash Development Association, Magnesium as a nutrient for crops and grass, 2017. PDA, Huntington, USA.

Prasad, S.V.S., S.B. Prasad, K. Verma, R.K. Mishra, V. Kumar and S. Singh. 2022. The role and significance of Magnesium in modern day research—A review. J. Magnes. Alloy. 10(1): 1–61.

Qadir, M., S. Schubert, J.D. Oster, G. Sposito, P.S. Minhas, S.A.M. Cheraghi et al. 2018. High-magnesium waters and soils: Emerging environmental and food security constraints. Sci. Total Environ. 642: 1108–1117.

Quamme, G.A., K.P. Schlingmann and M. Konrad. 2008. Mechanisms and disorders of magnesium metabolism. pp. 1747–1767. *In*: R.J. Alpern and S.C. Hebert [eds.]. Seldin and Giebisch's the Kidney. Academic Press, Cambridge, USA.

Rapant, S., V. Cvečková, K. Fajčíková, D. Sedláková and B. Stehlíková. 2017. Impact of calcium and magnesium in groundwater and drinking water on the health of inhabitants of the Slovak Republic. Int. J. Environ. Res. Pub. Health. 14(3): 278.

Rooney, M.R., K.D. Rudser, A. Alonso, L. Harnack, A.K. Saenger and P.L. Lutsey. 2020. Circulating ionized magnesium: Comparisons with circulating total magnesium and the response to magnesium supplementation in a randomized controlled trial. Nutrients. 12: 263.

[RSC] Royal Society of Chemistry, Magnesium, 2022. RSC, London, UK.

Saha, H., A. Harmoinen, A. Karvonen, J. Mustonen and A. Pasternack. 1998. Serum ionized versus total magnesium in patients with intestinal or liver disease. Clin. Chem. Lab. Med. 36(9): 715–718.

Schlingmann, K.P., S. Weber, M. Peters, L.N. Nejsum, H. Vitzthum, K. Klingel et al. 2002. Hypomagnesemia with secondary hypocalcemia is caused by mutations in TRPM6, a new member of the TRPM gene family. Nat. Genet. 31(2): 166–170.

Schlingmann, K.P., S. Waldegger, M. Konrad, V. Chubanov and T. Gudermann. 2007. TRPM6 and TRPM7—Gatekeepers of human magnesium metabolism. Biochim. Biophys. Acta Mol. Basis Dis. 1772(8): 813–821.

Schwalfenberg, G.K. and S.J. Genuis. 2017. The importance of magnesium in clinical healthcare. Scientifica. 2017: 4179326.

Sengupta, P. 2013. Potential health impacts of hard water. Int. J. Prev. Med. 4(8): 866–875.

Štofejová, L., J. Fazekaš and D. Fazekašová. 2021. Analysis of heavy metal content in soil and plants in the dumping ground of magnesite mining factory Jelšava-Lubeník (Slovakia). Sustainability. 13: 4508.

Swaminathan, R. 2003. Magnesium metabolism and its disorders. Clin. Biochem. Rev. 24(2): 47–66.

Tan, J. and S. Ramakrishna. 2021. Applications of magnesium and its alloys: A review. Appl. Sci. 11: 6861.

Tinawi, M. 2020. Disorders of magnesium metabolism: hypomagnesemia and hypermagnesemia. Arch. Clin. Biomed. Res. 4(3): 205–220.

UniProtKB 2022. https://www.uniprot.org/uniprot/Q8N6F1.

van Dam, R.A., A.C. Hogan, C.D. McCullough, M.A. Houston, C.L. Humphrey and A.J. Harford. 2010. Aquatic toxicity of magnesium sulphate, and the influence of calcium, in very low ionic concentration water. Environ. Toxicol. Chem. 29(2): 410–421.

Venkatesh, V., R.K.K. Singh and N.S. Devi. 2018. Critical limit of available magnesium for green gram in soils of Imphal West district, Manipur, India. Int. J. Chem. Stud. 6(4): 1092–1097.

Viering, D., J. de Baaij, S.B. Walsh, R. Kleta and D. Bockenhauer. 2017. Genetic causes of hypomagnesemia, a clinical overview. Pediatr. Nephrol. 32(7): 1123–1135.

Wang, L., P. Tai, C. Jia, X. Li, P. Li and X. Xiong. 2015. Magnesium contamination in soil at a magnesite mining region of liaoning province, China. Bull. Environ. Contam. Toxicol. 95(1): 90–96.

Wang, X.-Bn, X. Yan and X.-Y. Li. 2020. Environmental risks for application of magnesium slag to soils in China. J. Integr. Agric. 19(7): 1671–1679.

[WHO] World Health Organization, Calcium and Magnesium in Drinking-water, Public health significance, 2009. WHO, Geneva, Switzerland.

[WHO] World Health Organization, Magnesium oxide, 2021. WHO, Geneva, Switzerland.

Witkowski, M., J. Hubert and A. Mazur. 2011. Methods of assessment of magnesium status in humans: A systematic review. Magnes. Res. 24(4): 163–180.

Wolf, M.T. 2017. Inherited and acquired disorders of magnesium homeostasis. Curr. Opin. Pediatr. 29(2): 187–198.

Workinger, J.L., R.P. Doyle and J. Bortz. 2018. Challenges in the diagnosis of magnesium status. Nutrients. 10(9): 1202.

Woźniak, A., A. Wawrzyniak and A. Jeznach-Steinhagen. 2019. Hair as a biomarker to evaluate the intake of Iron, magnesium and zinc in children. J. Elem. 24(2): 727–738.

Zghoul, N., N. Alam-Eldin, I.T. Mak, B. Silver and W.B. Weglicki. 2018. Hypomagnesemia in diabetes patients: comparison of serum and intracellular measurement of responses to magnesium supplementation and its role in inflammation. Diabetes Metab. Syndr. Obes. 11: 389–400.

Zhan, J., T. Wallace, S. Butts, S. Cao, C. Weaver and N.G. Miller. 2020. Whole blood ionized magnesium, a novel nutrition biomarker for magnesium intake. Curr. Dev. 4(2): 1856.

Zhang, X., L.C. Del Gobbo, A. Hruby, A. Rosanoff, K. He, Q. Dai et al. 2016. The circulating concentration and 24-h urine excretion of magnesium dose- and time-dependently respond to oral magnesium supplementation in a meta-analysis of randomized controlled trials. J Nutr. 146(3): 595–602.

CHAPTER 13

Manganese

Sources of Exposure

Manganese (Mn) is the 12th most abundant metal on the Earth's crust, comprising 0.1%, and occurs in over 300 different minerals (Howe et al. 2004, Chiswell and Huang 2006). There are many environmental sources of Mn, which include eroded rocks, soils and decomposed plants (O'Neal and Zheng 2015). The major anthropogenic sources are municipal wastewater discharges, sewage sludge, mining and mineral processing, emissions from alloy, steel, iron (Fe) production, combustion of fossil fuels and emissions from the combustion of fuel additives (Chiswell and Huang 2006). Fertilizers, pesticides such as maneb and mancozeb, livestock feeding supplements, varnishes, pottery glazes, medical imaging contrast agents and water purification agents constitute additional anthropogenic sources. Mn dioxide and other Mn compounds are also used in products such as dry-cell batteries, glass and fireworks (Howe et al. 2004, Chiswell and Huang 2006, O'Neal and Zheng 2015, WHO 2021).

Air

Crustal rocks, soil erosion, ocean spray, forest fires, vegetation and volcanic activity are natural sources of the metal to the air, being the main anthropogenic sources the industrial emissions of ferroalloy production and Fe and steel foundries, power plants, coke ovens and the combustion of fossil fuels soils (Howe et al. 2004, WHO 2021). Methylcyclopentadienyl Mn Tricarbonyl (MMT) is an additive of nonlead gasoline as an octane enhancer in automobiles approved in several countries including the USA, Canada, Argentina, Australia, Bulgaria, France, Russia, New Zealand, China and in the European Union (O'Neal and Zheng 2015, Ikeda et al. 2017). Its use leads to increased airborne Mg in cities and areas along expressways, due to the release of Mn phosphates, sulphates and oxides into the air (Zayed 2001, O'Neal and Zheng 2015, Ikeda et al. 2017). The majority of Mg emitted from tailpipes of cars using MMT-containing gasoline are associated with particles smaller than 2.5 μm in size (PM2.5), which may get into the bloodstream and reach the blood-brain barrier (Li et al. 2022, Ikeda et al. 2017). In spite of MTT allowing the phasing out of leaded gasoline, MTT puts people at risk for excessive exposure to Mn (O'Neal and Zheng 2015, Evans and Masullo 2021).

In remote locations Mn values in air range between 0.5 and 14 ng/m³, rural areas present average concentrations of 40 ng/m³, and in urban areas the levels can vary from 65 to 166 ng/m³; near foundries annual averages Mn concentrations of 200 to 300 ng/m³ can be found in air and even surpass 500 ng/m³ near ferro- and silicomanganese industries (Espinosa et al. 2001, WHO 2021). Mn in particulate matter with less than or equal to 10 μm in diameter (PM10), was already estimated to be 6.68 ng/m³ (range 0.85–614 ng/m³) in urban areas. Exposure to Mn from air is generally several orders of magnitude less than exposure from diet, typically being around 0.04 ng/d , although this can vary depending on proximity to a Mn source (WHO 2021).

Workplace

Occupational exposure of Mn is a very relevant issue since is it linked to most reported cases of Mn intoxication, particularly in environments typically associated with an elevated atmospheric metal concentration; inhaled Mg is of greater concern because it bypasses the body's normal defence mechanisms (NIOSH 2103, Kim et al. 2015, Evans and Masullo 2021). Increased Mg-induced toxicity is observed among miners, industrial welders and farmers exposed to Mn-based pesticides such as maneb and mancozeb (Gyuri et al. 2015. Subjects involved in activities such as pipefitters, millwrights and dry-cell battery producers are also critically at risk (Cowan et al. 2009, NIOSH 2013, 2019, Baker et al. 2014, Evans and Masullo 2021).

Several agencies established occupational limits for Mg levels in the air: the National Institute for Occupational Safety and Health (NIOSH) sets an Immediately Dangerous to Life and Health (IDLH) concentration of 500 mg/m³, a Recommended Exposure Limit (REL) of 1 mg/m³ time-weighted average (TWA), and a 3 mg/m³ Short-Term Exposure Limit (STEL). The Occupational Safety and Health Administration (OSHA) Permissible Exposure Limit (PEL) is 5 mg/m³ (ceiling), while the American Conference of Governmental Industrial Hygienists (ACGIH) Threshold Limit Value (TLV) is 0.02 mg/m³ (TWA) (NIOSH 2013, OSHA 2021). However, the REL from NIOSH is under review since the agency considers that Mn exposure through activities such as welding is complicated by several factors, such the fact that welding fume is a complex mixture (NIOSH 2013). Additionally, there are variations in particle size depending on the industry. The largest particles which can be inhaled and deposited in the air passages of the extra thoracic region between the mouth, nose and the larynx are named, inhalable dust fraction; they consist of particles with an aerodynamic diameter up to 100 μm. Smaller particles which are able to reach the gas-exchange region of the lungs form the respirable dust fraction, the limit for entering the alveolar region being between 10 and 15 μm (Wippich et al. 2020). In this case, the European Scientific Committee on Occupational Exposure Limit Values (SCOEL) recommends the use of the respiratory fraction to measure the exposure to Mn, although an Occupational Exposure Limit (OEL) for the inhalable fraction is also obtained; these values are 0.2 mg/m³ (inhalable fraction) and 0.050

mg/m³ (respirable fraction) (SCOEL 2011, Visser et al. 2014). Yet, despite such concerns, welders exposed to airborne Mn at levels below federal occupational safety standards have exhibited neurological problems (Bhandari 2016).

Soil

The major pool of Mn to soils comes from crustal sources, along with direct atmospheric deposition, wash-off from plants and other surfaces, leaching from plant tissues and the shedding or excretion of material such as leaves, dead plants and animal material and animal excrement (Howe et al. 2004). Soils containing abundant Fe usually contain an appreciable amount of Mn (EPA 1995). Land disposal of Mg-containing wastes, mainly coal fly ash, is the principal anthropogenic source of Mg releases to soil. Notably, Mn deposition to soils from the use of MMT in gasoline estimated in two Canadian sites were considered insignificant when compared to natural background Mn levels (541 and 557 mg/kg) in these areas. The average background levels in soils range from around 40 to 900 mg/kg, with an estimated mean concentration of 330 mg/kg (ATSDR 2012).

Soil ingestion may result from putting in the mouth, contacting dirty hands or eating dropped food. Additionally, soil-pica is the recurrent ingestion of unusually high amounts of soil (i.e., on the order of 1,000–5,000 mg/d) usually by children and developmentally delayed individuals, may constitute a potential health concern. An estimated oral average intake of Mn for children from soils in the vicinity of a municipal solid waste incinerator was determined as approximately 0.0021 to 0.0032 mg/kg/d (ERG 2001, ATSDR 2012).

Water

Excessive Mn levels can be found in the aquatic environment in the presence of anoxic water carbon-rich soil, agricultural fertilizers and run-off from sites of human deposition of materials rich in the metal; sea disposal of mine tailings and liquor are relevant sources of Mn to the marine environment (Balzer 1982, Evans and Masullo 2021). Around the world the predominant sources of anthropogenic inputs are domestic wastewater and sewage sludge disposal (Balzer 1982).

Dissolved Mn in natural waters can range from 10 to > 10, 000 µg/L, but rarely exceeding 1,000 µg/L and being usually less than 200 µg/L (Howe et al. 2004). Concentrations of dissolved Mn of up to 4,400 µg/L have been recorded in waters receiving acid mine drainage, and in sediments from an urban lake receiving inputs from industrial and residential areas (Howe et al. 2004). Drinking water is a common source of concern for excessive exposure to Mn, with well-water specifically accruing relevant amounts of the metal via breakdown from adjacent rock beds (Evans and Masullo 2021). Low levels of Mn in source or treated water may also accumulate in the distribution system and periodically lead to high levels of Mn from the tap. Furthermore, Mn may not be adequately removed during treatment and since nearly all Fe ores contain Mn as a contaminant, it should be considered by water treatment plants in which Fe flocculation is used (Chiswell and Huang 2006, MDH 2021,

WHO 2021). Concentrations above 20 µg/L have led to complaints about staining of plumbing fixtures and laundry, water looking bad, smelling or tasting unpleasant . Mn can also create a brownish-black or black stains on toilets, showers, bathtubs, or sinks (MDH 2021, WHO 2021). Values less than 20 µg/L were previously calculated for drinking-water in the USA, but high levels are described for low- or middle-income countries such as Bangladesh, Myanmar, China, India and Costa Rica (WHO 2021). Exposure to excess levels of Mn in drinking water ($\geq$ 0.2 mg/L) may lead to neurological deficits in children, including poor school performance, impaired cognitive function, increased attenttion deficit/hyperactivit y disorder in adults the lowest observable adverse effect level of Mn in water is estimated by the Environmental Protection Agency (EPA) as 0.06 mg/kg/d or 4.2 mg/d for a 70 kg subject (Kim et al. 2015). In the USA, testing public water for Mn levels is not required although some community public water systems do it all the same; the current guidelines suggest a measured Mn level of fewer than 400 µg/L with a value lower than 100 µg/L considered safe for an infant who drinks tap water or drinks formula made with tap water (Evans and Masullo 2021, MDH 2021). The European Drinking Water Regulations 2014 have set a limit of 50 µg/L; however, values above this limit can affect the colour and the taste of the water deterring consumers from drinking it. Even so it is advised that water with Mn levels above 120 µg/L should not be consumed, this limit is to protect everyone in the population since is based on the most vulnerable such as babies in the womb, infants and young children (HSE 2019). The World Health Organization (WHO) sets a provisional health-based guideline value (pGV) of 80 µg/L for total Mn, based on identified health considerations for bottle-fed infants; again, infants being identified as the most susceptible population, this pGV is applicable to protect the general population as a whole (WHO 2021).

Food

Food is the most important source of Mn exposure for the general population and can be one or more orders of magnitude higher than the intake from drinking-water (even assuming a daily water intake of 2 L). However, the estimated total dietary intake of Mn is unlikely to pose a risk to healthy adults from both deficiency and excess (Finley and Davis 1999, Aschner and Erikson 2017, WHO 2021). Mn is available from a wide variety of foods. Plant sources have much higher Mn concentrations than animal sources, containing whole grains (wheat germ, oats and bran), rice and nuts (hazelnuts, almonds and pecans) the highest amounts of Mn; chocolate, tea, mussels, clams, legumes, fruit, leafy vegetables (spinach), seeds (flax, sesame, pumpkin, sunflower, and pine nuts) and spices (chili powder, cloves and saffron) are also rich in the metal (Aschner and Erikson 2017). Measurements of the Mn content in terrestrial plants range from 20 to 500 mg/kg, with blueberries regarded as Mn accumulators exhibiting foliar levels of more than 2,000–4,000 mg/kg. Shellfish were determined as having levels varying from 3 to 660 mg/kg d.w., while concentrations found in tissues of marine and freshwater fish range from < 0.2 to 19 mg/kg d.w.; however, concentrations above 100 mg/kg d.w. have been reported for fish in polluted surface waters (Howe et al. 2004).

Breastfed infants get their necessary nutritional requirements of Mn without the risk of toxicity, on the contrary those whose their main dietary intake is from infant formulas may be receiving higher concentrations of the metal. Particularly soy-based infant formulas contain higher levels of Mn than human breast milk or cow-based formulas; babies who drink such formulas had shown higher concentrations of Mn in hair samples than those who were breastfed (O'Neal and Zheng 2015, Evans and Masullo 2021, WHO 2021). The Food and Drug Administration (FDA) sets a minimum nutritional requirement of 5 µg/100 kcal for the amount of Mn in infant formulas but there is no maximum value, despite recommendations that infants should consume about 3 µg Mn/d for less than 6 mon. Since infants can drink up to a litre of formula a day, when the formula is prepared according to the manufacturer's instructions, they can consume from 32 to 51 µg/d of Mn, far exceeding this recommendation (O'Neal and Zheng 2015). The WHO/Food and Agriculture Organization of the United Nations Codex Committee and the Expert Panel of the Life Science Research Office have a different position, setting both minimum and maximum guidance levels of Mn for infant formula intended to be marketed as breast milk substitute. The minimum guidance level is 1 µg/100 kcal and the upper level 100 µg/100 kcal (67 µg/100 mL). An additional risk is the fact that some well-meaning but inadequately informed parents may perceive plant-based beverages, such as soy or rice beverages, as a good alternative to infant formula; however, Mn is found at high levels in both plants and calculated mean intakes from these beverages by infants up to 6 mon of age (assuming a complete substitution) approaches the Tolerable Upper Intake Level for 1 to 3 yr old subjects, may presenting an increased risk of adverse neurological effects (Cockell et al. 2004).

Regarding exposures in adults, exposure to Mn from food is estimated to vary between 2 and 6 mg/d , although higher values have been reported for people eating vegetarian diets, which may have Mn intakes as high as 10.9 mg/d (WHO 2021). Additional causes can lead to increased Mn exposure through diet: enhanced absorption as a consequence of dietary interactions, namely with Fe, and depressed Mn excretion as a consequence of biliary insufficiency (Finley and Davis 1999). The United States Food and Nutrition Board and the Institute of Medicine's Dietary Reference Intakes (DRI) cites ~2 mg/d as an adequate intake of Mn for adults (and 1.2–1.5 mg/d for children) (Aschner and Erikson 2017). The European Food Safety Authority (EFSA) proposes 3 mg/d for all adults, including pregnant and lactating women, 3 mg/d for adolescents, and 0.5 mg/d for children aged 1 to 3 yr (WHO 2021). The EPA concluded that an appropriate reference dose for oral Mn is 10 mg/d (0.14 mg/Kg/d) and considers a Tolerable Upper Intake Level of 9 to 11 mg/d for adults and 2–6 mg/d for children, which varies with the age (Chiswell and Huang 2006, Kim et al. 2015, Aschner and Erikson 2017). Considering the lack of information about the dietary intake of Mn, an interim guidance value of 0.16 mg/kg/d (based on the tolerable upper intake level for 70 kg adults of 11 mg Mn/d) was recommended by the Agency for Toxic Substances and Disease Registry (ATSDR) (Kim et al. 2015).

Mn present in many different forms in dietary supplements is a potential cause for concern. Not all multivitamin/mineral supplements contain Mn, but those that do typically provide 1.0 to 4.5 mg of the metal. Most supplements containing only Mn, or Mn with a few other nutrients, contain 5 to 20 mg (Aschner and Erikson 2017, NIH 2021). Parenteral Nutrition (PN) may also lead to excessive exposure to Mn due to the high bioavailability of the metal because normal regulatory mechanisms in the gastrointestinal tract are bypassed. Therefore, while PN administration is a potentially lifesaving therapy for patients who cannot tolerate enteral nutrition, at the same time could lead to a risk; such patients need to be closely monitored (Khan et al. 2020). Mn toxicity was reported in patients receiving long-term PN containing about 1 mg/da of parenteral Mn (adults) or more than 40 µg/kg/d (children) (Crossgrove and Zheng 2004). Patient-specific supplementation with Mn is difficult to accomplish in the clinical setting, since this trace element is routinely provided in a fixed-dose commercially available product which is added to PN preparations with other essential and trace vitamins and minerals. There have been significant variations on the recommended daily dosing for Mn supplementation by this route, although the American Society for Parenteral and Enteral Nutrition (ASPEN) guidelines recommend a daily dose of 55 µg/d of Mn for PN and the European Society for Parenteral and Enteral Nutrition (ESPEN) indicate that the recommended daily dose of Mn in adults should be up to 1 µg/kg/d (Reinert and Forbes 2021).

Drug abuse

The street drug "Bazooka" is another source of Mn exposure and consists of a cheaper cocaine-based product, which can be obtained at an early stage of the coca-leaf extraction process; during one of the steps of the extraction potassium permanganate is used and precipitates when added to the cocaine base (usually 70–90%) (Ensing 1985, ATSDR 2012, Kim et al. 2015). Potassium permanganate is also added as an oxidizing agent to ephedrone, which a psychostimulant drug that increases the release of catecholamines in the brain and induces behavioural effects similar to the ones induced by methamphetamine. Several Mn-induced Parkinsonism cases are reported among intravenous abusers in many areas of the world, and even after cessation of ephedrone use, some of the motor symptoms may continue to progress (Sikk et al. 2011, O'Neal and Zheng 2015). Since in these drug cocktails the final Mn concentration can be as high as 0.6 g/L, multiple injections per day can result in doses ranging from 60 to 180 mg/d , far exceeding the 0.1 mg Mn/d recommended levels as an intravenous supplement (O'Neal and Zheng 2015).

Essentiality and Toxicity

Mn is an essential element implicated in the metabolism of proteins, lipids and carbohydrates. Mn also plays unique roles that cannot be replaced by other metals, which include the activity of Mn-dependent enzymes such as arginase, agmatinase, glutamine synthetase and Mn superoxide dismutase (Balachandran et al. 2020, WHO 2021). Many of these enzymes play a role in several biological processes like bone formation, free radical defence, neurotransmitter synthesis and ammonia clearance in the brain (WHO 2021). Additionally, Mn is important for the immune function, biochemical regulation of energy consumption, coagulation and haemostatic function (Evans and Masullo 2021). The metal is also required for growth and development, including the development of the nervous system especially in early life (WHO 2021).

Deficiency

Evidence about Mn deficiency in humans is rare, and no specific Mn deficiency syndrome has been described so far. In animals this condition can lead to growth alterations, skeletal abnormalities, reproductive defects, ataxia and lipid and carbohydrate metabolism defects; effects such as hair colour change from black to red, slow nail growth and scaly dermatitis have also been observed, although only experimentally (Rondanelli et al. 2021).

Toxicity

Very little data exist on acute Mn toxicity, although a case of 12 oz. of MMT accidental ingestion is reported as leading to severe but reversible neurotoxicity, with seizures and altered mental status which 1 d after intubation disappeared (Nemanich et al. 2021). A case of ingestion of hydrated Mn sulphate for diet purposes (three tablespoons daily, total duration unknown) induced lethargy, vomiting, abdominal pain, profuse diarrhoea, liver failure, acute renal injury, acute respiratory distress, myocardial dysfunction, shock with lactic acidosis and death within 72 hr (WHO 2021).

Children are particularly susceptible to Mn toxicity given their decreased functionality of excretion mechanisms, increased ability of the gastrointestinal tract to absorb the metal and increased permeability of the blood-brain barrier to Mn, as compared with adults. In countries with high Mn levels in water sources, children are prone to deficiencies in IQ scores, memory, reasoning and general academic achievement (Evans and Masullo 2021). Increased infant mortality, namely in the first year of life, is described in Bangladesh due to exposure to Mn concentrations at or above standard values (O'Neal and Zheng 2015).

However, most of the reported cases of Mn toxicity occur in individuals repeatedly exposed to high concentrations of airborne Mn (> 5 mg/m^3) in occupational settings (Keen and Zidenberg-Cherr 2003, O'Neal and Zheng 2015, Evans and Masullo, 2021). Available evidence suggests that oxidative stress caused by Mn can induce selective brain damage leading to functional disability and impairment of motor functions (Sharma et al. 2021). Affected workers frequently show abnormal accumulations of Mn in a region of the brain known as the *globus pallidus*, which is an area in the brain that plays an important role in movement regulation (NIOSH 2103). The early onset of Mn intoxication is usually subtle and the symptoms of Mn intoxication, once established, usually become progressive and irreversible, reflecting permanent damage to neurological structures (Cowan et al. 2009). In its most severe form , the toxicosis is manifested by a permanent crippling neurological disorder of the extrapyramidal system, which is similar to Parkinson's disease. In its milder form, the toxicity is expressed by hyperirritability, violent acts, hallucinations, disturbances of libido and incoordination. The earlier symptoms, once established, can persist even after the Mn body burden returns to normal. Subtle signs of Mn toxicity, which included delayed reaction time, impaired motor coordination and impaired memory, were observed in workers exposed to airborne Mn concentrations lower than 1 mg/ m³ (Keen and Zidenberg-Cherr 2003). The fact is that even low-level occupational exposure, with air Mn concentrations at or below occupational standards, can be detrimental and neurochemical, neurobehavioral and neuroendocrine changes may occur before structural damage; this could occur at occupational exposures to welding fumes <0.2 mg/m3 (NIOSH 2103, O'Neal and Zheng 2015). Notably, manganism and parkinsonism are considered two different manifestations of Mn exposure, with each neurodegenerative manifestation depending on the intensity, duration and route of exposure. Manganism is the result of high acute exposure, whereas parkinsonism happens due to low chronic exposure (Tarale et al. 2016).

Damages in other systems of the human body may also be provoked by Mn, such as the reproductive and immune system dysfunction, nephritis, testicular damage, pancreatitis, lung disease and hepatic damage, but frequencies are unknown (Keen and Zidenberg-Cherr 2003). Animal and human data also suggests also that Mn exposure significantly alters cardiovascular function, although the exact mechanism of cardiac toxicity remains unknown (O'Neal and Zheng 2015). While there is a limited number of epidemiological data suggesting that high levels of Mn can result in an increased risk for colourectal and digestive tract cancers, most investigators do not consider Mn to be a carcinogen (Keen and Zidenberg-Cherr 2003).

Biomarkers

Biological monitoring is an important tool in the prevention of diseases related to those exposed to chemicals such as Mn on a regular basis. However, in occupational situations where exposures to Mn causing apprehension are likely to occur, biomarkers (BMs) are used, but no reliable one has been established to evaluate these exposures on an individual basis, or to evaluate the effects, as a complete scientific understanding of the mechanism of Mn induced toxicity remains undiscovered (Bahrami et al. 2015). As regards, to non-occupational settings, there is no strong evidence of a matrix that may serve as a valid BM of Mn exposure (Shilnikova et al. 2022). On the contrary, some authors have been defending the reliability of some of these same BMs.

Biomarkers of exposure

The measurement of Mn in biological fluids and tissues, such as blood, serum, plasma, urine, and less frequently hair, nails, cerebrospinal fluid, faeces and saliva, have been suggested as biomarkers (BM) of exposure, with blood and urinary Mn the most widely investigated (Flora 2014).

Manganese levels

Blood Blood Mn levels are often used as a bioindicator of exposure to Mn in epidemiological studies investigating Mn toxicity and are considered by some researchers (although not by others) as a useful proxy indicator of target organ Mn dose in occupational health; clinically, this BM poorly reflects the body burden of Mn and the disease status (Choi et al. 2005, Oulhote et al. 2014, Kim et al. 2015, Evans and Masullo 2021). Blood Mn concentration reflects only recent active exposure and while, while serving as a reasonable indicator of exposure on a group basis, appears be a modest indicator for distinguishing Mn-exposed subjects from control subjects at the individual level (Zheng et al. 2011). Another limitation of this BM is the significant discrepancy between blood half-life and tissues, which in blood is shorter (Zheng et al. 2011, O'Neal and Zheng 2015). A literature-based analysis explored the relationship between Mn levels in blood and Mn levels in air, over a number of exposure levels from different environments. There was a positive association between these parameters, when air concentrations were approximately 10 $\mu g/m^3$ or above, suggesting that there may be a degree of exposure above which Mn levels in the blood begin to act as an exposure BM for inhaled Mn. Possibly, below this concentration the Mn in the body is dominated by dietary Mn and causes only negligible changes in Mn blood levels (Baker et al. 2014). In spite of little data available on Mn levels in the blood of the general population, normal whole blood Mn levels are described as ranging from 7 to 12$\mu g/L$ (Flora 2014, Oulhote et al. 2014).

Most of the Mn content of whole blood is in the cellular components, containing erythrocytes 65% of the metal; this leads one to consider that erythrocytes represent a Mn compartment more akin to target tissues than serum, and may be a more sensitive indicator of the Mn body burden (Choi et al. 2005). The feasibility of using the Mn content in erythrocytes as a BM of Mn exposure was studied in a ferroalloy smelting factory, and workers exhibited Mn levels in erythrocytes higher than controls (Zheng et al. 2011); however, in a similar work, smelters did not exhibited significantly augmented levels of this BM, although erythrocytic Mn concentrations were significantly correlated with the palladial index obtained from Magnetic Resonance Imaging (MRI); this could indicate that increased erythrocytic Mn concentrations reflect Mn accumulation within the central nervous system (Zheng et al. 2011). Accordingly, Mn levels in erythrocytes (as well as in whole blood) were shown to reflect signal intensities of MRI better than Mn levels in plasma or urine in liver cirrhotic patients (Choi et al. 2005). The average concentration of the metal in these cells is documented as 22 ± 7.4 $\mu g/L$ (Maynar et al. 2020).

Normal ranges of 0.4–0.85 $\mu g/L$ in serum have been mentioned as raging 0.6 to 4.3 $\mu g/L$ (Cowan et al. 2009, Flora 2014). Elevated serum Mn concentration can be seen among active Mn-exposed welders, but without association with the years of employment. However, plasma Mn concentrations were also seen to begin declining although Mn exposure was still ongoing (Cowan et al. 2009, O'Neal and Zheng 2015). Therefore, the use of serum Mn levels as BM of exposure has proved to be difficult.

Urine Since Mn is mainly excreted in faeces via biliary excretion, and only in a very small proportion by urine, measuring urinary Mn is of little interest as BM of exposure (Flora 2014, Kim et al. 2015). Moreover, urinary Mn can be only indicative of recent exposures (only a few hours) because the metal can exit the body relatively quickly (Wu et al. 2017, WHO 2021). Little or no significant changes in urinary Mn have been observed in many studies and some authors have recommended abandoning Mn urinary levels as a BM of Mn exposure. Normal levels of Mn in urine range from 1 to 8 µg/L and in faeces varies widely among exposed workers, between 0.07 and 15.9 µg/g (Zheng et al. 2011, WHO 2021).

Saliva Saliva samples from groups of Mn-exposed welders were previously found to exhibit significantly higher Mn concentrations than controls, but the variation was only partly associated with the year of employment and had a much greater variation among tested subjects than in the plasma (Zheng et al. 2011). Another study involving persons with the same occupation, while the saliva Mn concentrations mirrored those of serum Mn levels, a fairly large variation in saliva Mn levels led the authors to not recommend the use of saliva Mn levels to assess Mn exposure (Flora 2014, O'Neal and Zheng 2015).

Hair and nails Hair and toenail Mn determinations have shown to be promising, even though the toxicokinetics of incorporation of Mn in these matrices are not yet fully elucidated (Ward et al. 2017). Hair may integrate or reflect exposures over longer timeframes than blood or urine (Eastman et al. 2013). The average value of Mn concentrations in the hair of exposed workers tend to increase with the year of employment, leading to believe that this BM can reflect Mn body burden (Zheng et al. 2011). Significant increases of Mn concentrations (> 300%) were also detected in axillary hair of workers in a dry-cell battery plant, when compared to a control population (Cowan et al. 2009). Studies in environmentally exposed children reported additionally that hair Mn (but not blood Mn) was a predictor of exposure and neurotoxic outcomes (Eastman et al. 2013). However, Mn concentrations in hair were already found to be significantly associated with Mn intake from water, but not with the dietary intake (Signes-Pastor et al. 2019, Kousa et al. 2021).

The mean values of Mn concentrations in hair are described as 0.121 µg/g, ranging 0.011 to 0.736 µg/g, but great differences in environmentally exposed subjects exist; to illustrate, Mn levels in hair samples of 32 children residing near a hazardous waste site, ranged widely (from 89.1 to 2,145.3 ppb) (ATSDR 2012, Eastman et al. 2013). Studies using hair samples have often used different methods for cleaning hair to resolve exogenous metal contamination prior to analysis, which may explain these differences (Jursa et al. 2018).

Nail Mn concentrations may reflect exposure over longer periods of time than hair, with correlations reported between Mn levels in toenails and individual exposure of 7 to 12 mon prior to the clipping of the toenails (Flora 2014, Ward et al. 2017). Besides, this BM seems to distinguish between subjects who are exposed on average above and below the ACGIH TLV (Ward et al. 2017). As regards to environmental exposures, maternal toenail samples were alleged to be a reliable BM of environmental exposure to Mn, and as reflecting exposure from drinking water, especially at higher Mn water concentrations (Signes-Pastor et al. 2019).

Bones The relatively long half-life of Mn in the skeletal system (8–9 yr) renders bone Mn concentrations an ideal indicator to assess the body burden (O'Neal and Zheng 2015). In this perspective, Neutron Activation Analysis (NAA) was already used as a testing system to quantify the Mn content in bones; results obtained from Mn quantification in the bones of hands suggested the viability of future studies using this methodology (Evans and Masullo 2021). Moreover, this is a non-invasive technique that allows a real-time quantification and can determine Mn concentrations as low as 0.5 mg/kg (O'Neal and Zheng 2015).

Other biomarkers

Ratio Mn/Fe Substantial evidence exists that Fe and Mn compete for absorption into mucosal cells because the Divalent Metal Transporter protein-1 (DMT1) is a transporter for both; increased absorption of Mn in Fe-deficient subjects can account for an observed fivefold increase in blood Mn. Additionally, transferrin (the major transport protein for Fe) has also been involved in the transport of Mn (Rahman et al. 2013). Since changes of Mn and Fe concentrations in the biological media rely on opposite directions, combining these two measurements into one could widen the differences between exposed and control subjects, thereby increasing the sensitivity of Mn-exposure assessment. In an occupational context, differences among controls, low or high exposure groups, were found to be higher after determining the ratio Mn/Fe (MIR) in erythrocytes or in the plasma, as compared with Mn concentrations alone in both biological samples. It was established that MIR was an effective BM for distinguishing Mn-exposed workers from the control subjects; additionally, correlations with airborne Mn levels suggested that the BM could reflect external Mn exposure (Cowan et al. 2009, Zheng et al. 2011).

Prolactin Dopamine is a hormone and neurotransmitter, with prolactin being an indirect indicator of dopaminergic function and homovanillic acid a downstream metabolite of dopamine; all three molecules have been tested as possible BMs for Mn exposure but with controversial results. As regards prolactin, increased levels have been observed in Mn-exposed male workers with previously manifested neurotoxicity's, whereas any significant alteration was found in workers engaged in dry-cell battery production. So far, the information on these molecules is limited and so far, inconclusive (Cowan et al. 2009, Zheng et al. 2011).

Magnetic Resonance Imaging The paramagnetic properties of Mn allow using non-invasive imaging techniques such as Magnetic Resonance Imaging (MRI), which can be used for pinpointing Mn accumulation in brain tissues (Zheng et al. 2011). MRI is based on the magnetization properties of atomic nuclei and in this technique, an external magnetic field is used to align the protons that are normally randomly oriented within the water nuclei of the tissue being examined. This alignment is then perturbed by introduction of an external radio frequency. As the nuclei return to their resting alignment through various relaxation processes, they emit radiofrequency energy. In MRI, tissues can be characterized by two different relaxation times, T1 (longitudinal relaxation time) and T2 (transverse relaxation time). The first one is a measure of the time taken for spinning protons to realign with the external magnetic field, while the second one is a measure of the time taken for spinning protons to lose phase coherence among the nuclei spinning perpendicular to the main field (Preston 2016). Deposition of Mn in the brain causes distinct MRI brain appearances with pronounced signals in changes the *globus pallidus*, which are characterized by hyperintensity on T1-weighted images and hypo intensity on T2-weighted images; leads to considering that MRI may be used as a non-invasive examination to detect Mn accumulation in the brain (Zheng et al. 2011, Li et al. 2014, Anagianni and Tuschl 2019). Accordingly, being the *striatum* and the *globus pallidus* the primary targets of Mn accumulation in the brain, brain MRI shows T1-weighted intensity abnormal signal enhancement in both brain areas *in vivo* and in asymptomatic occupationally Mn-exposed workers (Li et al. 2014, Lao et al. 2017). Researchers have also used the Pallidal Index (PI) as a semi-quantitative parameter to evaluate Mn accumulation in the brain, which is calculated as the signal intensity ratio of *globus pallidus* relative to the frontal white matter on T1-weighted intensity planes, multiplied by 100 (Li et al. 2014). PI has been used widely as a semi-quantitative indicator of brain Mn status in several human studies, discriminating between exposed and not exposed subjects, even when no clinical symptoms are evident (Zheng et al. 2011, O'Neal and Zheng 2015). However, limitations exist since values vary significantly between studies, and are not a proportional measure of the Mn concentration in the brain (Li et al. 2014). Some researchers believe that the PI value with reference on the neck muscle rather than the frontal cortex, reflect Mn accumulation in the brain more accurately. Additionally, PI calculated from high-resolution 3D T1-weighted MRI correlates better with air Mn concentrations and neurobehavioral performance indicators, when compared to the PI calculated from lower resolution images (Zheng et al. 2011).

Biomarkers of susceptibility

There are many factors which may influence a subject susceptibility to Mn exposures, including age, gender, ethnicity, pre-existing medical conditions and genetics.

Both younger and older subjects are more susceptible due to physiological features such as differences in absorption or excretion; women of all ethnicities are more prone to higher Mn burden due to sex-related metabolic differences in the regulation of Mn; the Asian population tend to accumulate significantly more Mn than non-Hispanic Caucasians or non-Hispanic Black individuals. Also patients with chronic liver disease, neurological disease or with Fe deficiency (implying enhanced Mn absorption in the gastrointestinal tract), have increased risk of developing Mn toxicity (Moreno et al. 2009, Oulhote et al. 2014, O'Neal and Zheng 2015, Rechtman et al. 2022).

As regards genetic or epigenetic markers, no reliable one has been validated to evaluate host susceptibility to the metal because a complete scientific understanding of their mechanism of toxicity remains undiscovered (Zheng et al. 2011).

Genetic biomarkers

Some contenders exist which include Single Nucleotide Polymorphisms (SNPs) in genes involved in Mn detoxification (Tarale et al. 2016).

CYP2D6 For instance, the human genome comprises several putatively functional protein-coding cytochrome P450 (CYP) genes; CYPs is the major enzyme family primarily found in liver cells (Shankar and Mehendale 2014, Tarale et al. 2016). Among these enzymes, CYP2D6 has shown the greatest impact on many xenobiotics due to its wide spectrum of genetic variants, from null alleles to several-fold gene amplification and its broad substrate selectivity (Kim et al. 2015). Namely, the SNP at the CYP2D6 (C > T 2850, rs16947) is associated with interindividual and interethnic differences in the metabolism and disposition of Mn, as evinced by some miners in central India, for whom this genotype could be involved in a faster metabolism of blood Mn, compared to wild and heterozygous genotypes (Vinayagamoorthy et al. 2010, Kim et al. 2015). Therefore, it has been accepted that possibly cytochrome P450 CYP2D6 can be considered as a susceptibility BM for identifying miners who are at a higher risk for developing manganism, although the mechanism through which their variants influence susceptibility to Mn is not well understood (Tarale et al. 2016).

SLC30A10 Mutations in the Solute Carrier Family 30 Member 10 Protein Coding gene (SLC30A10) are the first example of a genetic cause correlating neurological symptoms directly with Mn accumulation in humans; this was demonstrated in 2012 (Kim et al. 2015). The disease results from a loss of function of the protein encoded by SLC30A10, this protein being a member of the cation diffusion facilitator superfamily of metal transporters and an important and evolutionarily conserved Mn transporter (Quadri et al. 2015, Anagianni and Tuschl 2019, Mercadante et al. 2019). The protein is largely expressed in the liver and is present in neuronal cells of the *globus pallidus* under normal conditions and seems to be a key determinant of body Mn levels; implicated in regulating Mn export from cells, protecting against Mn-induced neurotoxicity (Kim et al. 2015, O'Neal and Zheng 2015, Anagianni and Tuschl 2019, Mercadante et al. 2019). Additional studies revealed its essentiality for hepatobiliary Mn excretion and significant contribution to intestinal Mn excretion. It is also known that elevated Mn levels increase the expression of SLC30A10 (Anagianni and Tuschl 2019, Liu et al. 2021). An autosomal-recessive mutation in this transport protein leads to an inherited hypermanganesemia. This results from its absence in the liver with ensuing impaired Mn efflux activity due to mislocalization of the transporter, and subsequent intracellular accumulation of the metal (Kim et al. 2015, Anagianni and Tuschl 2019). Results in a pleomorphic phenotype have included dystonia and adult-onset Parkinsonism and brain Mn excess, even without a history of environmental exposure (O'Neal and Zheng 2015, Anagianni and Tuschl 2019). Additionally, because Mn and Fe compete for binding at several transporters, it is not surprising that Fe stores are depleted in individuals with these SLC30A10 mutations, showing increased total Fe-binding capacity and low ferritin (Anagianni and Tuschl 2019).

Parkin/PARK2 Loss of functions of Parkin/PARK2 due to recessive mutations are reported to be responsible for early onset of Parkinsonism and its possible use as a BM for Mn susceptibility has also been studied (Tarale et al. 2016). The PARK2 gene encodes the Parkin protein, which is an E3 ubiquitin protein ligase thought to be involved both in protein degradation and in oxidative stress protection (Kim et al. 2015). Parkin-mediated ubiquitination promotes mitochondrial autophagy or degradation of target proteins, including the transporter DMT1 which plays a significant role in transporting Mn; since DMT1 is a substrate for Parkin, the last is involved in maintaining a steady state level of the transporter (Konovalova et al. 2015, Tarale et al. 2016, Fan et al. 2017, Zhong et al. 2020). Loss-of-function mutations in the PARK2 gene makes cells susceptible to elevated uptake of Mn via DMT1, whereas overexpression of PARK2 is shown to rescue dopaminergic neurons from cell death caused by Mn toxicity (Kim et al. 2015. Tarale et al. 2016, Fan et al. 2017). Workers occupationally exposed to Mn in their daily work were found to exhibit decreased expression of this gene, which may contribute (at least partially) to the mechanism of Mn-induced Parkinsonian disorder (Fan et al. 2017, Cao et al. 2022). Remarkably after cessation of an experimental Mn exposure *in vivo*, the amount of Park2 mRNA in the blood was observed to increase 1 mon after the recovery; after 5 mon blood and brain Mn levels returned to normal, rotarod activity recovered, and the level of Park2 mRNA in the blood and Park2/Parkin in the midbrain and *striatum* also returned to the normal (Cao et al. 2022). Hence, PARK2 mutations are proposed to be used as a new BM of susceptibility to Mn, while other researchers propose the use of Park2 mRNA in the blood as a novel BM for Mn exposure and recovery (Fan et al. 2017, Cao et al. 2022).

ATP13A2 ATP13A2 (also named PARK9) encodes a lysosomal P5-type ATPase that plays an important role in regulating cation homeostasis, and for which there is a possible involvement in the regulation of metal toxicity (Tan et al. 2011, Ugolino et al. 2019). ATP13A2 mutations were first associated to a very rare autosomal recessive juvenile Parkinsonism [the Kufor-Rakeb Syndrome (KRS) with dementia phenotype] but are also found in patients with various other types of Parkinsonism (Tan et al. 2011, Glykys and Sims 2017). ATP13A2 defective cells appear to impair α-synuclein degradation and would presumably lead to to brain α-synuclein accumulation and toxicity (Glykys and Sims 2017). Notably, while exposure to Mn is an environmental risk factor for Parkinson's Disease (PD), KRS pathogenic ATP13A2 cell mutants are known to be not able to protect cells from Mn induced cell death *in vitro*, through reducing intracellular Mn concentrations (Tan et al. 2011, Aboud et al. 2012). Additionally, taking into account that PD is characterized by loss of dopaminergic neurons, experiments with transgenic *C. elegans* overexpressing GFP-tagged human ATP13A2 protein in dopamine neurons were conducted. It was seen that the nematodes exhibited higher resistance to dopamine neuron degeneration after acute exposure to Mn as compared to controls (Ugolino et al. 2019). Additionally, the SNP rs2871776 G allele in ATP13A2 is identified as associated with an adverse effect of Mn on motor coordination, through the absence of a binding site for the transcription factor INSM1 that plays an important role in developing the central nervous system;

adult carriers of this genotype perform more poorly in motor coordination tests after Mn exposure (Kim et al. 2015).

Negative results Other gene polymorphisms that were already considered candidates for susceptibility BM due to their role in detoxification processes in which Mn mediated toxicity could be involved: glutathione transferases (GSTs) and NAD(P) H: quinone oxidoreductase (NQO1); however, studies with these genes led to negative results. GSTs are a family of multifunctional proteins that act as enzymes and as binding proteins in various detoxification processes and cell metabolism; the participation of these molecules in the detoxification of metals and products of oxidative stress is broadly reported, in the same way as oxidative stress has been implicated as a contributing mechanism by which Mn can be toxic to cells (Tsuchida and Sato 1992, Erikson et al. 2004, Kim et al. 2015). GSTM1 is one of these isozymes and its role in detoxifying reactive chemical species is played by catalyzing their conjugation to glutathione (GSH), which is the most abundant intracellular antioxidant found in many kinds of cells and tissues (Kim et al. 2015, Gad 2014). The isozyme GSTM1 has a common polymorphism (the null allele) that is present in about 50% of the white population and 30–70% in other racial/ethnic groups and causes lack of enzymatic activity (Landi et al. 2007). However, Mn-exposed miners with the GSTM1-null genotype show blood levels of Mn similar to those found in other genotypes leading to believe that GSTM1 polymorphisms possibly do not have an important role for Mn detoxification (Kim et al. 2015).

NAD(P)H: quinone oxidoreductase (NQO1), is a flavoenzyme that plays a role in protecting against quinones, protecting cells from oxidative damage (Nebert et al. 2002). Mn promotes dopamine auto-oxidation with formation of quinones and free radicals, which increases dopamine toxicity in high Mn-accumulating areas of the brain (Tarale et al. 2016, Harischandra et al. 2019). But again, several studies in occupational environments have led to the belief that NQO1 genetic polymorphism does not influence the accumulation of Mn (Zheng et al. 1999, 2002, Kim et al. 2015).

Epigenetic biomarkers

With regards to epigenetic alterations, Mn is known to increase the production of ROS in a catalytic fashion via redox cycling; in turn, oxidative DNA damage can interfere with the ability of methyltransferases to interact with DNA; these events result in a generalized altered methylation of cytosine residues at CpG sites (Kim et al. 2015, Tarale et al. 2017). Mn exposure leads to modifications in DNA methylation status of genes involved in several important biological pathways, and their perturbation may underlie neurotoxic mechanisms of the metal. These pathways include the regulation of neuronal differentiation and development, synaptic transmission, signal transduction, inflammation and programmed cell death; the affected genes are namely, PARK2 and PINK1, which are recognized as important contenders in the aetiology of PD and have a significant role in proteasomal degradation, protein kinase activity and dopamine biosynthesis. It is also suggested that epigenetic alterations of PINK1/

PARK2 could contribute to the Mn-mediated mitochondrial dysfunction (Tarale et al. 2017).

Studies with neuronal cultures revealed additionally complex Mn-mediated miRNA changes attributed to the fact that Mn can alter miRNA expression and release; some miRNAs demonstrate elevated expression while others are downregulated. To cite a few examples, miRNAs proinflammatory responses (such as TNFα and IL-6) downregulates, raising inflammation, whereas expression of hsa-miR-4306, which targets PARK9 (a suppressor of α-synuclein production and activity), increases it (Wallace et al. 2020).

Biomarkers of effect

Several biochemical and physiological effects of Mn exposure, based primarily on the hypothesized mechanisms of Mn toxicity, have been suggested as markers for monitoring Mn toxicity (Zheng et al. 2011).

Manganoproteins

Mn is a cofactor for manganoproteins which have been examined as potential markers. Glutamine Synthetase (GS) is one of them, selectively expressed in astroglial cells where the conversion of glutamate to glutamine is catalysed by GS preventing an increase of extracellular glutamate levels and glutamate-dependent overexcitation (Kim et al. 2015). Overexpression of GS mRNA reflects changes in intracellular levels of Mn, it is possible that Mn-potentiated cellular Fe overload while triggering oxidative stress, could inhibit the enzyme activity (which in turn stimulates its synthesis) (Flora 2014, Zheng et al. 2011). A study con ducted to indicate whether chronic Mn overload could affect the expression of GS *in vivo* showed a reduction of GS in astrocytes (Kim et al. 2015). However, several inconsistencies among results from different studies do not seem to support the use of GS as an indicator of Mn-induced toxic effect. Reports exist on increased GS levels in the hippocampus, but also decreased level in the hypothalamus on Mn inhalation; Mn-induced reduction of GS in cerebellum and decreased GS mRNA in the *striatum* are also described (Zheng et al. 2011).

Superoxide dismutase (Mn-SOD) is another manganoprotein, present in neurons, which regulates and detoxifies superoxide, an extremely powerful oxidant and a by-product of the cellular metabolism (Kim et al. 2015, Kitada et al. 2020). The role of Mn-SOD in preventing Mn toxicity is quite relevant, because on entering the brain the metal enhances the production of ROS toxicity (Martinez-Finley et al. 2013). As reported for GS, it is possible that the expression and activity of Mn-SOD are regulated by changes in the cellular levels of Mn; diminished Mn-SOD in neurons and neuroblastoma cells due to Mn overload, was shown earlier (Chtourou et al. 2011, Kim et al. 2015).

Metallothioneins

Metallothioneins (MTs) as metal-binding proteins have an important role as antioxidants. While it is reported that toxic concentrations of Mn decrease the gene expression level of MT, other studies challenged the interpretation of these results (Zheng et al. 2011, Kim et al. 2015). MT-1 expression was found to unexpectedly decrease in Mn-treated astrocytes, unlike its general increase in response to elevated intracellular divalent metals. Possibly increased intracellular Mn might diminish the content of the other divalent metals modulators of MT expression (such as zinc or copper) which exerts feedback inhibitions, thus decreasing MT mRNA expressions. Results from animal studies noted a significant decrease in MT mRNA in brain tissues after Mn exposure, which is in agreement in this interpretation. While, increased MT mRNA or any effect in brain MT levels of young rats exposed to Mn is also documented. Age and gender seem additionally to be substantial factors when assessing the neurotoxicity of Mn, as Mn-induced decrease of MT mRNA was already observed at the cerebellum, olfactory bulb and hippocampus in young male rats, but in a different way, such decreases were detected in the hypothalamus in the young female rats and in the hippocampus of senescent males (Madison et al. 2011, Zheng et al. 2011).

Divalent metal transporter 1

Mn is mostly found in the divalent form, being the Divalent Metal Transporter 1 (DMT1) the most important transporter of this metal (Kim et al. 2015, Wolff et al. 2018); DMT1 is largely present in the basal ganglia, which is a target area for both Parkinsonism and Mn toxicity (Ingrassia et al. 2012, Kwakye et al. 2015). Cultured choroidal epithelial cells challenged with Mn exhibit increased DMT1 levels, while increased DMT1 mRNA concentrations were Mn dose-dependent and exposure time-associated. It is plausible that Mn can act at transcriptional levels to increase DMT1 RNA production and/or RNA processing, or at the post-transcriptional level by blocking DMT1 mRNA degradation (Kim et al. 2015).

Serum Mn citrate

Serum Mn citrate has been suggested as a BM of elevated risk of Mn-dependent neurological disorders in occupational health (O'Neal and Zheng 2015). While in serum, Mn is mainly associated with proteins like Tf and only at slight amounts to small carriers such as citrate, in cerebrospinal fluid (CSF) Mn-citrate is identified as the most important Mn-species. When total serum Mn exceeds 1.6 µg/L, a positive linear relationship to Mn-citrate in CSF or serum can be found, allowing the use of the latest to estimate the Mn load in brain and CSF and hence, access an increased risk for internal Mn exposure of the brain (Michalke et al. 2015).

Imaging techniques

Magnetic resonance spectroscopy (MRS) is a useful technique to quantify neurochemical markers associated with Mn exposure and appear capable of detecting biochemical changes before full-blown manganism symptoms become evident. Studies using this technique found that in the thalamus and basal ganglia of Mn-exposed smelters, the levels of GABA nearly doubled even when the mean airborne Mn levels were below the occupational standard; this BM could indicate early metabolic or pathological change associated with low-level Mn exposure (O'Neal and Zheng 2015). N-acetyl-aspartate which is a molecule synthesized in neurons, and has been used as a neuronal integrity marker, were also found to be significantly decreased in the frontal cortex of exposed subjects; this change was concomitant with an 82% increase of GABA levels in the thalamus region. The authors proposed that the PI and the GABA level combination may be a powerful, non-invasive BM for both Mn exposure and pre-symptomatic Mn neurotoxicity (Zheng et al. 2011).

Positron Emission Tomography (PET) scans allow access the uptake of 6-[18F] fluorodopa by dopaminergic neurons, which are often significantly compromised in patients with Parkinsons . This approach has also been explored although so far, results for Mn toxicity studies have been inconsistent for distinguishing manganism from idiopathic PD (Zheng et al. 2011).

Final Remarks

This chapter described the relation of Mn with humans, as an essential metal, for which deficiency cases are not very common, but toxicity due to excessive exposures is likely to occur. Major concerns comprise chronic exposures in occupation, pesticides such as maneb and mancozeb, infant formula or parenteral nutrition, and possibly a near future the gasoline additive methylcyclopentadienyl Mn tricarbonyl. The brain is the primary target for Mn, with manganism resulting from high acute exposure, whereas Parkinsonism happens due to low chronic exposure. With more or less acceptance Mn levels in blood or their fractions, hair or nails have been used as exposure BMs. More recently the ratio Mn/Fe to access exposures or MRI for Mn accumulation in brain tissues, revealed to be very promising. Mutations with interest for susceptibility BMs are namely the ones found in cytochrome P450 CYP2D6 possibly may identify miners who are at higher risk for developing manganism, or in the SLC30A10 gene, which is the first example of a genetic cause correlating neurological symptoms directly with Mn accumulation in humans. Affected functions of PARK2 or PARK9 genes, earlier associated with Parkinsonism, have also been studied as possible BM for Mn susceptibility, as well as epigenetic alterations of PINK1/PARK2 genes Mn-mediated miRNA changes. Several biochemical and physiological effects of Mn exposure, based primarily on the hypothesized mechanisms of Mn toxicity, have been suggested as markers for monitoring Mn toxicity. These include manganoproteins (such as GS and Mn-SOD) and MT and DMT1 or serum Mn citrate. MRS is a novel useful technique to quantify neurochemical markers associated with Mn exposure and appear capable of detecting biochemical changes before full-blown manganism symptoms become evident.

However, the reliability of BMs for Mn is still controversial, with some authors defending the consistency of some of them, whereas others pointed that any reliable one has been established to evaluate the exposures on an individual basis or to evaluate the effects. As regards non-occupational settings, there is no conclusive evidence for any biological matrix as a valid BM of Mn exposure. Further research is required to introduce an appropriate BMs to prevent diseases in workers exposed to Mn, while a long way will be needed to find BMs suitable for non-occupationally exposed persons.

References

Aboud, A.A., A.M. Tidball, K.K. Kumar, M.D. Neely, K.C. Ess, K.M. Erikson et al. 2012. Genetic risk for Parkinson's disease correlates with alterations in neuronal manganese sensitivity between two human subjects. Neurotoxicology. 33(33): 1443–1449.

Anagianni, S. and K. Tuschl. 2019. Genetic disorders of manganese metabolism. Curr. Neurol. Neurosci. Rep. 9(6): 33.

Aschner, M. and K. Erikson. 2017. Manganese. Adv. Nutr. (Bethesda, Md.) 8(3): 520–521.

[ATSDR] Agency for Toxic Substances and Disease Registry 2012. Toxicological Profile for Manganese. Agency for Toxic Substances and Disease Registry, U.S. Department of Health and Human Services, Public Health Service. Atlanta, USA.

Bahrami, A., A. Jonidi Jafari, H. Asilian, M. Taghizadeh, A. Ardjmand and H. Akbari. 2015. Biomarkers of occupational manganese toxicity. Biosci. Biotechnol. Res. Asia 12(3): 2147–2156.

Baker, M.G., C.D. Simpson, B. Stover, L. Sheppard, H. Checkoway, B.A. Racette et al. 2014. Blood manganese as an exposure biomarker: state of the evidence. J. Occup. Environ. Hyg. 11(4): 210–217.

Balachandran, R.C., S. Mukhopadhyay, D. McBride, J. Veevers, F.E. Harrison, M. Aschner et al. 2020. Brain manganese and the balance between essential roles and neurotoxicity. J. Biol. Chem. 295(19): 6312–6329.

Balzer, W. 1982. On the distribution of iron and manganese at the sediment/water interface: Thermodynamic versus kinetic control. Geochim. Cosmochim. Acta. 46(7): 1153–1161.

Bhandari, T. 2016. Low levels of manganese in welding fumes linked to neurological problems. Current safety standards may not protect workers adequately. Washington University School of Medicine News Hub.

Bjørklund, G., M. Dadar, M. Peana, M.S. Rahaman and J. Aaseth. 2020. Interactions between iron and manganese in neurotoxicity. Arch. Toxicol. 94(3): 725–734.

Cao, Y.M., X.M. Fan, J. Xu, J. Liu and Q.Y. Fan. 2022. Manganese intoxication recovery and the expression changes of Park2/Parkin in rats. Neurochem. Res. 47(4): 897–906.

Chiswell, B. and S.-H.D. Huang. 2006. Manganese removal. pp. 172.–192. *In:* G. Newcombe and D. Dixon [eds.]. Interface Science and Technology, Interface Science in Drinking Water Treatment. Elsevier Ltd, Amsterdam, Netherlands.

Choi, Y., J.K. Park, N.H. Park, J.W. Shin, C.I. Yoo, C.R. Lee et al. 2005. Whole blood and red blood cell manganese reflected signal intensities of T1-weighted magnetic resonance images better than plasma manganese in liver cirrhotics. J. Occup. Health. 47(1): 68–73.

Chtourou, Y., K. Trabelsi, H. Fetoui, G. Mkannez, H. Kallel and N. Zeghal. 2011. Manganese induces oxidative stress, redox state unbalance and disrupts membrane bound ATPases on murine neuroblastoma cells *in vitro*: protective role of silymarin. Neurochem. Res. 36(8): 1546–1557.

Cockell, K.A., G. Bonacci and B. Belonje. 2004. Manganese content of soy or rice beverages is high in comparison to infant formulas. J. Am. Coll. Nutr. 23(2): 124–30.

Cowan, D.M., Q. Fan, Y. Zou, X. Shi, J. Chen, M. Aschner et al. 2009. Manganese exposure among smelting workers: blood manganese-iron ratio as a novel tool for manganese exposure assessment. Biomarkers. 14(1): 3–16.

Crossgrove, J. and W. Zheng. 2004. Manganese toxicity upon overexposure. NMR Biomed. 17(8): 544–553.

Eastman, R.R., T.P. Jursa, C. Benedetti, R.G. Lucchini and D.R. Smith. 2013. Hair as a biomarker of environmental manganese exposure. Environ. Sci. and Technol. 47(3): 1629–1637.

Ensing, J.G. 1985. Bazooka: Cocaine-base and manganese carbonate. J. Anal. Toxicol. 9(1): 45–46.

[EPA] Environmental Protection Agency. 1995. Proceedings Workshop on the Bioavaiability and Oral Toxicity of Manganese. Cincinnati, OH, US Environmental Protection Agency.

[ERG] Eastern Research Group, Inc. 2001. Soil-Pica. Summary Report for the Agency for Toxic Substances and Disease Registry, Division of Health Assessment and Consultation. Atlanta, USA.

Erikson, K.M., A.W. Dobson, D.C. Dorman and M. Aschner. 2004. Manganese exposure and induced oxidative stress in the rat brain. Sci. Total Environ. 334-335: 409–416.

Espinosa, A.J.F., M. Ternero-Rodriguez, F.J. Barragan de la Rosa and J.C. Jimenez-Sanchez. 2001. Size distribution of metals in urban aerosols in Seville (Spain). Atmos. Environ. 35(14): 2595–2601.

Evans, G.R. and L.N. Masullo. 2021. Manganese Toxicity. StatPearls. Treasure Island, Finland.

Fan, X., Y. Luo, Q. Fan and W. Zheng. 2017. Reduced expression of PARK2 in manganese-exposed smelting workers. Neurotoxicology 62: 258–264.

Finley, J.W. and C.D. Davis. 1999. Manganese deficiency and toxicity: Are high or low dietary amounts of manganese cause for concern? Biofactors 10(1): 15–24.

Flora, S.J.S. 2014. Metals. pp. 485–519. *In:* R.C. Gupta [ed.]. Biomarkers in Toxicology. Academic Press, Cambridge, USA.

Gad, S.C. 2014. Glutathione. pp. 751. *In:* P. Wexler. [ed.]. Encyclopedia of Toxicology. Academic Press, Cambridge, USA.

Glykys, J. and K.B. Sims. 2017. The neuronal ceroid lipofuscinosis disorders. pp. 390–404. *In:* K.F. Swaiman, S. Ashwal, D.M. Ferriero, N.F. Schor, R.S. Finkel, A.L. Gropman, P.L. Pearl and M.I. Shevell. [eds.]. Swaiman's Pediatric Neurology. Elsevier, Amsterdam, Netherlands.

Harischandra, D.S., G. Shivani, Z. Gary, J. Huajun, K. Arthi, A. Vellareddy et al. 2019. Manganese-induced neurotoxicity: new insights into the triad of protein misfolding, mitochondrial impairment, and neuroinflammation. Front. Neurosci. 13: 654.

Howe, P.D., H.M. Malcolm and S. Dobson. 2004. Manganese and its compounds: Environmental aspects. Centre for Ecology & Hydrology, Monks Wood, United Kingdom. World Health Organization, Geneva, Switzerland.

[HSE] HSE National Drinking Water Group. 2019. Manganese in Drinking Water. HSE National Drinking Water Group, Ireland.

Ikeda, K., T. Tachibana and Y. Manome. 2017. The applications, neurotoxicity, and related mechanisms of manganese-containing nanoparticles. pp. 205–225. *In:* X. Jiang and H. Gaoin [eds.]. Neurotoxicity of Nanomaterials and Nanomedicine. Academic Press, Cambridge, USA.

Ingrassia, R., A. Lanzillotta, I. Sarnico, M. Benarese, F. Blasi, L. Borgese et al. 2012 1B/ (−) IRE DMT1 Expression during Brain Ischemia Contributes to Cell Death Mediated by NF-κB/RelA Acetylation at Lys310. PLoS ONE. 7(5): e38019.

Jursa, T., C.R. Stein and D.R. Smith. 2018. Determinants of hair manganese, lead, cadmium and arsenic levels in environmentally exposed children. Toxics 6(2): 19.

Keen, C.L. and S. Zidenberg-Cherr. 2003. Manganese toxicity. pp. 3686–3691. *In:* B. Caballero [ed.]. Encyclopedia of Food Sciences and Nutrition. Academic Press, Cambridge, USA.

Khan, A., J. Hingre and A.S. Dhamoon. 2020. Manganese neurotoxicity as a complication of chronic total parenteral nutrition. Case Rep. Neurol. Med. 2020: 9484028.

Kim, G., H.-S. Lee, J.S. Bang, B. Kim, D. Ko and M. Yang. 2015. A current review for biological monitoring of manganese with exposure, susceptibility, and response biomarkers. J. Environ. Sci. Health, Part C, 33: 229–254.

Kitada, M., J. Xu, Y. Ogura, I. Monno and D. Koya. 2020. Manganese superoxide dismutase dysfunction and the pathogenesis of kidney disease. Front. Physiol. 11: 755.

Konovalova, E.V., O.M. Lopacheva, I.A. Grivennikov, O.S. Lebedeva, E.B. Dashinimaev, L.G. Khaspekov et al. 2015. Mutations in the Parkinson's Disease-Associated PARK2 Gene are accompanied by imbalance in programmed cell death systems. Acta Nature 7(4): 146–149.

Kousa, A., K. Loukola-Ruskeeniemi, T. Hatakka and M. Marjatta. 2021. High manganese and nickel concentrations in human hair and well water and low calcium concentration in blood serum in a pristine area with sulphide-rich bedrock. Environ. Geochem. Health 26.

Kwakye, G.F., M.M. Paoliello, S. Mukhopadhyay, A.B. Bowman and M. Aschner. 2015. Manganese-induced parkinsonism and parkinson's disease: shared and distinguishable features. Int. J. Environ. Res. Pub. Health. 12(7): 7519–7540.

Landi, S., F. Gemignani, M. Neri, R. Barale, S. Bonassi, F. Bottari et al. 2007. Polymorphisms of glutathione-S-transferase M1 and manganese superoxide dismutase are associated with the risk of malignant pleural mesothelioma. Int. J. Cancer. 120(12): 2739–2743.

Lao, Y., L.A. Dion, G. Gilbert, M.F. Bouchard, G. Rocha, Y. Wang et al. 2017. Mapping the basal ganglia alterations in children chronically exposed to manganese. Sci. Rep. 7: 41804.

Li, J., Y. Wang, K. Steenland, P. Liu, A. van Donkelaar, R.V. Martin et al. 2022. Long-term effects of PM2.5 components on incident dementia in the northeastern United States. The Innovation. 3(2): 100208.

Li, S.J., L. Jiang, X. Fu, S. Huang, Y.N. Huang, X.R. Li, J.W. Chen et al. 2014. Pallidal index as biomarker of manganese brain accumulation and associated with manganese levels in blood: A meta-analysis. PloS One. 9(4): e93900.

Liu, C., T. Jursa, M. Aschner, D.R. Smith and S. Mukhopadhyay. 2021. Up-regulation of the manganese transporter SLC30A10 by hypoxia-inducible factors defines a homeostatic response to manganese toxicity. Proc. Natl. Acad. Sci. 118(35): e2107673118.

Madison, J.L., M. Wegrzynowicz, M. Aschner and A.B. Bowman. 2011. Gender and manganese exposure interactions on mouse striatal neuron morphology. Neurotoxicology 32(6): 896–906.

Martinez-Finley, E.J., C.E. Gavin, M. Aschner and T.E. Gunter. 2013. Manganese neurotoxicity and the role of reactive oxygen species. Free Radic. Biol. Med. 62: 65–75.

Maynar, M., F.J. Grijota, J. Siquier-Coll, I. Bartolome, M.C. Robles and D. Muñoz. 2020. Erythrocyte concentrations of chromium, copper, manganese, molybdenum, selenium and zinc in subjects with different physical training levels. J. Int. Soc. Sports Nutr. 17: 35.

[MDH] Minnesota Department of Health. 2021. Manganese in Drinking Water. Minnesota Department of Health. Saint. Paul, USA.

Mercadante, C.J., M. Prajapati, H.L. Conboy, M.E. Dash, C. Herrera, M.A. Pettiglio et al. 2019. Manganese transporter Slc30a10 controls physiological manganese excretion and toxicity. J. Clin. Invest. 129(12): 5442–5461.

Michalke, B., L. Aslanoglou, M. Ochsenkühn-Petropoulou, B. Bergström, A. Berthele, M. Vinceti et al. 2015. An approach for manganese biomonitoring using a manganese carrier switch in serum from transferrin to citrate at slightly elevated manganese concentration. J. Trace Elem. Med. Biol. 32: 145–154.

Moreno, J.A., E.C. Yeomans, K.M. Streifel, B.L. Brattin, R.J. Taylor and R.B. Tjalkens. 2009. Age-dependent susceptibility to manganese-induced neurological dysfunction. Toxicological Sci. 112(2): 394–404.

Nebert, D., A.L. Roe, S.E. Vandale, E. Bingham and G.G. Oakley. 2002. NAD(P)H: quinone oxidoreductase (NQO1) polymorphism, exposure to benzene, and predisposition to disease: A Huge review. Genet. Med. 4: 62–70.

Nemanich, A., B. Chen and M. Valento. 2021. Toxic boost: Acute, reversible neurotoxicity after ingestion of methylcyclopentadienyl manganese tricarbonyl (MMT) mistaken for an energy drink. Am. J. Emerg. Med. 42: 261.e3–261.e5.

[NIH] National Institute of Health. 2021. Manganese Fact Sheet for Health Professionals. National Institute of Health, Office of Dietary Supplements, Bethesda, USA.

[NIOSH] National Institute of Occupational Safety and Health. 2013. Welding and Manganese. Department of Health and Human Services, Public Health Service, Centres for Disease Control and Prevention, National Institute for Occupational Safety and Health. Washington D. C., USA.

O'Neal, S.L. and W. Zheng. 2015. Manganese toxicity upon overexposure: A decade in review. Curr. Environ. Health Rep. 2(3): 315–328.

[OSHA] Occupational Safety and Health Administration. 2021. Adoption of Rules to Reduce Manganese Permissible Exposure Limit; Cross-References Welding Rules with Other Standards Including Confined Spaces. Occupational Safety and Health Administration, Washington, USA.

Oulhote, Y., D. Mergler and M.F. Bouchard. 2014. Sex- and age-differences in blood manganese levels in the U.S. general population: national health and nutrition examination survey 2011–2012. Environ. Health 13: 87.

Preston, D. 2016. Magnetic Resonance Imaging (MRI) of the Brain and Spine: Basics. D. C Preston. https://case.edu/med/neurology/NR/MRI%20Basics.htm. Assessed 22nd July 2022.

Quadri, M.L., M. Kamate, S. Sharma, S. Olgiati, J. Graafland, G.J. Breedveld et al. 2015. Manganese transport disorder: Novel SLC30A10 mutations and early phenotypes. Mov. Disord. 30(7): 996–1001.

Rahman, M., B. Rahman and N. Ahmed. 2013. High blood manganese in iron-deficient children in Karachi. Pub. Health Nutr. 16(9): 1677–1683.

Rechtman, E., E. Navarro, E. de Water, C.Y. Tang, P. Curtin, D.M. Papazaharias et al. 2022. Early life critical windows of susceptibility to manganese exposure and sex-specific changes in brain connectivity in late adolescence. Biol. Psychiatry Glob. Open Sci. In Press, Corrected Proof.

Reinert, J.P. and L.D. Forbes. 2021. Manganese toxicity associated with total parenteral nutrition: A review. J. Pharm. Technol. 37(5): 260–266.

Rondanelli, M., M.A. Faliva, G. Peroni, V. Infantino, C. Gasparri, G. Iannello et al. 2021. Essentiality of manganese for bone health: An overview and update. Nat. Prod. Commun. 16(5): 1–8.

[SCOEL] Scientific Committee on Occupational Exposure Limit Values. 2011. Recommendation from the Scientific Committee on Occupational Exposure Limits for manganese and inorganic manganese compounds. European Commission, Employment, Social Affairs and Inclusion, Scientific Committee on Occupational Exposure Limit Values. Brussels, Belgium.

Shankar, K. and H.M. Mehendale. 2014. Cytochrome P450. pp. 1125–1127. *In*: P. Wexler. [eds.]. Encyclopedia of Toxicology. Elsevier, Amsterdam, Netherlands.

Sharma, A., L. Feng, D. Muresanu, S.I. Sahib, Z.R. Tian, J.V. Lafuente et al. 2021. Manganese nanoparticles induce blood-brain barrier disruption, cerebral blood flow reduction, edema formation and brain pathology associated with cognitive and motor dysfunctions. Prog. Brain Res. 265: 385–406.

Shilnikova, N., N. Karyakina, N. Farhat, S. Ramoju, B. Cline, F. Momoli et al. 2022. Biomarkers of environmental manganese exposure. Crit. Rev. Toxicol. Ahead-of-print, 1–19.

Signes-Pastor, A.J., M.F. Bouchard, E. Baker, B.P. Jackson and M.R. Karagas. 2019. Toenail manganese as biomarker of drinking water exposure: a reliability study from a US pregnancy cohort. J. Expo. Sci. Environ. Epidemiol. 29(5): 648–654.

Sikk, K., S. Haldre, S.-M. Aquilonius and P. Taba. 2011. Manganese-induced parkinsonism due to ephedrone abuse. Parkinsons Dis. 2011: 865319.

Tan, J., T. Zhang, L. Jiang, J. Chi, D. Hu, Q. Pan et al. 2011. Regulation of intracellular manganese homeostasis by Kufor-Rakeb syndrome-associated ATP13A2 protein. J. Biol. Chem. 286(34): 29654–29662.

Tarale, P., T. Chakrabarti, S. Sivanesan, P. Naoghare, A. Bafana and K. Krishnamurthi. 2016. Potential role of epigenetic mechanism in manganese induced neurotoxicity. BioMed Res. Int. 2548792.

Tarale, P., S. Sivanesan, A.P. Daiwile, R. Stöger, A. Bafana, P.K. Naoghare et al. 2017. Global DNA methylation profiling of manganese-exposed human neuroblastoma SH-SY5Y cells reveals epigenetic alterations in Parkinson's disease-associated genes. Arch. Toxicol. 91(7): 2629–2641.

Tsuchida, S. and K. Sato. 1992. Glutathione transferases and cancer. Crit. Rev. Biochem. Mol. Biol. 27(4-5): 337–384.

Ugolino, J., K.M. Dziki, A. Kim, J.J. Wu, B.E. Vogel and M.J. Monteiro. 2019. Overexpression of human Atp13a2Isoform-1 protein protects cells against manganese and starvation-induced toxicity. PLoS One. 14(8): e0220849.

Venkataramani, V., T.R. Doeppner, D. Willkommen, C.M. Cahill, Y. Xin, G. Ye et al. 2018. Manganese causes neurotoxic iron accumulation via translational repression of amyloid precursor protein and H-Ferritin. J. Neurochem. 147(6): 831–848.

Vinayagamoorthy, N., K. Krishnamurthi, S.S. Devi, P.K. Naoghare, R. Biswas, A.R. Biswas et al. 2010. Genetic polymorphism of CYP2D6*2 C T 2850, GSTM1, NQO1 genes and their correlation with biomarkers in manganese miners of Central India. Chemosphere. 81: 1286–1291.

Visser, M.J., de Wit-Bos, L., N.G.M. Palmen and P.M.J. Bos. 2014. Overview of Occupational Exposure Limits within Europe. RIVM Letter report 2014-0151. National Institute for Public Health and the Environment. Bilthoven, Netherlands.

Wallace, D.R., Y.M. Taalab, S. Heinze, B. Tariba Lovaković, A. Pizent, E. Renieri et al. 2020. Toxic-metal-induced alteration in mirna expression profile as a proposed mechanism for disease development. Cells 9: 901.

Wang, X., G.J. Li and W. Zheng. 2006. Upregulation of DMT1 expression in choroidal epithelia of the blood-CSF barrier following manganese exposure *in vitro*. Brain Res. 1097(1): 1–10.

Ward, E.J., D.A. Edmondson, M.M. Nour, S. Snyder, F.S. Rosenthal and U. Dydak. 2017. Toenail manganese: A sensitive and specific biomarker of exposure to manganese in career welders. Ann. Work Expo. Health. 62(1): 101–111.

[WHO] World Health Organization. 2021. Manganese in drinking-water - Background document for development of WHO Guidelines for drinking-water quality. WHO/HEP/ECH/WSH/2021.5. World Health Organization, Geneva, Switzerland.

Wippich, C., J. Rissler, D. Koppisch and D. Breuer. 2020. Estimating respirable dust exposure from inhalable dust exposure. Ann. Work Expo. Health. 64(4): 430–444.

Wolff, N.A., M.D. Garrick, L. Zhao, L.M. Garrick, A.J. Ghio and F.A. Thévenod. 2018. A role for divalent metal transporter (DMT1) in mitochondrial uptake of iron and manganese. Sci. Rep. 8(1): 211.

Wu, C., J.G. Woo and N. Zhang. 2017. Association between urinary manganese and blood pressure: Results from National Health and Nutrition Examination Survey (NHANES), 2011–2014. PLoS ONE. 12(11): e0188145.

Zayed, J. 2001. Use of MMT in Canadian gasoline: health and environment issues. Am. J. Ind. Med. 39(4): 426–433.

Zheng, W., S.X. Fu, U. Dydak and D.M. Cowan. 2011. Biomarkers of manganese intoxication. Neurotoxicology 32(1): 1–8.

Zheng, Y.X., P. Chan, Z.F. Pan, N.N. Shi, Z.X. Wang, J. Pan et al. 2002. Polymorphism of metabolic genes and susceptibility to occupational chronic manganism. Biomarkers. 7(4): 337–346.

Zheng, Y., F. He, P. Chan, Z. Pan, Z. Wang, J. Pan et al. 1999. Genetic polymorphism and susceptibility to occupational chronic manganism: A case-control study. Zhonghua Yufang Yixue Zazhi/Chin. J. Prev. Med. 33(2): 78–80.

Zhong, Y., X. Li, X. Du, M. Bi, F. Ma, J. Xie et al. 2020. The S-nitrosylation of parkin attenuated the ubiquitination of divalent metal transporter 1 in MPP+-treated SH-SY5Y cells. Sci. Rep. 10(1): 15542.

CHAPTER 14

Zinc

Sources of Exposure

Zinc (Zn) is a metal that is normally found with an average of about 78 mg/kg on the Earth's crust, mainly as Zn oxide or sphalerite (ATSDR 2005a, Hussain et al. 2022, IDPH 2022). Most Zn finds its way into the environment as consequence of human activities, such as mining, smelting metals, steel production, as well as burning coal (IDPH 2022). Metallic Zn has many uses in industry, the most common being galvanization, which consists of coating steel and iron or other metals to prevent rust and corrosion. Metallic Zn is also used mixed with other metals to form alloys such as brass and bronze, to make pennies and also dry cell batteries. Powdered Zn is explosive and may burst into flames if stored in damp places (ATSDR 2005).

Air

Mining, purifying of Zn, lead and cadmium ores, steel production, coal burning and burning of wastes constitute activities that can increase Zn levels in the atmosphere (ATSDR 2005). The main emissions to air from Zn production are bound to dust with particles containing Zn may having up to 5 mm in size, where Zn occurs as sulphur dioxide, other sulphur compounds and acid mists, oxides of nitrogen and other nitrogen compounds (EEA 2019, Hussain et al. 2022).

Background air concentrations of Zn are generally < 1 µg/m³, typically between 0.010 and 0.1 µg/m³ in rural areas and may range from 0.1 to 1.7 µg/m³ in areas near cities; the average Zn concentration for a 1 yr period was already found to be 5 µg/m³ in an area close to an industrial source. The mean concentrations of Zn associated with particulate matter in ambient air in Canada were determined as 0.085 µg/m³, and in Finland, 0.17 µg/m³ (WHO 2003, ATSDR 2005).

Workplace

People are occupationally exposed to Zn in activities such as Zn mining, smelting and welding, and in the manufacture of brass, bronze or other Zn-containing alloys, galvanized metals, machine parts, rubber, paint, linoleum, oilcloths, batteries, some kinds of glass and ceramics and dyes (ATSDR 2005). The main sources are fumes generated during thermal and chemical processes; these are mainly Zn oxide fume formed during high-temperature procedures, and dust generated during the mechanical processing of Zn-containing materials (Pakulska and Czerczak 2017). The effects of exposures depend on the particle size; the critical effect of acute exposure to a respirable fraction is "metal fume fever", whereas impaired lung function and asthma symptoms are the main effects of exposure to the inhalable fraction (Pakulska and Czerczak 2017). Therefore, occupational exposure limits are established separately for the respirable and the inhalable fractions, although some authors defend that exposure to finely divided Zn oxide dust can produce symptoms similar to those for metal fume fever (Cooper 2008, NIOSH 2011, Pakulska and Czerczak 2017). The Occupational Safety and Health Administration (OSHA) Permissible Exposure Limit (PEL) is 5 mg/m^3 for Zn oxide (dusts and fumes) in the workplace air during an 8-hr workday, in a 40-hr work week (NJDHSS 2007, Plum et al. 2010). The National Institute for Occupational Safety and Health (NIOSH) recommends that the level of Zn oxide in a workplace air should not exceed an average of 1 mg/m^3 over a 10-hr period of a 40-hr work week, while the American Conference of Governmental Industrial Hygienists (ACGIH) sets a limit of 10 mg/m^3 as an 8-hr Time Weighted Average (TWA) for Zn oxide, when measured as total dust (ATSDR 2005, NIOSH 2011). The inhalation of Zn oxide nanoparticles (ZnO-NP) may occur at factories that produce or use nanomaterials during production and is a major concern. Nanomaterials are able to cross biological membranes and access cells, tissues and organs, while larger-sized particles normally cannot (Ramakrishna and Rao 2011). It is seen that the inhalation of ZnO-NPs at concentrations well below occupational exposure limits (such as the one proposed by the German Institute for Occupational Safety and Health, which is 40,000 particles/cm^3) causes inflammation (Czyżowska and Barbasz 2022, IFA 2022).

Soil

The natural introduction of Zn in soils is attributed to parent rock erosion and weathering, and it is known that the Zn content is comparatively greater in heavier soils than in lighter ones; atmospheric deposition through volcanic ash, forest fires are additional natural sources of Zn to soils. Anthropogenic sources include the combustion of fossil fuels, galvanization, tyre and railing rust, motor oil, cement, tar production and hydraulic fluid, as well as disposal of Zn wastes from metal manufacturing industries and coal ash from electric utilities; fertilizers and municipal sludges applied to cropland soils can also contribute also to increase the levels of Zn in the soil (ATSDR 2005, Hussain et al. 2022).

A typical concentration of Zn in unfertilized and uncontaminated soil ranges from 10 to 300 mg/kg. Along with lead, Zn is identified as one of the two major soil pollutants in peri urban areas (Noulas et al. 2018, Hong et al. 2019, Hussain et al. 2022). In Europe, limit values for concentrations of Zn in soil are 150 to 300 mg/kg d.w., whereas limit values for Zn concentrations in sludge for use in agriculture are from 2,500 to 4,000 mg/kg (EC 2020).

Water

Normal levels of Zn in surface water and groundwater do not exceed 0.01 and 0.05 mg/L, respectively (WHO 2003). However, industrial waste streams from Zn and other metal manufacturing, domestic wastewater, run-off from soil mine drainage, municipal wastes, urban runoff, coal-fired power stations, burning of waste materials containing Zn, all constitute sources of water pollution with the metal (ATSDR 2005, Noulas et al. 2018). Zn may leach to groundwater by some mineral fertilizers and old galvanized metal pipes which can be dissolved by acidic waters, as the groundwater is often extremely polluted near mines of sulphide minerals (Noulas et al. 2018). A study to assess groundwater contamination in an area adjoining a Zn smelter effluent stream showed that wells within an 80 m vicinity, had concentrations above the permissible limits for drinking purposes, and that they were even not suitable for irrigation (Garg and Totawat 2004, Sankhla et al. 2019).

Drinking water may provide about 10% of the Zn daily intake but may contain high levels of Zn if they are stored in metal containers (in the same way as other beverages) or flow through pipes that have been coated with Zn to resist rust (WHO 2003, ATSDR 2005, Sankhla et al. 2019). Drinking water containing Zn at 5 mg/L may have a milky appearance, and even when the concentrations are above 3 mg/L tends to be opalescent, develops a greasy film when boiled, and has an undesirable astringent taste (WHO 2003, Noulas et al. 2018). In a Finnish survey of public water supplies, the median Zn content was below 20 µg/L; other surveys found higher concentrations, the highest being 1.1 mg/L (WHO 2003).

Levels to protect freshwater aquatic organisms are according to the Environmental Protection Agency (EPA, 2005) below 120 µg/L, and the recommended maximum content in irrigation water is 2 mg/L to prevent pollution of water aquifers and toxicity to many plant species. In Europe, threshold values ranging from 60 to 5,000 µg/L among countries, are settled for groundwater (ATSDR 2005, EC 2006, Noulas et al. 2018, Li et al. 2019).

Food

Exposure of the general population to Zn is primarily by ingestion, and foods are a good source of Zn for safe consumption by humans it is also considered that naturally occurring Zn from food does not appear to cause toxicity (Hussain et al. 2022). The metal is widespread in usually consumed foods, while being higher in those of animal origin, particularly some sea foods such as oysters; the richest common dietary source is red meat (ATSDR 2005, Wolf et al. 2022). Meats, fish, poultry may contain 29 mg/L of Zn, with leafy vegetables containing approximately 2 mg/L (ATSDR 2005, Noulas et al. 2018, Maret and Sandstead 2006). People in a poverty situation tend to consume less frequently foods in which Zn is more bioavailable Zn, and intake of foods rich in phytate and other Zn-binding ligands, such as whole grains and legumes (Maret and Sandstead 2006, Wolf et al. 2022). The average daily intake of Zn from food is 0.07–0.23 mg/kg b.w./day (corresponding to 5.2–16.2 mg/day when assuming a 70-kg average body weight), the Food and Nutrition Board'

Recommended Dietary Allowance (RDA) 11 mg/d for men and 8 mg/d for women; extra dietary levels of Zn are advised for women during pregnancy and lactation, which are 11–12 mg/d y and 12–13 mg/d y, respectively. Owing to their lower average body weights, infants are recommended to have a Zn intake of 2–3 mg/d y, and children 5–9 mg/d y (ATSDR 2005, Bodar et al. 2005, Plum et al. 2010). The Joint FAO/WHO committee on food additives propose a Daily Dietary Requirement (DDR) for Zn of 0.3 mg/kg b.w. and a Provisional Maximum Tolerable Daily Intake (PTDI) of 1.0 mg/kg b.w. (Noulas et al. 2018). No acute-duration oral Minimal Risk Level (MRL) ($\leq$14 d) is obtained for Zn, but MRLs for intermediate-duration oral exposure (15–364 d) or chronic-duration oral exposure ($\geq$1 yr) are settled, both as 0.3 mg/kg/d (ATSDR 2005b).

Zn supplementation provides beneficial effects, namely on the immune system, but excessive intake can produce adverse health effects. This problem is currently increasing at an alarming rate attributed to lack of awareness by consumers of this type of product ; moreover, the amount of Zn absorbed from supplements is greater than the absorption from diet (Ryu and Ademir 2020, Hussain et al. 2022). Even though it is not necessarily mandatory for fulfilling dietary Zn requirements, the consumption of these products is widespread. Excessive Zn intake is frequently associated with copper (Cu) deficiency, particularly due to the consumption of large doses over extended periods of time (Plum et al. 2010, Hussain et al. 2022). Such deficiency seems to be caused by the competitive absorption of Zn and Cu within enterocytes, mediated by metallothionein (MT) whose expression is upregulated by high dietary Zn content. As MT binds Cu with higher affinity than Zn, available Cu ions are bound by MT and the resulting complex is excreted, lowering Cu in the body (Plum et al. 2010).

Ceramic and glass containers with food can also be a source of Zn; quantifiable amounts of Zn were already found in flat and deep ceramic dishes (Dong et al. 2014, Mania et al. 2018). Regarding Zn nanoparticles, due to smaller particle size and larger surface area, nanomaterial may penetrate through the network structure of packaging materials and diffuse into food more easily; on consumption of these foods, nanoparticles transfer to the surrounding tissues or organs and break through the blood brain barrier and could cause potential food safety issues (Liu et al. 2016). Yet, current regulations for the food additive ZnO do not include particle size, nor the percentage of nanoparticles (Youn and Choi 2002). Literature reports indicate that oral exposure to ZnO-NPs leads to serious damage to organs, such as the liver, kidneys or lungs (Czyżowska and Barbasz 2022). Some other studies have confirmed the release of ZnO from polypropylene food containers into three kinds of food-simulating solutions, within a range of 0.15–0.56 µg/L, and are considered security risks (Liu et al. 2016). Moreover, the amounts needed to achieve the antibacterial activity of ZnO-NPs are sufficient to have a negative impact on living organisms (Czyżowska and Barbasz 2022). It is estimated that approximately 400 companies around the world have plans for introducing nanotechnology in foods and food packaging (Paidari et al. 2021).

Consumer products

People using excessive denture adhesive for ill-fitting dentures are at maximum risk of hyperZnemia, which generally occurs with the use of large amounts over years; problems such as bone marrow suppression and polyneuropathy, which can result in numbness and paraesthesia of the extremities, loss of balance and walking problems, are described (Singh et al. 2015).

Zn is also used in several other consumer products which include buckets, nails, gutters, paints, drugstore products, such as baby care ointment, gargle and eye drops, as well as in various cosmetics (eye shadow, sunscreen, deodorant, dandruff or shampoos); ZnO (together with titanium oxide) is the most common ingredient in personal care products (Odar et al. 2005, Czyżowska and Barbasz 2022). Recently, it was shown that small amounts of ZnO from solar filters penetrate through the protective layers of the skin and are seen in the blood and urine of the consumers. Since other studies stated that there was no evidence of ZnO-NP penetration into living epidermis or its cytotoxicity, such different outcomes might be related to the individual variability of skin features and/or to the overall skin condition of the subjects (Czyżowska and Barbasz 2022).

Essentiality and Toxicity

Zn is an essential metal, meaning that humans should obtain the metal from diets. Due to several important roles played by Zn in the organism, its deficiency can lead to health disorders. On the other hand, excess Zn in the body is toxic depending the effects on the dose and exposure route.

Zn is a part of different proteins, acts as a cofactor or coenzyme of more than 300 enzymes, playing a role in the activity and regulation of various enzymes, proteins, DNA and DNA binding proteins, in immunity, in the production of hormones and their receptors and in cell metabolism; Zn also aids in suppressing the generation of toxic Reactive Oxygen Species (ROS) (Hussain et al. 2022).

Deficiency

Dietary factors reducing the availability of Zn are the most common causes of Zn deficiency. Both nutritional and inherited Zn deficiency produce similar symptoms with a broad range of pathologies produced, which is attributed to the large number of physiological processes for which Zn is required (Ackland and Michalczyk 2006). Human Zn deficiency was first reported in 1961, and as mentioned earlier, is currently a potential widespread problem in both developing and industrialized nations. The most severe inherited forms of Zn deficiency is the rare autosomal recessive metabolic disorder known as acrodermatitis enteropathica, which may lead to growth retardation, impaired immune function and multiple skin or gastrointestinal lesions; signs and symptoms in infancy could include diarrhoea, mood changes, anorexia and neurological disturbance (WHO 2003, Nistor et al. 2016). Acquired severe Zn deficiency may occur in patients receiving total parental nutrition without supplementation of Zn, after excessive alcohol ingestion and due to iatrogenic causes (e.g., treatment with histidine or penicillamine) (WHO 2003, Hussain et al. 2022). Zn deficiency also occurs secondary to other diseases that impair intestinal absorption and/or increase intestinal loss of Zn, such as sprue, cystic fibrosis and other intestinal malabsorption syndromes; chronic increased urinary Zn loss may occur in some renal diseases or liver cirrhosis (Maret and Sandstead 2006).

Clinical manifestations of moderate Zn deficiency are mainly found in patients with low dietary Zn intake, alcohol abuse, malabsorption, chronic renal disease and chronic debilitation. Symptoms include growth retardation, hypogonadism in men, skin changes, poor appetite, mental lethargy, delayed wound healing, taste abnormalities and abnormal dark adaptation (Plum et al. 2010).

Acute toxicity

Probably Zn homeostasis mechanisms allow efficient handling of an excess of orally ingested Zn; the oral LD50 for Zn is close to 3 g/kg b.w., which is more than 10-fold higher than the dose for cadmium and 50-fold higher than the dose for mercury (Plum et al. 2010). Therefore, there is a low possibility of Zn toxicity being fatal, although the prognosis largely depends on how quickly the patient receives treatment (Agnew and Slesinger 2022). Acute toxicity appears from the ingestion of excessive amounts of Zn salts, accidentally or deliberately or due to emetic or dietary supplementation. But one of the most common causes of acute Zn toxicity is the over-consumption of dietary Zn supplements (WHO 2003, Hussain et al. 2022). Food poisoning attributed to the use of galvanized Zn containers in food preparation has also been reported (WHO 2003). Accidental intravenous dosage of > 7 g of Zn over a 60-hr period is reported to cause death; hematemesis may occur due to Zn direct caustic effects as well as renal injury, liver necrosis, thrombocytopenia and coagulopathy (WHO 2003, Agnew and Slesinger 2022).

At lower doses, symptoms usually do not become evident until ingestions exceed approximately 1 to 2 g of Zn, though some authors have stated that vomiting usually occurs after oral consumption of more than 500 mg of Zn sulphate and that gastrointestinal distress may occur with 50 mg (WHO 2003, Ryu and Aydemir 2020, Agnew and Slesinger 2022). The immediate symptoms include abdominal pain, nausea, and vomiting, and additional effects include lethargy, anaemia and dizziness (Plum et al. 2010).

Toxicity can vary in severity depending on the specific compound involved and the duration of exposure when Zn is inhaled. For example, smoke bombs containing Zn chloride can cause chest pain, airway irritation and even an Acute Respiratory Distress Syndrome (ARDS)-like clinical picture, with pulmonary fibrosis as long-term sequelae; this type of exposure occurs primarily in soldiers and armed forces (Agnew and Slesinger 2022, Hussain et al. 2022). Metal workers exposed to metals including oxides of Zn have been reported to suffer from metal fume fever. Symptoms can include flu-like symptoms with nausea, cough, headache, fever and gastrointestinal and muscle pain, generally, appearing a few hours after acute exposure (Wallig and Keenan 2013). Usually metal fume fever is not life threatening and the respiratory effects disappear within 1 to 4 d (Plum et al. 2010).

Zn overload may also appear from a group of inherited disorders of Zn metabolism, which can provoke baseline Zn plasma levels above 3 mg/L. Such concentrations are more than three times the physiological level and exceed the amount normally found in serum after Zn intoxication; symptoms can include anaemia, growth failure and systemic inflammation, and quite interestingly resemble Zn deficiency more than chronic or acute intoxication. This can be explained by excessive binding of Zn to serum proteins, such as albumin or overexpression of the Zn-binding S100 protein calprotectin; consequently, large amounts of bound Zn in serum potentially deplete the biologically available Zn (Sankhla et al. 2019).

Chronic toxicity

Chronic ingestion of excessive Zn may result in effects of the bone marrow and neurologic alterations which include dizziness, pain, headache and tiredness (Sankhla et al. 2019, Hussain et al. 2022). Cu deficiency is the major consequence of chronic ingestion of Zn, which may occur after Zn therapy (150–405 mg/d) for coeliac disease, sickle cell anaemia or acrodermatitis enteropathica (WHO 2003). Cu deficiency can lead to sideroblastic anaemia, granulocytopenia and myelodysplastic syndrome; a sensorimotor polyneuropathy syndrome has also been observed with elevated levels of Zn and Cu deficiency, but the factual cause of this syndrome remains unknown (Hussain et al. 2022). Additionally, Zn supplementation of healthy adults with 20 times the recommended dietary allowance for 6 wk resulted in the impairment of various immune responses. (WHO 2003, Plum et al. 2010). Currently EPA classifies Zn and their compounds as not listed to human carcinogenicity (group D) (EPA 2005). Zn is possibly involved in the pathogenesis and progression of prostate malignancy, possibly not due to direct carcinogenicity of Zn, but because its induced immunosuppression increases the incidence of cancer (Prasad 2008, Plum et al. 2010). Such concerns warrant further investigation, as some studies suggest that high intraprostatic Zn levels may protect against prostate carcinogenesis, whereas others suggest the opposite (Leitzmann et al. 2003).

Biomarkers

Since both Zn deficiency and overload are likely to occur in several conditions, and may result in serious health consequences, biomarkers (BMs) are important tools in the biomonitoring of Zn status and consequent effects, and in the identification of susceptible groups of persons.

Biomarkers of exposure

Zinc levels

As there are no Zn stores in the organism and that the homeostatic control of Zn is complex, the search for a reliable indicator for Zn status has been challenging. There are gaps regarding accurate assessment of Zn status, especially in mild to moderate deficiency, and the results are conflicting and inconsistent; moreover, most concerns have been addressed to assess Zn deficiency and much less to assess excess and/or toxicity (Freitas et al. 2017). Some authors consider that so far, assessments of Zn intake and mineral bioavailability are the best methods for estimating Zn concentration at both individual and population levels (Gibson et al. 2008, Freitas et al. 2017).

Blood Only 1% of total Zn in the body is present in circulating blood, with a concentration on whole blood of 4 to 8 µg/mL (King et al. 2015, Freitas et al. 2017), while a study that aimed to establish a reference value of childbearing women in China proposed a range of 4 to 7 µg/mL (Zhang et al. 2021).

Zn concentrations in various cell types such as erythrocytes, have been investigated for their potential to indicate Zn status, since this would provide an assessment over a longer time compared to that of the rapidly turning over plasma pool (Gibson et al. 2008); the Zn content in erythrocytes is normally 8–14 µg/g w. w. or 10–11 mg/10^{10} cells (King et al. 2015). Mixed results have been reported from human Zn-depletion/repletion studies or prolonged high Zn supplements, without consistent changes in erythrocytic Zn levels; besides, in some of them, during the phase of Zn depletion , no response by erythrocyte concentrations have been noted while there was evidence of functional Zn deficiency (Gibson et al. 2008, Lowe et al. 2009, King et al. 2015). The Biomarkers of Nutrition for Development (BOND) Zn Expert Panel classifies this BM as "not useful." (Gibson et al. 2008). Red cell membrane Zn seems to be sensitive to Zn depletion, but the complexity of sample preparation is likely to exclude the widespread use of this BM (Hambidge 2003).

Leukocytes contain up to 25 times more Zn than erythrocytes, which leads to the belief that they are more sensitive to changes in Zn nutrition than erythrocytes; generally, mixed white blood cells contain 75 µg Zn/10^{10} cells. But again, during an experimental phase of Zn depletion , no response by leucocyte Zn concentrations was detected (King et al. 2015). However, in a similar study, Zn concentrations in lymphocytes and granulocytes were observed to decrease after 20 wk of a Zn-deficient diet, which led to consider that this measurement could be useful, despite little sensitive (Prasad 2017). Some constrains are the fact that sample preparation is complex and that there are not established reference values for these matrices (Hambidge 2003, Gibson et al. 2008).

Systematic laboratory analyses have revealed that there are no significant differences in Zn concentrations between the plasma and serum (Poddalgoda et al. 2019). Until now, the measurement of serum or plasma Zn concentration is the only biochemical indicator recommended by WHO/UNICEF/IAEA/IZnG and by the EURopean micronutrient RECommendations Aligned (EURRECA) Network of Excellence to assess the Zn status in populations, and to measure population-level exposure to Zn supplementation; therefore, this BM is the most frequently used for evaluating the likelihood of Zn deficiency (Maret and Sandstead 2006, Gibson et al. 2008, Wessells et al. 2021). Such measures not only reflect recent Zn intake but also respond rapidly (within 2–5 d) to short-term changes in Zn intake via supplementation and persist for the duration of this period regardless of the initial concentrations of plasma Zn (Wessells et al. 2020). Nevertheless, although an available reference range for serum Zn levels could serve as a useful indicator for clinical decision making, homeostatic regulation to maintain a constant Zn concentration in plasma is a limitation; furthermore, age, gender, ethnicity, geochemical factors and altitude of the resident population , have an influence in fixing reference values (Poddalgoda et al. 2019, Barman et al. 2020). The levels of Zn in healthy adults are approximately 1 µg/mL in serum, and several efforts have been developed to define reference values according with specific geographic areas and gender. A range of 0.6–1.2 mg/L (0.59–1.25 mg/L for male and 0.5–1.0 mg/L for female, with significant differences) was estimated in Bangladesh; respectively, 9.6–31.6, 8.9–29.9, and 9.3–30.8 µmol/L were estimated for men, women and for the entire population, in Iran; and 0.75–1.77 mg/L are reported for Chinese childbearing women (ATSDR 2005b, Ghasemi et al. 2012, Barman et al. 2020, Zhang et al. 2021). Additional confounders include fasting, diurnal and circadian rhythms, infection and inflammation, exercise, stress or trauma and tissue catabolism during starvation which can result in the release of Zn into circulation temporarily increasing plasma Zn levels; the normal empirical lower limit for fasting is 10.7 µmol/L (700 µg/L) of Zn in plasma, and a value of 7.65 µmol/L is proposed to indicate severe Zn deficiency (Maret and Sandstead 2006, Lowe et al. 2009, Silveira and Falco 2015, Poddalgoda et al. 2019, Wessells et al. 2020). On an individual basis, serum/plasma Zn concentrations seem to be sensitive enough to detect serious deficiencies, such as in untreated acrodermatitis enteropathica, but present constraints in identifying marginal deficiencies; the measure can also be useful to assess a population's response to Zn deficiency and Zn intervention (supplementation) (Hambidge 2003, Freitas et al. 2017, Poddalgoda et al. 2019). Moreover, in cases of high intakes of Zn generally achieved by using Zn supplements, which can elevate plasma Zn concentrations, this BM may have the ability to detect potential toxicity (Hambidge 2003). On the other hand, severe criticism regarding this BM is there, an example being Maret and Sandstead (2006) who stated that plasma/ serum Zn concentrations are convenient indices, but physiologically insensitive, while Silveira and Falco (2015) went even further in considering that this BM is the least sensitive indicator of Zn status. The fact is, that contrary to above mentioned results, many inconsistent results are also reported . Plasma Zn concentrations were already observed to be maintained within the accepted normal range for several weeks to months, even though diets provided only 2.6–3.6 mg/d (40–55 mmol/d), which are amounts of Zn insufficient for neurobiological function; concentrations were also reported as may (or may not) decrease with Zn deficiency, and may (or may not) increase with toxicity (Maret and Sandstead 2006, Sandstead 2015, Silveira and Falco 2015). The use of population distribution curves could improve the interpretation of Zn concentrations in the plasma, as individual fluctuations would be levelled out. On the other hand, such an approach could result in a loss of detail, precision and sensitivity of the BM, giving only population trends (Lowe et al. 2009).

Urine and faeces Urinary excretion of Zn accounts for around 15% of the daily losses and when a diet is low in Zn, the amount of Zn excreted through this way is reduced by 96% (Wieringa et al. 2015). Urine Zn levels are not an established BM of Zn status but can be potentially useful as a BM, as urinary Zn excretion rates decline with severe dietary Zn restriction and may reflect dietary Zn intake (or absorption) over a wide range of intake. Other researchers defend that the effectiveness of urinary Zn concentration as a BM of intake during depletion is inconclusive due to insufficient data (Hambidge 2003, Wieringa et al. 2015, Poddalgoda et al. 2019, Phiri et al. 2021). However, this BM is also identified as an effective quantitative BM during Zn supplementation studies, with reports of an approximate five-fold increase in urinary elimination after supplementation (Poddalgoda et al. 2019, Phiri et al. 2021). Supplementation trials with intakes of 15–25, 26–50 or 51–100 mg/d , already resulted in a statistically significant increase of urinary Zn excretion, in spite of only 15–25 and the 26–50 mg/d subgroups included enough studies to state this marker as useful (Lowe et al. 2009). In general terms, there is not enough data on urinary Zn to make strong recommendations on the validity and usefulness of this BM as an indicator for Zn status (Wieringa et al. 2015). A limitation is the fact that since excretion of Zn in urine varies diurnally due to differences in hydration status, 24-hr urine collection would be preferred relatively to spot urine collection, which is impractical in large cross-sectional surveys (Phiri et al. 2021).The levels of Zn in healthy adults are approximately 0.5 mg/g creatinine in urine, while with severe Zn restriction, the renal excretion in 24 hr can reduce to 200 μg (Solomons 2003, ATSDR 2005b, Sandstead 2015) Conversely, adults chronically consuming 10–15 mg of dietary Zn can excrete daily about 500 μg of Zn (Solomons 2003).

Excretion of Zn through faeces is regulated by changes in recent Zn absorption and Zn status, with high levels of Zn in these samples considered as may be indicative of recent exposure (Hambidge 2003, ATSDR 2005b). This BMs was used in studies centred on the evaluation of Zn absorption and endogenous faecal Zn losses in toddlers at risk for enteric dysfunction (Mondal et al. 2019). Such measurements, however, require tracer techniques and metabolic collections and are only applicable as a special research tool (Hambidge 2003).

Hair and nails The concentrations of Zn in hair and nails were recently classified by the BOND Zn expert panel as potential and emerging BMs; they are thought to reflect the amount of Zn in the blood at the time of integument synthesis (Wessells et al. 2020).

Low hair Zn concentrations in childhood can reflect chronic suboptimal Zn status, when the confounding effect of protein-energy malnutrition is absent and have been associated with other indices of sub-optimal Zn status; these include impaired taste acuity, low growth percentiles and high dietary phytate: Zn molar ratios or little consumption of meat (Hambidge 2003, Gibson et al. 2008). Whether hair Zn is a valid index of chronic suboptimal Zn status among adults is uncertain and more studies are required (Gibson et al. 2008). Following Zn supplementation, hair

Zn concentrations have been shown to increase despite exhibiting an inconsistent response among infants and young children (Fleming et al. 2020, Wessells et al. 2020). Since hair Zn levels varies with age, sex, season, hair growth rate and possibly hair colour and other hair cosmetic products, these confounders must be taken into account when using this BM (Gibson et al. 2008).

Zn in the hair of healthy children have been determined as 112.55 µg/g (ranging 88.3–136.0 µg/g) with a cut-off value < 70 µg/g for those deemed to be Zn deficient; a range of 0.52 ± 0.3 µg/mL was found in healthy adult controls (Lowe 2016).

In the same way as hair, nail is a keratin-based tissue, and thus, it is expected that nail Zn concentration might be a useful marker of Zn status. However, in most investigations' plasma and nail (as well as hair) Zn concentrations were not correlated, possibly because they reflect Zn exposure or status over different time frames with different sensitivities. On the contrary, the use of this BM might be supported by the results of one recent study, where Zn concentration in nails indicated Zn nutritional status (which was assessed by dietary measures). Moreover, the earlier lack of sensitive measurement methods seems to be surpassed by a recent promising portable X-Ray Fluorescence (XRF) technique to assess Zn concentrations in nail clippings (Fleming et al. 2020, Wessells et al. 2020).

Until now and in general terms, hair Zn is classified as a potential BM and nail Zn as an emerging-BM of Zn exposure by the BOND Zn Expert Panel; however, both are not recommended to be used as single assessment indicators of Zn status. More data based on standardized procedures are needed to confirm their sensitivity and specificity, and to establish reference limits indicative of Zn deficiency (IZnG 2018).

Molecular biomarkers

A more appropriate approach to address the status of Zn is the assessment of mechanisms used to control Zn homeostasis in response to fluctuations in Zn intake (Hennigar et al. 2016).

Enzymes Several Zn metalloenzymes have been investigated in the plasma, erythrocytes, erythrocyte membranes or in specific cell types (Gibson et al. 2008, King et al. 2015).

While the obtained results have been inconsistent, alkaline phosphatase (ALP) activity in serum, erythrocytes or erythrocyte membranes has been most frequently studied and used as a BM of Zn status (Gibson et al. 2008, Ray et al. 2017). ALP is released from the liver and bone, and in small amounts from the intestine, but is present more than 80% in serum. This enzyme is a major regulator of bone mineralization and in the liver and is important for breaking down proteins; the activity of ALP is usually measured as a part of liver function test (Szulc et al. 2013, Lowe et al. 2021). More recently some attention has been addressed to know the conditions associated with decreases in its activity and deficiencies in micronutrients like Zn appears to be important causes of low ALP activity (Ray et al. 2017). But while it is accepted that in severe Zn deficiency serum ALP activity decreases, this may not occur in response to changes in dietary Zn intake in populations with marginal Zn deficiency (Cho et al. 2007, Bui et al. 2013). In six studies on the response of plasma ALP activity after Zn supplementation, no consistent effect of Zn intake on overall plasma ALP activity was evident (King et al. 2015). Inversely, Zn supplementation at supraphysiological doses increased parameters of bone formation in healthy men, including ALP levels and such increases were noted for both serum total ALP and bone specific ALP (Peretz et al. 2001). Therefore, the results obtained so far lack consistency. Since ALP exists as three different isozymes (intestinal, placental and liver/kidney/bone), ALP in circulation is a mixture of these isoforms, and examining each isoform would yield more sensitive and consistent data (King et al. 2015).

The activity of ecto purine 5'nuecleotidase in plasma or lymphocytes may be promising and warrants further study, despite a few inconsistent results obtained so far. In some cases, associations were observed between plasma ecto purine 5'nuecleotidase levels and changes in Zn intakes or Zn depletion-repletion (Gibson et al. 2008). Prasad (2020) defends that this enzyme is a very sensitive BM of human marginal Zn deficiency and that is even more sensitive for Zn deficiency than Zn in plasma or in peripheral blood cells. Gibson et al. (2008) stated that when this BM is used in community settings where mild Zn deficiency is more likely to exist, the response is less consistent, possibly due to large between-subject variation.

Metallothioneins and Zinc transporters The gene expression of metallothionein (MT) and ZIP families of Zn transporters were highlighted in a BOND review as emerging BMs of Zn status (Lowe 2016).

MT are metal storage proteins found in most tissues and MT-bound Zn is thought to be the predominant form of MT (Gibson et al. 2008, Hennigar et al. 2016). Their circulating levels in plasma respond to changes in Zn intake, although they can also increase in response to infection and stress (Gibson et al. 2008). In leukocytes, MT expression decreases in response to Zn depletion and increases in response to Zn supplementation, in a dose-dependent manner and at the earliest time point measured. A systematic review highlighted MT expression in leukocytes as more reliable and sensitive to changes in dietary Zn than other indicators (including Zn plasma levels) and had potential to become a preferred BM to assess Zn status both in clinical and field settings (Hennigar et al. 2016). When taking into account factors affecting MT expression (inflammation agents, free radicals and pharmacologic agents), a combination of serum Zn and serum MT to assess Zn nutriture is recommended. When using both BMs, a low level of serum Zn and MT can indicate a decrease in Zn intake, whereas low serum Zn concentrations plus elevated levels of MT can indicate redistribution of tissue Zn in response to infection or stress (Gibson et al. 2008).

MT monocyte mRNA also have promising potential, since their levels markedly reduce with mild experimental dietary Zn restriction and increase rapidly with Zn supplementation; the measuring method, reverse transcriptase–polymerase chain reaction, is also interesting for epidemiologic studies (Hambidge 2003).

Zn transporters can be divided into two distinct families: the Zrt-, Irt-like Protein (ZIP) and the Zn transporter (ZnT, SLC30A) (Hennigar et al. 2016). ZnT and ZIP proteins appear to have opposite roles in cellular Zn homeostasis: ZnT reduces intracellular cytoplasmic Zn by promoting Zn efflux from cells or into intracellular vesicles, whereas ZIP transporters increase intracellular cytoplasmic Zn by promoting extracellular and, probably , vesicular Zn transport into cytoplasm (Liuzzi and Cousins 2004). Hence, it is likely that when dietary Zn intakes increase, the expression of the ZnT genes would be upregulated, whereas the ZIP genes would be downregulated (Andree et al. 2004). In accordance, the expression of Zn transporters genes in leukocyte subsets have been shown to be directly proportional to the availability of Zn *in vitro* (Ryu et al. 2011). Nevertheless, there is some lack of consistency on Zn transporter expression data, since there are a large number of Zn transporters (Hennigar et al. 2016). Additionally, several factors affect Zn transporter expression, such as inflammation, free radicals and pharmacologic agents.

It is conceivable that the use of a sole molecular indicator to assess Zn status is not sufficient; rather a combination of MT and Zn transporter expression in leukocytes (perhaps used in conjunction with plasma Zn) may provide a more comprehensive view of Zn status. This approach would minimize the influence of confounders on any single variable (Hennigar et al. 2016).

Transcripts of blood cell genes associated with Zn homeostasis also hold the potential to indicate the dietary Zn status (Ryu et al. 2011). MT transcripts are upregulated by Zn and other metals through the binding of metal ions to metal-responsive transcription factor-1, which in turn is able to bind to the metal regulatory elements in the promoters of the MT genes (Andree et al. 2004). Additionally, supplementation with Zn already resulted in a significant decrease of ZIP2 and ZIP3 mRNA (Ryu et al. 2011).

Fatty acid metabolism Zn status may affect Fatty Acid (FA) metabolism because it acts as a cofactor in FA desaturases; its activity is sensitive to early stage of Zn deficiency (Knez et al. 2016, Chimhashu et al. 2018). Being the delta 6-catalyzed step required for conversion of Linoleic Acid (LA) to dihomo-γ-linolenic acid (DGLA), an elevation in the LA: DGLA ratio seems to be a sensitive marker for Zn deficiency, as demonstrated earlier in a chicken model. Results found in humans are encouraging as they showed that concentrations of DGLA are decreased and LA: DGLA ratio is increased in people with lower dietary Zn intake, possibly due to reduced activity of delta 6 desaturases (Knez et al. 2016).

Biomarkers of susceptibility

No BMs of susceptibility to Zn toxicity are validated, although some factors are known to affect individual susceptibility for the metal.

Plasma Zn decrement seem to be strongly predicted by age, and is very probably more dependent on physio pathological changes occurring with ageing, than on nutritional intake; these changes may involve the dysregulation of Zn transporters expression and Zn homeostasis (Giacconi et al. 2017).

Genetic background seems to be quite relevant in Zn status. An interesting investigation with twins, aimed to elucidate to what extent Zn concentrations in erythrocytes were influenced by genetic or by environmental factors, led to conclude that 20% of the variation in Zn concentration is due to genetic factors (Day et al. 2017).

Acrodermatitis enteropathica

The most well-known genetic condition relates to mutations in the SLC39A4 gene coding for the intestinal transporter ZIP4, which cause the earlier mentioned acrodermatitis enteropathica (Kienast et al. 2007, Day et al. 2017) The disease is characterized by an inability to absorb Zn resulting in systemic deficiency of the metal. Under normal conditions ZIP4 would play a role in augmenting intracellular Zn availability, through promoting extracellular Zn uptake and vesicular release into the cytoplasm; the reason that the transporter does not function normally in acrodermatitis enteropathica, is the consequence of an incorrect enteral Zn absorption (WHO 2003, Li et al. 2007). Skin lesions appear as dry, scaly erythematous plaques on the face or the anogenital area and the upper lip is usually spared. Nail changes may be present, and the hair becomes brittle, dry and lustreless. In profound deficiency, diffuse alopecia may be seen but several other effects occur; these include neuropsychological disturbances with irritability, weight loss, reduced immune function, diarrhoea, lethargy, anorexia, growth retardation, anaemia, amenorrhea, perinatal morbidity, hypogonadism and eye abnormalities including conjunctivitis, blepharitis, corneal opacities and photophobia (Jagadeesan and Feroze 2022). Acrodermatitis enteropathica can be lethal in the absence of treatment, but with adequate Zn supplementation the prognosis is excellent, with an expected response rate of 100% (WHO 2003, Jagadeesan and Feroze 2022).

Breast milk

There is another inherited form of Zn deficiency, which manifests in premature breastfed infants, and less common in term babies. These infants exhibit symptoms characteristic of nutritional Zn deficiency; they suffer from dermatitis, diarrhoea, alopecia, loss of appetite, impaired immune function and neuropsychiatric changes (Ackland and Michalczyk 2006, Kienast et al. 2007). Such conditions are a consequence of reduced levels of Zn in the maternal milk, which are less than 40% that of normal milk at a matched week of lactation. Notably, these low Zn levels found in milk are not attributed to a maternal Zn deficiency, but rather to a condition which predisposes mothers to produce Zn-deficient breast milk, and this condition is inherited. No mutations have been discovered so far, but very few data suggest that defects in ZnT5 and/or possibly ZnT6 may underlie these mothers' disorder (Ackland and Michalczyk 2006).

Mitochondrial citrate transporter

The mitochondrial citrate transporter is apparently involved in mitochondrial Zn influx, being coded by the gene SLC20A3. The influence of the single nucleotide polymorphism (SNPs) rs11126936, in susceptibility to Zn was also investigated (Eide 2006, Day et al. 2017). Zn serum concentrations are significantly lower in carriers of the C allele of rs11126936, when compared to T carriers and in addition, the CC genotype is more frequently observed in subjects with low Zn serum. Significant association was also seen between another SNP, rs73924411, in the same gene and Zn concentrations in relation to cognitive impairment scores (Day et al. 2017).

Heat shock proteins

Heat Shock Proteins (HSPs) have also been investigated; they are a family of chaperones involved in the folding and correct functions of proteins and its most representative member is Hsp70, whose expression increases after stressor stimuli, to protect cells from damage (Giacconi et al. 2014). The accumulation of Hsp70 *in vitro* induced by a sublethal shock of Zn (and Cu) was shown to provide a protective mechanism against metal cytotoxicity (Urani et al. 2001). Zn can induce Hsp70 production both *in vitro* and *in vivo*, playing a role in cells protection against inflammation induced by the Tumour Necrosis Factor (TNF)-α (whose levels constitute an inflammatory marker) (Zelová and Hošek 2013, Giacconi et al. 2014). Accordingly, it is described that Zn deficiency promotes inflammation, whereas an adequate intracellular Zn ion bioavailability downregulates TNF-α (Giacconi et al. 2014). Some researchers tested the association of an Hsp70 polymorphism (+1267 A/G) with various parameters of Zn status, but unfortunately no association was found. In a different manner, a TNF-α (-308 G/A SNP) polymorphism can modulate TNF-α production and Zn status in chronic or acute illness, while not in healthy ageing people; a close association of this genetic variant with reduced Zn plasma concentrations in old patients affected by pulmonary infections was already observed (Giacconi et al. 2014).

Biomarkers of effect

As with susceptibility BMs, so far, no effect BM are validated for Zn deficiency nor toxicity. A multi-omics study (omics will be outlined with more detail in the next chapter) developed very recently to identify candidate BMs, revealed numerous altered proteins and metabolites whose changes induce pathway alterations in Zn-deficient rats; these molecules can account for most manifestations of Zn deficiency and offer the potential for Zn deficiency diagnosis. To highlight, decreased organ coefficient of thymus, thymosin, IL-1, IL-6 and interferon-g was found in this study; ATP metabolic processes were inhibited by Zn deficiency and included biosynthesis of amino acids, glycolysis/gluconeogenesis, pyruvate metabolism the TCA cycle and oxidative phosphorylation, which were all decreased in the liver; decreased insulin and increased glucagon implied high blood glucose. Some selected BMs were tested further, in two Zn-deficient populations: in a Zn supplementation trial and in a Zn-deficient population; only glutathione sulfotransferase omega-1 (GSTO1) was considered reliable in these human groups (Wang et al. 2021).

Other parameters exploring Zn toxicity mechanism(s) were used in a study focused on biogenic Zn NPs exposure; these were all unspecific BMs and including oxidative stress parameters such as malondialdehyde concentration and glutathione content and the activity of antioxidant enzymes such as superoxide dismutase and catalase; caspase-3 expression was also evaluated (Salimi et al. 2019).

Additionally, and since Zn is a cofactor or structural component for proteins involved in DNA damage repair, a deficiency of this metal has deleterious effect on DNA integrity, resulting in an increase in the number of strand breaks. There is evidence that in Zn deficient individuals, there is measurable reduction DNA damage following Zn supplementation, irrespective of a concurrent increase in plasma Zn levels (Lowe 2016).

Final Remarks

This chapter presented Zn as an essential metal for which under a conservative estimate 25% of the world's population is at risk of Zn deficiency. On the other hand, Zn is known to pollute several environmental parts constituting many concerns for human health, exposures due to occupation and unsupervised supplementation; the inhalation of Zn as nanoparticles as well as its use in food packages are also causes for apprehension. Acute toxic effects vary according with the route of exposure, if ingested or inhaled, while Cu deficiency is the major consequence of the chronic ingestion of Zn, with ensuing blood and neurological effects. Since the homeostatic control of Zn is complex, the search for a reliable indicator for Zn status has been challenging, with gaps regarding accurate assessment of Zn status, especially in mild to moderate deficiency. Moreover, most concerns have been addressed to assess Zn deficiency and much less to assess excess and/or toxicity. Until now, the measurement of serum or plasma Zn concentration is the only indicator recommended by experts to assess the Zn status and measure exposure to Zn supplementation, while hair and nail concentrations, expression of MT and ZIP families of Zn transporters, have been considered potential and emerging BMs. Most known genetic Zn deficiencies are acrodermatitis enteropathica and inherited conditions predisposing mothers to produce Zn-deficient breast milk. Genetic polymorphism which could be explored as susceptibility BMs include mutations in genes coding for the intestinal transporter ZIP4. Any effect BM is validated for Zn deficiency or toxicity, although a multi-omics study developed very recently to identify candidate BMs, revealed numerous altered proteins and metabolites. The study and development of these potential BMs will be vital to improve Zn population and individuals monitorization and improve the assessment of Zn toxicity and their effects, which need greater advances when compared with deficiencies assessments.

References

Ackland, M.L. and A. Michalczyk. 2006. Zinc deficiency and its inherited disorders—A review. Genes Nutr. 1(1): 41–50.

Agnew, U.M. and T.L. Slesinger. 2022. Zinc Toxicity. StatPearls. Treasure Island, Finland.

Andree, K.B., J. Kim, C.P. Kirschke, J.P. Gregg, H.Y. Paik, H. Joung et al. 2004. Investigation of lymphocyte gene expression for use as biomarkers for zinc status in humans. J. Nutr. 134(7): 1716–1723.

[ATSDR] Agency for Toxic Substances and Disease Registry 2005. Toxicological Profile for Zinc. U.D. Department of Health and Human services, Public Health Service, Agency for Toxic Substances and Disease Registry, Atlanta, USA.

[ATSDR] Agency for Toxic Substances and Disease Registry. 2005b. ToxGuideTM for Zinc Zn CAS# 7440-66-6. Agency for Toxic Substances and Disease Registry, Atlanta, USA.

Barman, N., M. Salwa, D. Ghosh, M.W. Rahman, M.N. Uddin and M.A. Haque. 2020. Reference value for serum zinc level of adult population in Bangladesh. Electron. J. Int. Fed. Clin. Chem. Lab. Med. 31(2): 117–124.

Bodar, C.W., M.E. Pronk and D.T. Sijm. 2005 The European Union risk assessment on zinc and zinc compounds: the process and the facts. Integr. Environ. Assess. Manag. 1(4): 301–319.

Bui, V.Q., J. Marcinkevage, U. Ramakrishnan, R.C. Flores-Ayala, M. Ramirez-Zea, S. Villalpando et al. 2013. Associations among dietary zinc intakes and biomarkers of zinc status before and after a zinc supplementation program in Guatemalan schoolchildren. Food Nutr. Bull. 34(2): 143–150.

Chimhashu, T., L. Malan, J. Baumgartner, P.J. van Jaarsveld, V. Galetti, D. Moretti et al. 2018. Sensitivity of fatty acid desaturation and elongation to plasma zinc concentration: A randomised controlled trial in Beninese children. Br. J. Nutr. 119(6): 610–619.

Cho, Y.E., R.A. Lomeda, S.H. Ryu, H.Y. Sohn, H.I. Shin, J.H. Beattie et al. 2007. Zinc deficiency negatively affects alkaline phosphatase and the concentration of Ca, Mg and P in rats. Nutr. Res. Pract. 1(2): 113–9.

Cooper, R.G. 2008. Zinc toxicology following particulate inhalation. Indian. J. Occup. Environ. Med. 12(1): 10–3.

Czyżowska, A. and A. Barbasz. 2022. A review: Zinc oxide nanoparticles—friends or enemies? Int. J. Environ. Health Res. 32(4): 885–901.

Day, K.J., M.M. Adamski, A.L. Dordevic and C. Murgia. 2017. Genetic variations as modifying factors to dietary zinc requirements—a systematic review. Nutrients. 9(2): 148.

Dong, Z., L. Lu, Z. Liu, Y. Tang and J. Wang. 2014. Migration of toxic metals from ceramic food packaging materials into acid food simulants. Math. Probl. Eng. 2014: 759018.

[EC] European Commission. 2006. ANNEX 3 to the Commission Staff Working Document accompanying the Report from the Commission in accordance with Article 3.7 of the Groundwater Directive 2006/118/EC on the establishment of groundwater threshold values Information on the Groundwater Threshold Values of the Member States. European Commission, Brussels, Belgium.

[EC] European Commission. 2020. No L 181/6 Official Journal of the European Communities 4.7.8. COUNCIL DIRECTIVE of 12 June 1986 on the protection of the environment, and in particular of the soil, when sewage sludge is used in agriculture (86/278/EEC). European Commission, Brussels, Belgium.

[EEA] European Environmental Agency. 2019. Zinc production Manufacture of basic precious and non-ferrous metals, EMEP/EEA air pollutant emission inventory guidebook. European Environmental Agency, Washington D.C., USA.

Eide, D.J. 2006. Zinc transporters and the cellular trafficking of zinc. Biochim. Biophys. Acta. 1763: 711–722.

[EPA] Environmental Protection Agency. 2005. Toxicological review of zinc and compounds (cas no. 7440-66-6). Environmental Protection Agency, Washington D.C., USA.

Fleming, D.E.B., S.L. Crook, C.T. Evans, M.N. Nader, M. Atia, J.M.T. Hicks et al. 2020. Portable X-ray fluorescence of zinc applied to human toenail clippings. J. Trace Elem. Med. Biol. 62: 126603.

Freitas, E.P., A.T. Cunha, S.L. Aquino, L.F. Pedrosa, S.C. Lima, J.G. Lima et al. 2017. Zinc status biomarkers and cardiometabolic risk factors in metabolic syndrome: A case control study. Nutrients. 9(2): 175.

Garg, V.K. and K.L. Totawat. 2004. Ground water contamination in the area adjoining zinc smelter effluent stream. J. Environ. Sci. Eng. 46(1): 61–64.

Ghasemi, A., S. Zahediasl, F. Hosseini-Esfahani and F. Azizi. 2012. Reference values for serum zinc concentration and prevalence of zinc deficiency in adult iranian subjects. Biol. Trace Elem. Res. 149(3): 307–314.

Giacconi, R., L.M. Costarelli, F. Malavolta, R. Piacenza, N. Galeazzi, A. Gasparini et al. 2014. Association among 1267 A/G HSP70-2, 2308 G/A TNF-a polymorphisms and pro-inflammatory plasma mediators in old ZincAge population. Biogerontology 15: 65–79.

Giacconi, R., L. Costarelli, F. Piacenza, A. Basso, L. Rink, E. Mariani et al. 2017. Main biomarkers associated with age-related plasma zinc decrease and copper/zinc ratio in healthy elderly from ZincAge study. Eur. J. Nutr. 56(8): 2457–2466.

Gibson, R., S. Hess, C. Hotz and K. Brown. 2008. Indicators of zinc status at the population level: A review of the evidence. Br. J. Nutr. 99(S3): S14–S23.

Hambidge, M. 2003. Biomarkers of trace mineral intake and status. J. Nutr. 133(3): 948S–955S.

Hennigar, S.R., A.M. Kelley and J.P. McClung. 2016. Metallothionein and zinc transporter expression in circulating human blood cells as biomarkers of zinc status: A systematic review. Adv. Nutr. 7(4): 735–746.

Hong, Y., R. Shen, H. Cheng, Y. Chen, Y. Zhang, Y. Liu et al. 2019. Estimating lead and zinc concentrations in peri-urban agricultural soils through reflectance spectroscopy: Effects of fractional-order derivative and random forest. Sci. Total Environ. 651(2): 1969–1982.

Hussain, S., M. Khan, T.M.M. Sheikh, M.Z. Mumtaz, T.A. Chohan, S. Shamim et al. 2022. Zinc essentiality, toxicity, and its bacterial bioremediation: A comprehensive insight. Front. Microbiol. 13: 900740.

[IDPH] Illinois Department of Public Health. 2022. Zinc fact sheet. Illinois Department of Public Health, Chicago, USA.

[IFA] Institute for Occupational Safety and Health. 2022. Nanoparticles at the workplace - Criteria for assessment of the effectiveness of protective measures. Institute for Occupational Safety and Health, Washington D.C., USA.

[IZiNCG] International Zinc Nutrition Consultative Group. 2018. Assessing population zinc exposure with hair or nail zinc. IZiNCG Technical Brief, 8. International Zinc Nutrition Consultative Group, Oakland, USA.

Jagadeesan, S. and F. Kaliyadan. 2022. Acrodermatitis Enteropathica. StatPearls Treasure Island, Finland.

Kienast, A., B. Roth, C. Bossier, C. Hojabri and P.H. Hoeger. 2007. Zinc-deficiency dermatitis in breast-fed infants. Eur. J. Pediatr. 166(3): 189–194.

King, J.C., K.H. Brown, R.S. Gibson, N.F. Krebs, N.M. Lowe, J.H. Siekmann et al. 2015. Biomarkers of Nutrition for Development (BOND)-Zinc review. J. Nutr. 146(4): 858S–885S.

Knez, M., C.R. James Stangoulis, M. Zec, J. Debeljak-Martacic, Z. Pavlovic, M. Gurinovic et al. 2016. An initial evaluation of newly proposed biomarker of zinc status in humans—linoleic acid: dihomo-γ-linolenic acid (LA: DGLA) ratio. Clin. Nutr. 15: 85–92.

Li, M., Y. Zhang, Z. Liu, U. Bharadwaj, H. Wang, X. Wang et al. 2007. Aberrant expression of zinc transporter ZIP4 (SLC39A4) significantly contributes to human pancreatic cancer pathogenesis and progression. Proc. Natl. Acad. Sci. USA 104(47): 18636–18641.

Li, X.F., P.F. Wang, C.L. Feng, D.Q. Liu, J.K. Chen and F.C. Wu. 2019. Acute toxicity and hazardous concentrations of zinc to native freshwater organisms under different pH values in China. Bull. Environ. Contam. Toxicol. 103(1): 120–126.

Leitzmann, M.F., M.J. Stampfer, K. Wu, G.A. Colditz, W.C. Willett and E.L. Giovannucci. 2003. Zinc supplement use and risk of prostate cancer. J. Natl. Cancer. Inst. 95(13): 1004–1007.

Liu, J., J. Hu, M. Liu, G. Cao, J. Gao and Y. Luo. 2016. Migration and characterization of nano-zinc oxide from polypropylene food containers. Am. J. Food Technol. 11: 159–164.

Liuzzi, J.P. and R.J. Cousins. 2004. Mammalian zinc transporters. An. Rev. Nutr. 24(1): 151–172.

Lowe, D., T. Sanvictores and S. John. 2021. Alkaline Phosphatase. StatPearls. Treasure Island, Finland.

Lowe, N.M. 2016. Assessing zinc status in humans. Curr. Opin. Clin. Nutr. Metab. Care 19(5): 321-327.

Lowe, N.M., K. Fekete and T. Decsi. 2009. Methods of assessment of zinc status in humans: A systematic review. Am. J. Clin. Nutr. 89(6): 2040S–2051S.

Mania, M., T. Szynal, M. Rebeniak and J. Postupolski. 2018. Exposure assessment to lead, cadmium, zinc and copper released from ceramic and glass wares intended to come into contact with food. Rocz. Panstw. Zakl. Hig. 69(4): 405–411.

Maret, W. and H.H. Sandstead. 2006. Zinc requirements and the risks and benefits of zinc supplementation. J. Trace Elem. Med. Biol. 20(1): 3–18.

Mondal, P., J.M. Long, J.E. Westcott, M.M. Islam, M. Ahmed, M. Mahfuz et al. 2019. Zinc absorption and endogenous fecal zinc losses in bangladeshi toddlers at risk for environmental enteric dysfunction. J. Pediatr. Gastroenterol. Nutr. 68(6): 874–879.

[NIOSH] National Institute for Occupational Safety and Health. 2011. Zinc oxide dust. National Institute for Occupational Safety and Health, Washington D.C., USA.

Nistor, N., L. Ciontu, O.E. Frasinariu, V.V. Lupu, A. Ignat and V. Streanga. 2016. Acrodermatitis enteropathica: A case report. Medicine (Baltimore). 95(20): e3553.

[NJDHSS] New Jersey Department of Health and Senior Services. 2007. Zinc oxide. New Jersey Department of Health and Senior Services, Trenton, USA.

Noulas, C., M. Tziouvalekas and T. Karyotis. 2018. Zinc in soils, water and food crops. J. Trace Elem. Med. Biol. 49(2018): 252–260.

[OSHA] Occupational Safety and Health Administration. 2022. Occupational Chemical Database Zinc oxide, dust and fume. Occupational Safety and Health Administration, Washington D.C., USA.

Paidari, S., R. Tahergorabi, E.S. Anari, A.M. Nafchi, N. Zamindar and M. Goli. 2021. Migration of various nanoparticles into food samples: A review. Foods 10: 2114.

Pakulska, D. and S. Czerczak. 2017. Zagrożenia zdrowotne wynikające z narażenia na cynk i jego związki nieorganiczne w przemyśle [Health hazards resulting from exposure to zinc and its inorganic compounds in industry]. Med Pr. 68(6): 779–794.

Peretz, A., T. Papadopoulos, D. Willems, A. Hotimsky, N. Michiels, V. Siderova et al. 2001. Zinc supplementation increases bone alkaline phosphatase in healthy men. J. Trace Elem. Med. Biol. 15(2-3): 175–178.

Phiri, F.P., E.L. Ander, R.M. Lark, E.J.M. Joy, A.A. Kalimbira, P.S. Suchdev et al. 2021. Spatial analysis of urine zinc (Zn) concentration for women of reproductive age and school age children in Malawi. Environ. Geochem. Health 43: 259–271.

Plum, L.M., L. Rink and H. Haase. 2010. The essential toxin: impact of zinc on human health. Int. J. Environ. Res. Public Health. 7(4): 1342–1365.

Poddalgoda, D., K. Macey and S. Hancock. 2019. Derivation of biomonitoring equivalents (BE values) for zinc. Reg. Toxicol. Pharmacol. 106: 178–186.

Prasad, A.S. 2008. Zinc in human health: Effect of zinc on immune cells. Mol. Med. 14(5-6): 353–357.

Prasad, A.S. 2017. Discovery of zinc for human health and biomarkers of zinc deficiency. pp. 241–259. *In*: J.F. Collins [ed.]. Molecular, Genetic, and Nutritional Aspects of Major and Trace Minerals. Academic Press, Cambridge, USA.

Prasad, A.S. 2020. Lessons learned from experimental human model of zinc deficiency. J. Immunol. Res. 2020: 9207279.

Ramakrishna, D. and P. Rao. 2011. Nanoparticles: Is toxicity a concern? Electron. J. Int. Fed. Clin. Chem. Lab. Med. CC. 22(4): 92–101.

Ray, C.S., B. Singh, I. Jena, S. Behera and S. Ray. 2017. Low Alkaline Phosphatase (ALP) In Adult Population an Indicator of Zinc (Zn) and Magnesium (Mg) Deficiency. Curr. Res. Nutr. Food Sci. 5(3).

Ryu, M.-S. and T.B. Aydemir. 2020. Zinc. *In*: B.P. Marriott, D.F. Birt, V. Stalling and A.A. Yates [eds.]. Present Knowledge in Nutrition. Academic Press, Cambridge, USA.

Ryu, M.-S., B. Langkamp-Henken, S.-M. Chang, M.N. Shankar and R.J. Cousins. 2011. Genomic analysis, cytokine expression, and microRNA profiling reveal biomarkers of human dietary zinc depletion and homeostasis. Appl. Biol. Sci. 108(52): 20970–20975.

Salimi, A., R. Hamid-Reza, H. Forootanfar, E. Jafari, A. Ameri and M. Shakibaie. 2019. Toxicity of microwave-assisted biosynthesized zinc nanoparticles in mice: a preliminary study. Artif. Cells Nanomed. Biotechnol. 47(1): 1846–1858.

Sandstead, H.H. 2015. Zinc. pp. 1369–1385. *In*: G.F. Nordberg, B.A. Fowler and M. Nordberg [eds.]. Handbook on the Toxicology of Metals. Academic Press, Cambridge, USA.

Sandstead, H.H. 2015. Plasma/Serum zinc, specific metals. *In*: G.F. Nordberg, B.A. Fowler, M. Nordberg and L.T. Friberg [eds.]. Handbook on the Toxicology of Metals. Academic Press, Cambridge, USA.

Sankhla, M.S., R. Kumar and L. Prasad. 2019. Zinc impurity in drinking water and its toxic effect on human health. Indian J. Forensic Med. Toxicol. 17(4): 84–87.

Silveira, E.A. and M.O. Falco. 2015. Nutritional Treatment Approach for ART-Naïve HIV-Infected Children in Health of HIV Infected People. pp. 291–302. *In*: R.R. Watson [ed.]. Health of HIV infected people. Food, nutrition and lifestyle with Antiretroviral drugs. Academic Press, Cambridge, USA.

Singh, V.D., S.K. Misra, V. Singh, V. Misra, P. K. Yadav and A. Chaturvedi. 2015. Denture Adhesive and Zinc Toxicity. Int. J. Oral Health Med. Res. 2(2): 111–112.

Solomons, N.W. 2003. Zinc. pp. 6272–6277. *In*: B. Caballero [ed.]. Encyclopedia of Food Sciences and Nutrition, Academic Press, Cambridge, USA.

Szulc, P. and D.C. Bauer. 2013. Biochemical markers of bone turnover in osteoporosis. pp. 297–306. *In*: C.J. Rosen [ed.]. Primer on the Metabolic Bone Diseases and Disorders of Mineral Metabolism. American Society for Bone and Mineral Research. John Wiley and Sons, Hoboken, USA.

Urani, C., P. Melchioretto, F. Morazzoni, C. Canevali and M. Camatini. 2001. Copper and zinc uptake and hsp70 expression in HepG2 cells. Toxicol. *In Vitro*. 15(4-5): 497–502.

Wallig, M.A. and K.P. Keenan. 2013. Safety Assessment including current and emerging issues in toxicologic pathology. pp. 1051–1073. *In*: W.M. Haschek, C.G. Rousseaux and M.A. Wallig [eds.]. Haschek and Rousseaux's Handbook of Toxicologic Pathology. Elsevier, Amsterdam, Netherlands.

Wang, M., L. Fan, W. Wei, C. Sun, Y. Li, F. Wang et al. 2021. Integrated multi-omics uncovers reliable potential biomarkers and adverse effects of zinc deficiency. Clin. Nutr. 40(5): 2683–2696.

Wessells, K.R., K.H. Brown, C.D. Arnold, M.A. Barffour, G.-M. Hinnouho, D.W. Killilea et al. 2021. Plasma and nail zinc concentrations, but not hair zinc, respond positively to two different forms of preventive zinc supplementation in young laotian children: A randomized controlled trial. Biol. Trace Elem. Res. 199: 442–452.

[WHO] World Health Organization. 2003. Zinc in Drinking-water. Background document for development of WHO Guidelines for Drinking-water Quality. World Health Organization, Geneva, Switzerland.

Wieringa, F.T., M.A. Dijkhuizen, M. Fiorentino, A. Laillou and J. Berger. 2015. Determination of zinc status in humans: which indicator should we use? Nutrients. 7(5): 3252–3263.

Wolf, J., H.H. Sandstead and L. Rink. 2022. Zinc. pp. 963–984. *In*: G.F. Nordberg and M. Costa [eds.]. Handbook on the Toxicology of Metals. Academic Press, Cambridge, USA.

Youn, S.-M. and S.-J. Choi. 2002. Food additive zinc oxide nanoparticles: Dissolution, interaction, fate, cytotoxicity, and oral toxicity. Int. J. Mol. Sci. 23: 6074.

Zelová, H. and J. Hošek. 2013. TNF-α signalling and inflammation: interactions between old acquaintances. Inflamm. Res. 62(7): 641–651.

Zhang, H., Y. Cao, Q. Man, Y. Li, J. Lu and L. Yang. 2021. Study on reference range of zinc, copper and copper/zinc ratio in childbearing women of China. Nutrients 13(3): 946.

Part IV
Novel Trends in Metals Biomarkers

CHAPTER 15

Biomarkers of Metal Mixtures

Exposure to Metal Mixtures

Large quantities of different chemicals are released on Earth finding their way into several environmental systems. Therefore, they do not exist individually but are present as joint mixtures, which constitute the real-life scenario of human exposures. Metals are a major source of environmental pollution caused by anthropogenic activities, which among all pollutants, have received paramount attention due to their toxic nature; in the same way as other chemicals, metals exist as mixtures in the environment (Heysab et al. 2016, Masindi and Muedi 2018, Anyanwu et al. 2018). Owing to their wide dispersion in soil, water, air, dust, human food chain and manufacturing products, and the fact that they do not degrade readily in the environment, bioaccumulate and biomagnify, people are generally exposed to a cocktail of mixtures of metal pollutants throughout their lifetime (EPA 2007, Wu et al. 2016, Kortenkamp 2009, Wang et al. 2018). However, most toxicity tests have been conducted using individual metals neglecting the potential effect of metal mixtures, especially at very low concentrations, which are grossly underestimated (Wu et al. 2016). Exposures to metal mixtures may result in effects that can withdraw from the summation of effects of single metals (as will be depicted next), such exposures are sometimes detrimental to the organism, even when metals are present at concentrations lower than No Observable Effect Concentrations (NOEC) (Kortenkamp 2009, Wang et al. 2018).

Air and workplace

Air pollution is a major cause of premature death and disease. In Europe, air pollution is considered the single largest environmental health risk and according to the Global Burden of Disease Study, 4.9 million deaths and 1.4 billion Disability-Adjusted Life Years (DALYs) in 2017 were attributed to this environmental problem (EEA 2021, Potter et al. 2021). Particulate Matter (PM) is a major component of air pollution, with 307,000 of premature deaths attributed to chronic exposure to fine PM in EU Member States (although decreased by 33% in 2019 when compared 2005) (EEA 2021). The surface of PM, particularly PM2.5, can carry several adsorbed chemicals which include metals, such as metal ions, non-metallic anions and their inorganic derivatives (Di Marco et al. 2020, Potter et al. 2021). Concentration of metals in PM depends on the nature of the environment, but high levels of adsorbed toxic metals are reported, with concentrations ranging 30 to 35 µg/m³ (EEA 2022, Popoola et al. 2018, Potter et al. 2021). Manganese (Mn), zinc (Zn), iron (Fe), cadmium (Cd), copper (Cu), arsenic (As), barium (Ba), lead (Pb), aluminium (Al) and nickel (Ni) are among the most found metals in these particles; the presence of chromium (Cr), potassium, calcium (Ca), vanadium (V) and strontium (Sr) is also described (Popoola et al. 2018, Potter et al. 2021). The Enrichment Factor (EF) is a widely used metric for determining how much the presence of an element in a sampling media has increased relative to average natural abundance due to human activity; EFs for Zn, Cu, Pb and Cd in air are calculated to be 2.1, 6.1, 11.7 and 1.1, respectively (Popoola et al. 2018, Bern et al. 2019).

In occupational settings, industrial processes can result in exposure to a number of metals simultaneously and/or consecutively (Omrane et al. 2018). Mercury (Hg), Pb, Cd, Ni, Cr, Mn, As, antimony (Sb), Zn, Cu, cobalt (Co), V and beryllium (Be) are used in the industry and are known to cause adverse health effects both as metal and as metallic compounds (ILO 2017).

Soil and water

In soils from contaminated sites the most common metals, in order of abundance, are Pb Cr, As, Zn, Cd, Cu and Hg. All of them are capable of reducing crop production, bioaccumulate and biomagnificate in the food chain, also posing the risk of superficial and groundwater contamination (Wuana and Okieimen 2011). In surface water bodies from several places around the world, the average concentrations of Cr, Mn, Fe, Co, Ni, As and Cd are well above the maximum allowed values for drinking water (Zamora-Ledezma et al. 2021). International organizations continue to set standards using results from individual metals studies, following the assumption that there is a minimal interaction among them; they consider that even if there are interactions, the degree will not exceed the safety factors applied (Wu et al. 2016, Babich et al. 2021). In fact, effects of exposure to multiple metal contaminants through the drinking water are still being studied. Results already available evince that exposure to chemical mixtures with their components at concentrations below Maximum Concentration Limits (MCLs) have biological consequences (Babich et al. 2021). To exemplify, an experimental study involving the administration of a mixture of eight drinking water-containing metals affected the structural integrity of several organs at 10 and 100 times that each metal alone (Wu et al. 2016).

Food

The global assessment of mixtures toxicity is mentioned as deserving particular attention and as indispensable for a more realistic risk assessment of food contaminants. In western countries, most contaminants of concern are detected at low doses in foodstuff, although this does not mean that there are no effects resulting from their interaction in mixtures (Kopp et al. 2020).

Risk assessment of metal mixtures

A general opinion is increasingly expressed by several authors that current regulatory limits for individual metals from the different sources of exposure, might not sufficiently protect people against the toxicity of metal mixtures (Wang et al. 2018). Pb, Hg, Cd and As are ranked among priority metals that are of public health importance as they have a high degree of toxicity and are the most environmentally abundant metal/metalloids (Wu et al. 2016).

The criteria for selection of metal mixtures in risk assessment include the possibility that they occur as a chemical mixture being part of a chemical product, if they are produced and emitted together from an industrial process or that occur together in the same environmental compartment or in the human body. Other classifying criteria is based on similarities in the chemical structure or derived from mechanistic examinations. Alternative proposals recommend a move towards, establishing grouping criteria by focusing on common adverse outcomes, with less emphasis on the similarity of mechanisms. This is based on the recognition that biological effects can be similar, although details of toxicokinetic (TK) and toxicological mechanisms, may differ greatly in many respects. Attention should be given on those metals having the most potential impact on human health, and not rely simply on the concentrations of the compounds in the mixture, taking note of their expected contribution to relevant endpoints of toxicity (ATSDR 2004, Kortenkamp 2009).

Metal mixtures interactions

As mentioned, traditional toxicological studies and human health risk assessments have focused primarily on the impacts of individual metals, using methods such as toxic equivalency factors and the hazard index; such methods assume that mixture toxicity could be predicted by the sum total of the individual component toxicities (Heysab et al. 2016, Wang et al. 2018). When no interactions exist, the concept of Concentration Additions (CA) can be applied and assuming that components of mixtures exhibit similar modes of action in sub-lethal or lethal effects; they can be regarded as dilutions of one and another since they act on the same target and biochemical pathway. The Independent Action (IA) is a different concept which assumes that completely distinct and independent mode of actions is presented by components in a mixture (Bopp 2015, Wu et al. 2016). However, when there are chemical interactions, it can be inferred that the combined action deviate from both CA and IA concepts (Wu et al. 2016). Such is the case of many metal species, which due to their chemistry interacting with each other as well as with other chemicals and biological structures, rather than simply exerting their own toxic effects (Heysab et al.2016). This implies that some or all individual components of a mixture may influence each other's toxicity and the risk of these interactions has been significantly underestimated by the traditional methods (Heysab et al.2016, Hernández et al. 2017).

Interactions outside the organism

Even before passing into the organism, the joint toxicity of metals is linked with several interaction processes occurring in different environmental media (Anyanwu et al. 2018). These affect their bioavailability, that is, the amount of chemical made available for the organism when introduced through intake, breathing, injection or contact with the skin (Eggleton and Thomas 2004). For example, in soils, metals interact with particles (organic matter, inorganic particles and nanoparticles) a process which affect the concentration of metals that can be absorbed by an organism. In water, during Cu interactions with humic substances there is competitive effect of Fe or Al binding (Domingos et al. 2015, Wu et al. 2016).

Interactions within the organism

Within the organism, metals in mixtures have competitive interactions with macromolecule/transporters because of their functional similarities and the fact that they are usually transported and eliminated through many shared cellular mechanisms. Namely, toxic metals have significant interactions with essential metals (e.g., Fe, Mn, Ca) which influence essential metal status in the human body (Karri et al. 2016).

Interaction refers to the combined effect of two or more chemicals and can be stronger (e.g., synergistic, potentiating) or weaker (e.g., antagonistic) than would be expected based on dose/concentration addition or response addition (DGHC 2012). More specifically, a synergistic effect occurs when the combined effect of two chemicals is greater than the sum of the effects of each chemical given alone; potentiation, is a form of synergism and occurs when the toxicity of a chemical on a certain tissue or organ system is enhanced when given together with another chemical that does not have toxic effects on the same tissue or organ system; inversely, an antagonistic effect occurs when the combined effect of two chemicals is less than the sum of each chemical given alone (DVFA 2003). It should be noted that interactions may vary according to the relative dose, route(s), timing and duration of exposure (and includes the biological persistence of the mixture components) and biological target(s) (DGHC 2012). Generally, the metals' interactions may appear during the following related processes: TK and toxicodynamics (TD) (Feng et al. 2018). TK refers to the processes of Absorption, Distribution, Metabolism/biotransformation and Excretion (ADME) of toxics in relation to time (Gupta 2016). TD refers to the quantitative description of the effects of a toxicant on a biological system, and include a range of endpoints and products, from the molecular level, to cells, tissues, organ systems and life-history traits (Gehring and Merwe 2014).

Toxicokinetic interactions

TK interactions are a common cause of deviations from additive effects (DGHC 2012).

Absorption During absorption, the ingestion of metals in combination is known as could increase or decrease the absorption of individual metals in the digestive tract (Mitchell et al. 2011). Such events occur due to competitive interactions among metals for surface transporters, could change the rate of intake of metals in the mixture and affect the toxic outcome of each metal in the mixture (Feng et al. 2018). A known example is the Divalent Metal Transporter (DMT)-1 which transfer as many as eight metals, and which was demonstrated in assays that Cd, Cu and Pb inhibit Mn transport (Garrick et al. 2003). It is also recognized that marginal mineral deficiencies in Ca and Fe can enhance the body burden of Cd from the diet, whereas natural competitors of Cd (such as Zn) contained in foods can minimize Cd absorption (Cui et al. 2005). Exposure to metal mixtures may also modify the permeability of biological membranes, since metals which damage membranes allow the uptake of other chemicals into intracellular compartments or inversely, enable their loss from cells, leading to a reduction in intracellular concentrations (Feng et al. 2018).

Metabolism Metabolism/biotransformation is the process by which substances that enter the body are changed from hydrophobic to hydrophilic chemicals to facilitate their excretion, usually generating products with few or no toxicological effects, although there are occasions where harmful toxicants are formed instead (bioactivation) (Dekant 2009, Gerba 2019). Metabolic interactions are events which occur when chemicals modify the metabolism of other mixture components; when metal mixtures enter an organism, they can induce, interact or inhibit a range of responses and metabolic pathways associated with detoxification or bioactivation (DGHC 2012, Wu et al. 2016). These include the engagement of phase I cytochrome P450 (CYPs) enzymes, antioxidant defence enzymes, metallothioneins (MT), among other endogenous compounds.

The CYP enzyme family is a relevant example since it is generally known as the first-phase response to exposure of toxicants in humans and catalyze a number of oxidation and reduction reactions (Wu et al. 2016, Sharaf et al. 2019, Hakkola et al. 2020, Esteves et al. 2021). Since chemicals with CYP activity may be inhibitors, inducers or substrates for a specific CYP enzymatic pathway, they can alter the metabolism of other chemicals; the inhibition of an enzymatic pathway of CYP may cause decreased clearance and increased accumulation of harmful chemicals metabolized by the same pathway, resulting in toxicity or the inverse when the pathway is induced (McDonnell and Dang 2013, Gräns 2015). CYPs are mostly found to be inhibited by metals (Cao et al. 2017). Concomitantly, oxidative stress is a fundamental molecular mechanism underlying metal-induced toxicity, for which the activity of enzymatic systems, including catalase, superoxide dismutase, glutathione S transferase and the non-enzymatic antioxidant reduced glutathione, are critical to counteract this harmful condition (Chen et al. 2018, Darma et al. 2021). Multiple metal mixtures have been shown to induce more profound toxicity effects on antioxidant defence system than the components of the mixture alone (Wang and Fowler 2008, Darma et al. 2021). In addition, while the role of the metal-specific stress proteins MTs is central for the intracellular regulation of metals (such as Cu, Zn and Cd), increased MT synthesis is associated with a higher capacity for binding these metals and protection against metal toxicity. It is demonstrated that the co-exposure to Pb and As induce higher levels of MT protein than the exposure to each one alone (Roesijadi 1994, Wang and Fowler 2008). The regulation role of MT on metal mixtures interaction is demonstrated *in vitro* and *in vivo*, with the induction of MT by one metal could impact the effects of another metal (Wang and Fowler 2008).

Excretion Regarding excretion and urinary elimination by the kidneys plays an important role in regulating the plasma levels of both xenobiotic and essential metals with potential for toxicity when in excess. DMT1, which is largely expressed in the renal tissue and can transport both metals and consequently, when the body is contaminated with xenobiotic metals, they enter the renal cells with concomitant decrease of the essential trace element entry due to competition events. Decreased clearance of essential metals can result in their accumulation to excessive concentrations and ensuing toxicity (Barbier et al. 2005, Abouhamed et al. 2006). To illustrate, a transport system exists involved in Zn excretion in the kidneys' proximal tubules, where Zn can be transported complexed with cysteine or histidine via a sodium-amino acid cotransporter; toxic metals, such as Hg and Cd, can bind these amino acids competing with Zn (Barbier et al. 2005). Additionally, metal-induced kidney damage is a condition that can interfere with metal mixtures handling by the organism; needed, multi-metal mixtures may act together and pose an additive and/ or synergistic effect on kidney function. The amount of oxidation reactions occurring in mitochondria turns the kidney particularly vulnerable to damage caused by oxidative stress; most metals are redox-active being As, Cd and Pb well-established nephrotoxicants, at least at higher doses (Chen et al. 2018, Moody et al. 2020, Zhou et al. 2021). Reduced renal function is additionally associated with lower urinary metals, leading to increased metal body burden (Orr and Bridges 2017, Jin et al. 2018).

Toxicodynamic interactions

Several multiple metal interactions in TD are documented. Exposure to low dose metal mixtures can affect the homeostasis of both toxic and essential metals in tissues, ionic mimicry being a relevant phenomenon in such events; it refers to the ability of a cationic form of a toxic metal to mimic an essential element or a cationic species of an element (Bridges and Zalups 2005, Cobbina et al. 2015). To give a few examples, Pb and As administrated as a mixture was found augmented in all organs during a study, and this was attributed to their ability to replace essential metals like Ca, Fe and Zn (Cobbina et al. 2015). Cationic species of certain toxic metals, such as Cd, can use ion channels (in particular Ca channels) to gain access into target cells (Bridges and Zalups 2005).

Neurotoxicity Pb uptake through the Blood Brain Barrier (BBB) can disrupt Ca transport mechanisms and promote the activation of stress responsive Mitogen-Activated Protein (MAP) kinases, which plays an important role in regulating apoptosis (Wada and Penninger 2004, Rai et al. 2010). Additionally, mechanisms of induced cognitive dysfunction by the mixture of Pb, Cd, As and Hg involve sharing routes to the brain. It was proposed that additive/synergetic effects may appear due to common binding affinity for the N-methyl-D-aspartate NMDA receptor (Pb, As, Hg), the Na+ − K+ ATP-ase pump (Cd and Hg), biological Ca (Pb, Cd, Hg) and the glutamate neurotransmitter (Pb and Hg) (Karri et al. 2016).

Since intercellular Reactive Oxygen Species (ROS) is an important effect induced by the joint toxicity of multi-metal mixtures, which was confirmed on experimental administration of a mixture of eight metals in drinking water, while observing the intercellular ROS generation, and as well apoptosis and cell cycle arrest at the S phase (Zhou et al. 2016). The mixture of As, Cd and Pb was also proved to induce synergistic toxicity in astrocytes raising ROS and intracellular Ca ion; increased apoptosis by proximal activation of extra cellular signal-regulated kinase (ERK) signalling and downstream activation of Jun N-terminal Kinase (JNK) pathways were also observed (Rai et al. 2010).

Carcinogenicity Other examples of joint effects are the influences of mixed combinations of metals on carcinogenic processes, which can be protected by inhibiting tumour formation, but can show a positive interaction for an increased incidence of cancer. Carcinogenic metals can interfere in the biochemical function of Zn, Ca and magnesium (Mg) ions in their role to maintain the integrity and proliferation of DNA; Zn-mimicry by Cd is a known case (Madden 2003). Upregulation of AP-1 and NF-κB transcription factors (both involved in the apoptosis process) on multi-metal exposure may also promote carcinogenesis (Zhou et al. 2016). It has also been documented that mixtures of Pb, As and Cd, result in *in vitro* miRNA and mRNA expression profiles which predict the disruption of cellular death, growth, proliferation and inflammatory response and cancer (Wu et al. 2016). Conversely, carcinogenic and other toxic effects of certain metals (such as Ni, Cd, Be, and Pb) can be partially or totally prevented in some cases by several other metals. Mg may protect against Cd-induced testicular tumours, while Mn can inhibit Ni-induced carcinogenesis in the muscle and kidneys (Rodriguez and Kasprzak 1989).

Other effects Effects of metals interactions on many other adverse outcomes are described, such as the ones observed on experimental co-exposure to As, Pb and Cd which may aggravate liver dysfunction, more than in single metal treated groups (Huang et al. 2021). Epidemiological potential positive association between exposure to metal mixtures and cardiovascular disease (and cancer mortality) have been documented, mainly on a large-scale study in the U.S. general population; in this study, blood metal concentrations of Pb, Cd and Hg, and urinary Ba, Cd, Co, cesium, molybdenum, Pb, Sb, titanium, tungsten and uranium, were determined (Duan et al. 2020). It is also reported that co-exposure of Pb and Cd increases the association between Cd and renal biomarkers (BM), and that in occupational conditions, co-exposure to As, Cd and Pb is demonstrated to provoke explicit impairment of peripheral nerves (Wu et al. 2016, Koszewicz et al. 2021).

So far, studies on metal mixture toxicity are widespread and inconsistent. Mixture effects are hard to predict as all potential outcomes have been observed, and interactions can be conflicting among different experiments and may also be dependent on concentration. Moreover, acute tests, which are performed more frequently, do not account for metal interactions taking place during longer-term detoxification, which lead to results varied from chronic tests (Anyanwu et al. 2018).

Biomarkers

Human exposure to environmental chemicals is most accurately characterized as exposure to mixtures of these agents. Evaluating these interactions is essential for risk assessment, with further studies on interactions of metal/metalloid mixtures utilizing biomarker (BM) endpoints highly warranted (Wang and Fowler 2008).

Multi-media biomarkers

The recognition that exposure to metal mixtures is the real-life scenario have led to many studies on mixtures using BMs of exposure for each metal but relying on a traditional approach of determining the concentration of each metal in a single sample, such as blood or urine (Levin-Schwartz et al. 2021). Limiting factors such as financial cost and methodologic challenges has contributed for mixture studies based on BMs in a unified human specimen (Ashrap et al. 2021). However, each metal has a unique toxicokinetics, which implies that different metals are distributed differently across different compartments. Therefore, no single medium can fully capture the toxicokinetic profile for all the chemicals in a mixture, reflecting overall human exposure accurately (Ashrap et al. 2021, Levin-Schwartz et al. 2021). A proposed solution is to combine exposure data across different media to derive integrated estimates of each chemical's internal concentration, a concept which is formalized as a Multi-Media Biomarker (MMB) (Ashrap et al. 2021, Levin-Schwartz et al. 2021). The preference for either blood or urine concentrations as a better indicator, vary across metals as differences between As and Pb are only an example. Inorganic As is cleared from the blood within a few hours and is excreted primarily in urine (about 70%) a few days after ingestion and accordingly, several epidemiological studies indicate that urinary As determinations are useful to provide indications of internal dose (ATSDR 2011, Marchiset-Ferlay et al. 2012). In a different way about 99% of Pb is found in erythrocytes, their levels in blood the most used BM of Pb exposure, both in the general population and at occupational exposures; it is generally accepted that this BM reflects both recent exposure and the body-burden. Pb excretion occurs through urine, but also through faeces and despite several attempts to use this BM to surrogate Pb blood levels to indicate exposure, its reliability is still an ongoing debate (Fukui et al. 1999, Sallsten et al. 2022).

An additional factor that should be taken into account while choosing a biological sample for each metal in a MMB approach. Since each medium depicts BM levels in a particular body part , it is important to have a previous understanding of how the presence of a particular metal in a specific media can impact a specific health outcome being studied (Ashrap et al. 2021). The possibility of capturing different exposure sources and pathways is an advantage attributed to MMB determinations, but the external source must be also considered while selecting matrices (Ashrap et al. 2021). A study revealed that while higher soil and outdoor dust Mn accounted for most of nail Mn concentrations, higher air and soil Mn accounted for most of saliva levels (Butler et al. 2019).

Biomarkers of susceptibility

Studies on genes polymorphism focused on genes involved in metal metabolism, metal transport, in structures that allow metals to gain access to cells or protecting cells from their harmful effects, are certainly a way to find susceptibility BMs for metal mixtures. So far results from human studies are still conflicting, although some BMs are pointed as promising to protect people for whom genetics makes them more susceptible to toxicities induced by metal mixtures (Wirth and Mijal 2010).

Metallothioneins

MTs as Zn-ion-binding proteins are involved in defence mechanisms against oxidative damage and inflammatory stress (Alvarez et al. 2015). The most broadly MT isoform expressed in human tissues is MT2A, whose expression is modulated by a wide variety of metals (and by oxidative stress). Under conditions of excess Zn, MT acts as a Zn chelator, but when oxidative stress is elevated MT behaves as a scavenger of ROS (Park et al. 2018). Zn induce MT expression activating a Metal Responsive Element (MRE)-binding transcription factor-1 (MTF-1), which binds to MRE regions and initializes the gene transcription. Other metals, such as Cd or Cu, also possess the ability to induce the transcription of the MT gene, although via a different pathway (Klaasseni and Lehman-McKeeman 1989, Kayaaltı et al. 2011, Herbert 2021). The expression of MTs is encoded by a multigene family of linked genes and can be influenced by Single Nucleotide Polymorphisms (SNPs) in these genes. Twenty-four SNPs have been identified in the MT2A gene with and incidence of about 1% in different population groups. The SNPs rs28366003 and the rs10636 are among the most studied ones, with associations detected between metal levels in the body and these SNPs. The rs28366003 (MT2A −5A/G) SNP is an A/G substitution that occurs in the 5'-Untranslated Region (UTR-5), a core promoter region of the MT2A gene. This substitution can affect MT transcription, namely reducing Cd-induced transcription, and therefore affecting the element concentrations in the body and adversely affect health; a marginal contribution of this SNP to higher placental Cd and Pb, maternal Pb and cord blood Cd concentrations are also reported (Sekovanić et al. 2020). Epidemiological studies showed that subjects with the GG genotype (genotypic frequency of 0.6%) had lower Zn levels and higher Cd and Pb concentrations in blood samples, than AA and AG genotypes carriers. Since the order of binding affinity of MTs to metals is Cd > Pb > Cu > Hg > Zn > Ag > Ni > Co, MTs are capable of binding to Cd and Pb much more than Zn. In this regard, a higher sensitivity for metal-induced toxic outcomes of MT2A − 5 GG genotype subjects is proposed to be due to lower MT expression, and the consequence is that low expressed MT previously binds to Cd and Pb, compared with Zn; adverse effects may also appear because adequate Zn levels would be important as component of biomembranes and a number of enzymes (Kayaaltı et al. 2011). Less information is reported for associations between the SNP rs10636 (MT2A +838G/C) polymorphism and susceptibility to metal toxicity. But it is described that Hg levels are lower in the urine of subjects with the CC genotype, when compared with those with the GG genotype. C allele carriers were also found to have lower concentrations of Cd, Cu and Zn in urine, Pb in blood, Fe in plasma and higher Zn and Cu levels in red blood cells (Sekovanić et al. 2020).

Enzymes

Paraoxonase 1 (PON1) is an antioxidant enzyme synthesized predominantly in the liver which circulates in blood plasma. The enzyme is transported in association with High-Density Lipoproteins (HDL) exerting a protective effect against Low-Density Lipoprotein (LDL) oxidation, which attributes to PON1 anti-atherogenic properties; interestingly, serum PON1 also plays a major role in the detoxication of specific organophosphorus compounds such as the pesticides paraoxon/parathion (Furlong et al. 2000, Lopes et al. 2017). The PON1 gene is located on the long arm of chromosome 7 between q21 and q22, and has two amino acid polymorphisms, one at position 55 (methionine/leucine, M/L) and the other at position 192 (arginine/ glutamine, R/Q) (Kamal et al. 2011). PON1 enzymatic activity is mostly determined by the Q192R polymorphism, which results in three genotypes possibilities, RR, RQ and QQ (Lopes et al. 2017). Several metals interfere with the protective function of PON1 causing significant inhibition of its activity, both *in vitro* and *in vivo* (Kamal et al. 2011). The individual susceptibility to environmental toxicants in relationship to PON1 gene polymorphism, is demonstrated in workers from a Pb-acid factory carrying the R192 allele. RR genotype carriers were found to have lower PON1 activity than those with the QR genotype, exhibiting higher levels of Pb and Cd than the other genotype carriers, and were more likely to develop atherosclerosis (Lopes et al. 2017).

Delta aminolaevulinic acid dehydratase (ALAD) is a cytoplasmic enzyme that catalyze the second step of the heme biosynthesis pathway, that is, the condensation of two molecules of delta-aminolaevulinic acid into porphobilinogen; ALAD is a Zn-dependent enzyme (Bernard and Lauwerys 1987). *In vitro* experiments have shown that ALAD can be activated or inhibited by several other metal ions including Pb, As, Cd, Hg, silver and Cu, which possibly bind to thiol groups of allosteric sites and, according to their structure, provoke allosteric transitions which activate or inactivate ALAD (Thompson et al. 1977, Bernard and Lauwerys 1987). ALAD is coded by an autosomal gene, being a relatively common variant the 177G>C (rs1800435) in the exon 4 of ALAD. This variant is characterized by a C > G allele change, which substitutes asparagine with lysine on residue 59, producing two codominant alleles, ALAD1 and ALAD2, in three genotypes, ALAD 1-1, ALAD 1-2 and ALAD 2-2 (Shaik and Jamil 2008, Mohamadkhani et al. 2019, Perini et al. 2021). In European and American populations, the ALAD2 allele have a prevalence of about 10 to 20% (Daniell et al. 1997). There is evidence that ALAD polymorphism may influence metal toxicokinetics, due to the fact that the mentioned lysine-asparagine variation creates a different functional isozyme with a distinct binding affinity for metals (Perini et al. 2021). Carriers of the ALAD2 (C allele) are known since a long time to be prone to having higher blood Pb concentrations than the frequent ALAD1 (G allele) carriers, rendering them to be more susceptible to Pb poisoning (Daniell et al. 1997, Wirth and Mijal 2010, Mohamadkhani et al. 2019). It was proposed that the amino acid change caused by the ALAD 177 C > G polymorphism result in a more negatively charged ALAD isozyme, which makes it more attracted to metals, such as Pb. Additionally, the ALAD enzyme interact with other metals besides Pb and as Pb and Hg have similar atomic radii, it is postulated that Hg is capable of binding to the same sites in ALAD. This was in agreement, with children who were ALAD2 carriers and were chronically exposed to Hg, and were observed to exhibit higher Hg concentrations in blood. These genotypes could contribute to changes in Hg half-life time, harmful effects and higher susceptibility to Hg-induced neurological disorders (Perini et al. 2021). Studies on ALAD polymorphisms and co-exposure to Pb and Hg are certainly warranted to elucidate the value of these susceptibility BM in protecting these risk groups.

Biomarkers of effect

Multifactorial mechanisms are involved in metal toxicity and among them, metal-induced effects are largely related to their oxidative state and reactivity with endogenous compounds. Redox-active metals, such as Fe, Cu and Cr, undergo redox cycling, whereas redox-inactive metals, such as Pb, Cd, Hg and others deplete cells' major antioxidants, particularly thiol-containing antioxidants and enzymes. Thus, through different routes, both redox-active or redox-inactive metals may cause an increase in production of ROS, leading to oxidative stress (Ercal et al. 2001, Das and Roychoudhury 2014, Ranjbar et al. 2014). Since the induction of oxidative stress is a relevant toxic mechanism shared by most metals, a number of oxidative stress BMs have been used in epidemiological studies to assess metal mixtures effects. It is accepted that estimates of oxidative stress among multi-metal exposures helps to distinguish people with higher priority for monitoring (Zendehdel et al. 2014).

Cell membranes damage

Malondialdehyde (MDA) is formed via the degradation process of cell membranes, where polyunsaturated fatty acids react with ROS, with MDA being a final product of lipid peroxidation; the levels of urinary MDA have been extensively used to evaluate metal-induced oxidative stress, in general and in occupationally exposed populations (Lopes et al. 2017). This effect BM was already used in an occupational study enlisting manual metal arc welding workers, who were exposed to metal mixtures (Cd, Cr, Pb, Ni, Mg and other metals). In the range of metals exposure, the induction of oxidative stress for the exposed group was observed by increased urine MDA concentrations (Zendehdel et al. 2014).

DNA and RNA damage

A major cause of DNA damage is oxidative stress, including these damages, abasic sites, strand breaks and base modifications (Gonzalez-Hunt et al. 2018). In DNA, 8-oxoGuanine (8-oxoG) is more easily oxidized than any of the four natural nucleobases. Therefore, a widely considered marker of early biological effects of ROS on DNA is 8-oxo-7,8-dihydroguanine (8-oxo-dG), which is the product of two-electron oxidation of guanine residues (Misiaszek et al. 2005, Loft et al. 2012, Domingo-Relloso et al. 2019). Studies on PM2.5 exposure have already led to finding that 8-oxodG concentration in lymphocytes was significantly associated with the exposure to the mixture of V, Cr, Fe, Ni, Cu and platinum that was present in these particles (Sørensen et al. 2005).

Although RNA is more prone than DNA to be damaged by metals, it has not been a major focus in investigating the consequences of oxidative stress (Kournoutou et al. 2017). Oxidative modification to RNA results in disturbances of the translational process and impairment of protein synthesis, which can cause cell deterioration or even death (Kong and Lin 2010). In a group of workers of a metal carpentry industry, urinary concentrations of 16 different metal exposure BMs were determined. While metal concentrations found in these workers were well below occupational exposure limit values, Ba, Hg, Pb and Sr were correlated with the RNA oxidative stress BM, 8-oxo-7, 8-dihydroguanosine (8-oxoGuo). This BM was able to discriminate exposed workers from controls with a high level of specificity and sensitivity (Buonaurio et al. 2021).

Antioxidant defence

To counteract oxidative stress several antioxidant defences that exist in the body, for instance, reduced glutathione (GSH) which provides reduced equivalents for the enzymatic reaction mediated by glutathione peroxidase (GPx) resulted in oxidized glutathione (GSSG). Hence, under oxidative stress conditions GSSG accumulate and the ratio GSSG/GSH increases; therefore, the ratio GSSG/GSH is considered a marker of oxidative stress at the cellular cytoplasm. Urinary GSSG/GSH ratio was previously used to investigate exposure to metal mixtures (which was assessed by the measurement of urinary Sb, Ba, Cd, Cr, Co, Cu, Mo, V and Zn). The principal component analysis showed a positive association between both non-essential and essential metals, with this effect BM (Domingo-Relloso et al. 2019).

Heme synthesis

Heme synthesis is a biochemical pathway requiring a number of steps, substrates and enzymes (Fig. 7); therefore, any deficiency in an enzyme or substrate of this pathway leads to accumulation of their intermediates in blood, tissues and urine leading to a clinically significant outcome named porphyria (Ogun et al. 2022). Metals are capable of impairing various aspects of the heme synthesis: gene expression, Fe integration into protoporphyrin IX and enzyme activity (Schauder et al. 2010). Namely, Pb interferes with heme synthesis by inhibiting several enzymes, in particular ALAD and ferrochelatase, thereby decreasing heme biosynthesis with ensuing increase of the rate-limiting enzyme for this biosynthesis pathway, d-aminolaevulinic synthetase (ALAS). As a consequence, urinary porphyrins, coproporphyrin and delta-aminolaevulinic acid (ALA) increase, as well as blood and plasma ALA and erythrocyte protoporphyrin (Roney et al. 2011). Although the associated levels of porphyrinuria are lower than those seen with Pb, it has been proved that Hg, As and other metal exposures can also affect heme synthesis (Daniell et al. 1997). A BM of Hg exposure is described as characterized by increased urinary concentrations of pentacarboxyporphyrin and coproporphyrin and the atypical keto-isocoproporphyrin, based on selective interference of the metal with the fifth (uroporphyrinogen decarboxylase) and sixth (coproporphyrinogen oxidase) enzymes of the heme biosynthetic pathway; such patterns of porphyrins urinary excretion were found in dentists exposed to Hg (Daniell et al. 1997, Heyer et al. 2006). In turn, As interferes with the activities of ALAS, porphobilinogen deaminase, uroporphyrinogen III synthase, uroporphyrinogen decarboxylase, coproporphyrinogen oxidase, ferrochelatase and heme oxygenase. An epidemiological study involving subjects exposed to As via drinking water, showed that despite no increase in urinary porphyrin excretion was found in exposed individuals, an inversion of coproporphyrin/uroporphyrin ratio could be observed (García-Vargas and Hernández-Zavala 1996).

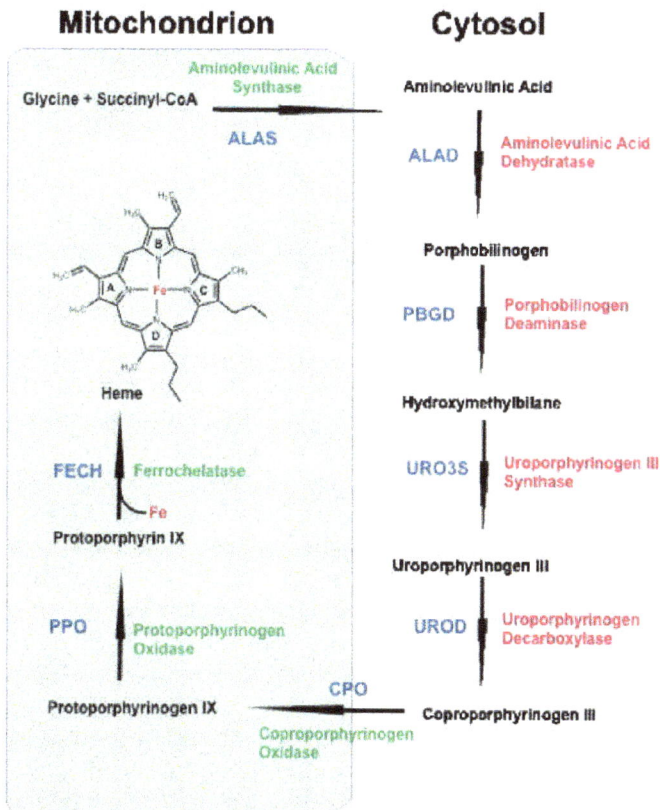

Figure 7. Heme biosynthesis pathway (Ajioka et al. 2006).

Additionally, *in vitro* and *in vivo* studies demonstrated that a number of other metals were porphyrinogenic, including aluminium, Cd, Co, gallium arsenide and others (Daniell et al. 1997). In addition, while intracellular heme levels are affected by the activity of heme oxygenase (whose activity is partially Fe-dependent) increased activity of this enzyme resulting in a depletion of cellular haemoproteins and can also be induced by Co, Cr, Mn, Fe, Cu, Zn and Pb (Garnica 1981)

Hence, the utility of porphyrins as BM of metal exposures is based largely on the properties of metals to selectively alter porphyrinogen metabolism in target tissues by mechanisms which lead to metal-specific changes in urinary porphyrin excretion patterns (Woods 1995). Hence, the measurement of changes in heme precursor excretion offer potential biological markers of effect which can be used for detecting harmful effects of specific chemical exposures, while their biological effects are still preclinical and potentially reversible (Levin-Schwartz et al. 2021). The high degree of correlation between excretion of specific porphyrins in the urine and other ultrastructural/biochemical alterations in organelles, such as the mitochondrion, indicates the use of porphyrinurias in detecting early stages of cell injury. Most importantly, metal-induced disturbances in this pathway have also proved to be useful for examining the interactions between metals under mixture exposure conditions. Several studies in the last 20 yr have used specific metal porphyrinuria patterns as BMs of exposure to metal mixtures (Fowler 2001). An example is addressing metal environmental toxicity in autistic children, using porphyrins excretion as an effect BM (Nataf et al. 2006).

Final Remarks

This chapter augmented that metals do not exist individually in the environment, but are present as joint mixtures, which comprise real-life scenario of human exposures. Exposures to metal mixtures may result in effects that can withdraw from the total effects of single metals, and such exposures are sometimes detrimental to the organism, even when metals are present at concentrations lower than no observable effect concentrations. In fact, metal species due to their chemistry interact with each other as well as with other chemicals and biological structures at TK and TD levels, rather than simply exerting their own toxic effect; some interaction, such as synergistic or enhancing can lead to stronger effects than would be expected on the basis of dose/concentration addition or response addition. Therefore, risk assessment based on current regulatory limits for individual metals (the traditional approach) might not sufficiently protect people against the toxicity of metal mixtures. Since no single medium to assess exposures can fully reproduce the TK profile for all the chemicals in a mixture, combined exposure data across different media to obtain integrated estimates of each chemical's internal concentration, a is a novel concept known as MMB. Studies on genes polymorphism focused on genes involved in metal metabolism, metal transport, in structures that allow metals to gain access to cells or protecting cells from their harmful effects, have been performed to find susceptibility BMs for metal mixtures, with single nucleotide polymorphisms in genes coding MTs, paraxonase or ALAD pointed as promising candidates. The induction of oxidative stress, a relevant toxic mechanism shared by most metals, several oxidative stress BMs have been used in epidemiological studies to assess metal mixtures effects; markers of lipid oxidation (MDA), DNA or RNA damage (8-oxo-7,8-dihydroguanine) or antioxidant defences depletion (ratio GSSG/GSH) have been used for such purposes. Additionally, metals are known to selectively alter porphyrinogen metabolism in target tissues by mechanisms which lead to metal-specific changes in urinary porphyrin excretion patterns. Hence, the measurement of changes in heme precursor excretion offer potential BMs of effects induced by metal mixtures.

Evaluating metal interactions is essential for risk assessment, and further studies on interactions on metal/metalloid mixtures using BMs are highly warranted to access real life scenarios of exposure.

References

Abouhamed, M., J. Gburek, W. Liu, B. Torchalski, A. Wilhelm, N.A. Wolff et al. 2006. Divalent metal transporter 1 in the kidney proximal tubule is expressed in late endosomes/lysosomal membranes: Implications for renal handling of protein-metal complexes. Am. J. Physiol. Renal Physiol. 290(6): 1525–1533.

Ajioka, R.S., J.D. Phillips and J.P. Kushner. 2006. Biosynthesis of heme in mammals. Biochim. Biophys. Acta - Mol. Cell Res. 1763(7): 723–736.

Alvarez, L., H. Gonzalez-Iglesias, C. Petrash, M. Garcia, M. Petrash, A. Sanz-Medel et al. 2015. Metallomics of the human lens: A focus on the zinc-metallothionein system. Investig. Ophthalmol. Vis. Sci. 56(7): 5576.

Anyanwu, B.O., A.N. Ezejiofor, Z.N. Igweze and O.E. Orisakwe. 2018. Heavy metal mixture exposure and effects in developing nations: An update. Toxics. 6(4): 65.

Ashrap, P., D.J. Watkins, B. Mukherjee, Z. Rosario-Pabón, C.M. Vélez-Vega, A. Alshawabkeh et al. 2021. Performance of urine, blood, and integrated metal biomarkers in relation to birth outcomes in a mixture setting. Environ. Res. 200: 111435.

[ATSDR] Agency for Toxic Substances and Disease Registry, U.S. Department of Health and Human Sciences. 2004. Interaction profile for: Lead, Manganese, Zinc and Copper. Agency for Toxic Substances and Disease Registry, U.S. Department of Health and Human Sciences, Atlanta, USA.

[ATSDR] Agency for Toxic Substances and Disease Registry. 2011. As Toxicity. What is the Biologic Fate of As in the Body? Agency for Toxic Substances and Disease Registry, U.S. Department of Health and Human Sciences, Atlanta, USA.

Babich, R., E. Craig, A. Muscat, J. Disney, A. Farrell, L. Silka et al. 2021. Defining drinking water metal contaminant mixture risk by coupling zebrafish behavioral analysis with citizen science. Sci. Rep. 11: 17303.

Barbier, O., G. Jacquillet, M. Tauc, M. Cougnon and P. Poujeol. 2005. Effect of heavy metals on, and handling by, the kidney. Nephron. Physiol. 99(4): 105–110.

Bern, C.R., K. Walton-Day and D.L. Naftz. 2019. Improved enrichment factor calculations through principal component analysis: Examples from soils near breccia pipe uranium mines, Arizona, USA. Environ. Poll. 248: 90–100.

Bernard, A. and R. Lauwerys. 1987. Metal-induced alterations of delta-aminolevulinic acid dehydratase. Ann. N. Y. Acad. Sci. 514: 41–47.

Bopp, S., E. Berggren, A. Kienzler, S. van der Linden and A. Worth A. 2015. Scientific methodologies for the combined effects of chemicals—a survey and literature review. EUR 27471. Luxembourg (Luxembourg): Publications Office of the European Union; 2015. JRC97522.

Bridges, C.C. and R.K. Zalups. 2005. Molecular and ionic mimicry and the transport of toxic metals. Toxicol. Appl. Pharmacol. 204(3): 274–308.

Buonaurio, F., M.L. Astolfi, D. Pigini, G. Tranfo, S. Canepari, A. Pietroiusti et al. 2021 Oxidative Stress biomarkers in urine of metal carpentry workers can be diagnostic for occupational exposure to low level of welding fumes from associated metals. Cancers (Basel). 13(13): 3167.

Butler, L., C. Gennings, M. Peli, L. Borgese, D. Placidi, N. Zimmerman et al. 2019. Assessing the contributions of metals in environmental media to exposure biomarkers in a region of ferroalloy industry. J. Expo. Environ. Epidemiol. 29: 674–687.

Cao, X., R. Bia and Y. Song. 2017. Toxic responses of cytochrome P450 sub-enzyme activities to heavy metals exposure in soil and correlation with their bioaccumulation in Eisenia fetida. Ecotoxicol. Environ. Saf. 144: 158–165.

Chen, P., J. Bornhorst, M.D. Neely and D.S. Avila. 2018. Mechanisms and disease pathogenesis underlying metal-induced oxidative stress. Oxid. Med. Cell. Longev. 2018: 7612172.

Cobbina, S.J., Y. Chen, Z. Zhou, X. Wu, W. Feng, W. Wang et al. 2015. Low concentration toxic metal mixture interactions: Effects on essential and non-essential metals in brain, liver, and kidneys of mice on sub-chronic exposure. Chemosphere. 132: 79–86.

Cui, Y., Y.G. Zhu, R. Zhai, Y. Huang, Y. Qiu and J. Liang. 2005. Exposure to metal mixtures and human health impacts in a contaminated area in Nanning, China. Environ. Int. 31(6): 784–790.

Daniell, W.E., H.L. Stockbridge, R.F. Labbe, J.S. Woods, K.E. Anderson, D.M. Bissell et al. 1997. Environmental chemical exposures and disturbances of heme synthesis. Environ. Health Perspect. 105(11): 37–53.

Das, K. and A. Roychoudhury. 2014. Reactive oxygen species (ROS) and response of antioxidants as ROS-scavengers during environmental stress in plants. Front. Environ. Sci. 02(52).

Dekant, W. 2009. The role of biotransformation and bioactivation in toxicity. EXS. 99: 57–86.

[DGHC] Directorate-General for Health and Consumers. 2012. Toxicity and Assessment of Chemical Mixtures. European Commission, Directorate-General for Health and Consumers, Brussels, Belgium.

Di Marco, V., A. Tapparo, D. Badocco, S. D'Aronco, P. Pastore and C. Giorio. 2020. Metal ion release from fine particulate matter sampled in the Po valley to an aqueous solution mimicking fog water: Kinetics and solubility. Aerosol Air Qual. Res. 20: 720–729.

Domingo-Relloso, A., M. Grau-Perez, L. Briongos-Figuero, J.L. Gomez-Ariza, T. Garcia-Barrera, A. Dueñas-Laita et al. 2019. The association of urine metals and metal mixtures with cardiovascular incidence in an adult population from Spain: the Hortega Follow-Up Study. Int. J. Epidemiol. 48(6): 1839–1849.

Domingos, R.F., A. Gélabert, S. Carreira, A. Cordeiro, Y. Sivry and M.F. Benedetti. 2015. Metals in the aquatic environment—interactions and implications for the speciation and bioavailability: A critical overview. Aquat. Geochem. 21: 231–257.

Duan, W., C. Xu, Q. Liu, J. Xu, Z. Weng, X. Zhang et al. 2020. Levels of a mixture of heavy metals in blood and urine and all-cause, cardiovascular disease and cancer mortality: A population-based cohort study. Environ. Pollut. 263(Pt A): 114630.

[DVFA] Danish Veterinary and Food Administration. 2003. Combined Actions and Interactions of Chemicals in Mixtures—The Toxicological Effects of Exposure to Mixtures of Industrial and Environmental Chemicals. The Danish Veterinary and Food Administration, Glostrup, Denmark.

[EEA] European Environmental Agency. 2021. Health impacts of air pollution in Europe. European Environmental Agency, Copenhagen, Denmark.

Eggleton, J.D. and K.V. Thomas. 2004. A review of factors affecting the release and bioavailability of contaminants during sediment disturbance events. Environ. Int. 30(7): 973–980.

[EPA] Environmental Protection Agency. 2007. Framework for Metals Risk Assessment. Environmental Protection Agency, Washington, USA.

Ercal, N., H. Gurer-Orhan and N. Aykin-Burns. 2001. Toxic metals and oxidative stress Part I: Mechanisms involved in metal-induced oxidative damage. Curr. Top. Med. Chem. 1(6): 529–539.

Esteves, F., J. Rueff and M. Kranendonk. 2021. The central role of cytochrome p450 in xenobiotic metabolism-a brief review on a fascinating enzyme family. J. Xenobiot. 11(3): 94–114.

Feng, J., Y. Gao, Y. Ji and L. Zhu. 2018. Quantifying the interactions among metal mixtures in toxicodynamic process with generalized linear model. J. Hazard. Mater. 345: 97–106.

Fowler, B.A. 2001. Porphyrinurias induced by mercury and other metals. Toxicol. Sci. 61(2): 197–198.

Fukui, Y., M. Miki, H. Ukai, S. Okamoto, S. Takada, K. Higashikawa et al. 1999. Urinary lead as a possible surrogate of blood lead among workers occupationally exposed to lead. Int. Arch. Occup. Environ. Health. 72(8): 516–520.

Furlong, C.E., W.F. Li, V.H. Brophy, G.P. Jarvik, R.J. Richter, D.M. Shih et al. 2000. The PON1 gene and detoxication. Neurotoxicology. 21(4): 581–587.

García-Vargas, G.G. and A. Hernández-Zavala. 1996. Urinary porphyrins and heme biosynthetic enzyme activities measured by HPLC in as toxicity. Biomed Chromatogr. 10(6): 278–284.

Garnica, A.D. 1981. Trace metals and hemoglobin metabolism. Ann. Clin. Lab. Sci. 11(3): 220–228.

Garrick, M.D., K.G. Dolan, C. Horbinski, A.J. Ghio, D. Higgins, M. Porubcin et al. 2003. DMT1: A mammalian transporter for multiple metals. BioMetals 16: 41–54.

Gehring, R. and D. van der Merwe. 2014. Toxicokinetic-toxicodynamic modelling. pp. 149–153. *In*: R.C. Gupta [ed.]. Biomarkers in Toxicology. Academic Press, Cambridge, USA.

Gerba, C.P. 2019. Environmental toxicology. pp. 511–540. *In*: M.L. Brusseau, I.L. Pepper and C.P. Gerba [eds.]. Environmental and Pollution Science. Academic Press, Cambridge, USA.

Gonzalez-Hunt, C.P., M. Wadhwa and L.H. Sanders. 2018. DNA damage by oxidative stress: Measurement strategies for two genomes. Curr. Opin. Toxicol. 7: 87–94.

Gräns, J. 2015. Chemical mixtures and interactions with detoxification mechanisms and biomarker responses in fish. PhD thesis. Department of Biological and Environmental Science, University of Gothenburg, Sweden.

Gupta, P.K. 2016. Fundamentals of Toxicology: Essential Concepts and Applications. Academic Press, Cambridge, USA.

Hakkola, J., J. Hukkanen, M. Turpeinen and O. Pelkonen. 2020. Inhibition and induction of CYP enzymes in humans: An update. Arch. Toxicol. 94: 3671–3722.

Herbert, M. 2021. Metallothionein genes research literatures. Rep. Opinion 13(7): 23–160.

Hernández, A.F., F. Gil and M. Lacasaña. 2017. Toxicological interactions of pesticide mixtures: An update. Arch. Toxicol. 91(10): 3211–3223.

Hernández-Zavala, A., L.M. Del Razo, G.G. García-Vargas, C. Aguilar, V.H. Borja, A. Albores et al. 1999. Altered activity of heme biosynthesis pathway enzymes in individuals chronically exposed to As in Mexico. Arch. Toxicol. 73(2): 90–95.

Heyer, N.J., A.C. Bittner Jr., D. Echeverria and J.S. Woods. 2006. A cascade analysis of the interaction of mercury and coproporphyrinogen oxidase (CPOX) polymorphism on the heme biosynthetic pathway and porphyrin production. Toxicol. Lett. 161(2): 159–66.

Heysab, K.A., F. Richard, M. Shoreb, G. Pereirab, K.C. Jonesa and F.L. Martin. 2016. Risk assessment of environmental mixture effects. RSC Adv. 6: 47844–47857.

Huang, R., H. Pan, M. Zhou, J. Jin, Z. Ju, G. Ren et al. 2021. Potential liver damage due to co-exposure to As, Cd, and Pb in mining areas: Association analysis and research trends from a Chinese perspective. Environ. Res. 201: 111598.

[ILO] International Labour Organisation. 2017. Metals. United Nations, International Labour Organisation, Geneva, Switzerland.

Jin, R., X. Zhu, M.J. Shrubsole, C. Yu, Z. Xia and Q. Dai. 2018. Associations of renal function with urinary excretion of metals: Evidence from NHANES 2003–2012. Environ. Int. 121(2): 1355–1362.

Kamal, M., Fathy, M.M., Taher, E., Hasan, M. and M. Tolba. 2011. Assessment of the role of paraoxonase gene polymorphism (Q192R) and paraoxonase activity in the susceptibility to atherosclerosis among lead-exposed workers. Ann. Saudi. Med. 31(5): 481–487.

Karri, V., M. Schuhmacher and V. Kumar. 2016. Heavy metals (Pb, Cd, As and MeHg) as risk factors for cognitive dysfunction: A general review of metal mixture mechanism in brain. Environ. Toxicol. Pharmacol. 48: 203–213.

Kayaaltı, Z., V. Aliyev and T. Söylemezoğlu. 2011. The potential effect of metallothionein 2A-5A/G single nucleotide polymorphism on blood cadmium, lead, zinc and copper levels. Toxicol. Appl. Pharmacol. 256(1): 1–7.

Klaasseni, C.D. and L.D. Lehman-McKeeman. 1989. Induction of metallothionein. J. Med. Toxicol. 8(7): 1315–1321.

Kong, Q. and C.L. Lin. 2010. Oxidative damage to RNA: Mechanisms, consequences, and diseases. Cell. Mol. Life Sci. 67(11): 1817–1829.

Kopp, B., P. Sanders, I. Alassane-Kpembi, V. Fessard, D. Zalko and L. Le Hégarat. 2020. Synergic toxic effects of food contaminant mixtures in human cells. Mutagenesis. 35(5): 415–424.

Kortenkamp, F. 2009. State of the Art Report on Mixture Toxicity—Final Report. UE Commission, Brussels, Belgium.

Koszewicz, M., K. Markowska, M. Waliszewska-Prosol, R. Poreba, P. Gac, A. Szymanska-Chabowska et al. 2021. The impact of chronic co-exposure to different heavy metals on small fibers of peripheral nerves. A study of metal industry workers. J. Occup. Med. Toxicol. 16: 12.

Kournoutou, G.G., P.C. Giannopoulou, E. Sazakli, M. Leotsinidis and D.L. Kalpaxis. 2017. Oxidative damage of 18S and 5S ribosomal RNA in digestive gland of mussels exposed to trace metals. Aquat. Toxicol. 192: 136–147.

Levin-Schwartz, Y., M.D. Politis, C. Gennings, M. Tamayo-Ortiz, D. Flores, C. Amarasiriwardena et al. 2021. Nephrotoxic metal mixtures and preadolescent kidney function. Children. 8: 673.

Loft, S., P. Danielsen, M. Løhr, K. Jantzen, J.G. Hemmingsen, M. Roursgaard et al. 2012. Urinary excretion of 8-oxo-7,8-dihydroguanine as biomarker of oxidative damage to DNA. Arch. Biochem. Biophys. 518(2): 142–150.

Lopes, A.C.B.A., M.R. Urbano, A. Souza-Nogueira, G.H. Oliveira-Paula, A.P. Michelin, M.F.H. Carvalho et al. 2017. Association of lead, cadmium and mercury with paraoxonase 1 activity and malondialdehyde in a general population in Southern Brazil. Environ. Res. 156: 674–682.

Madden, E.F. 2003. The role of combined metal interactions in metal carcinogenesis: A review. Rev. Environ. Health. 18(2): 91–109.

Marchiset-Ferlay, N., C. Savanovitch and M.-P. Sauvant-Rochat. 2012. What is the best biomarker to assess As exposure via drinking water? Environ. Int. 39: 150–171.

Masindi, V. and K.L. Muedi. 2018. Environmental contamination by heavy metals. *In*: H. El-Din, M. Saleh and R.F. Aglan. [eds.]. Heavy Metals. IntechOpen, London, UK.

McDonnell, A.M. and C.H. Dang. 2013. Basic review of the cytochrome p450 system. J. Adv. Pract. Oncol. 4(4): 263–268.

Misiaszek, R., Y. Uvaydov, C. Crean, N.E. Geacintov and V. Shafirovich. 2005. Combination reactions of superoxide with 8-Oxo-7,8-dihydroguanine Radicals in DNA: kinetics and end products. J. Biol. Chem. 280(8): 6293–6300.

Mitchell, E., S. Frisbie and B. Sarkar. 2011. Exposure to multiple metals from groundwater—A global crisis: Geology, climate change, health effects, testing, and mitigation. Metallomics. 3: 874–908.

Mohamadkhani, A., M. Pourasgari, M. Saveh, H. Fazli, P. Shahnazari and H. Poustchi. 2019. Association of delta-aminolevulinic acid dehydratase gene variant with serum level of alanine aminotransferase. Hepat. Mon. 19(8): e94664.

Moody, E.C., E. Colicino, R.O. Wright, E. Mupere, E.G. Jaramillo, C. Amarasiriwardena et al. 2020. Environmental exposure to metal mixtures and linear growth in healthy Ugandan children. PLoS ONE. 15(5): e0233108.

Nataf, R., C. Skorupka, L. Amet, A. Lam, A. Springbett and R. Lathe. 2006. Porphyrinuria in childhood autistic disorder: Implications for environmental toxicity. Toxicol. Appl. Pharmacol. 214(2006): 99–108.

Ogun, A.S., N.V. Joy and M. Valentine. 2022. Biochemistry, Heme Synthesis. *In*: StatPearls. Treasure Island, Finland.

Omrane, F., I. Gargouri, M. Khadhraoui, B. Elleuch, D. Zmirou-Navier. 2018. Risk assessment of occupational exposure to heavy metal mixtures: a study protocol. BMC Public Health. 18(1): 314.

Orr, S.E. and C.C. Bridges. 2017. Chronic kidney disease and exposure to nephrotoxic metals. Int. J. Mol. Sci. 18(5): 1039.

Park, Y., J. Zhang and L. Cai. 2018. Reappraisal of metallothionein: Clinical implications for patients with diabetes mellitus. J. Diabetes. 10(3): 213–231.

Perini, J.A., M.C. Silva, A.C.S. Vasconcellos, P.V.S. Viana, M.O. Lima, I.M. Jesus et al. 2021. Genetic Polymorphism of Delta Aminolevulinic Acid Dehydratase (ALAD) Gene and Symptoms of Chronic Mercury Exposure in Munduruku Indigenous Children within the Brazilian Amazon. Int. J. Environ. Res. Public Health. 18(16): 8746.

Popoola, L.T., S.A. Adebanjo and B.K. Adeoye. 2018. Assessment of atmospheric particulate matter and heavy metals: A critical review. Int. J. Environ. Sci. Technol. 15: 935–948.

Potter, N.A., G.Y. Meltzer, O.N. Avenbuan, A. Raja and J.T. Zelikoff. 2021. Particulate matter and associated metals: A link with neurotoxicity and mental health. Atmosphere. 12: 425.

Rai, A., S.K. Maurya, P. Khare, A. Srivastava and S. Bandyopadhyay. 2010. Characterization of developmental neurotoxicity of As, Cd, and Pb mixture: Synergistic action of metal mixture in glial and neuronal functions. Toxicol. Sci. 118(2): 586–601.

Ranjbar, A., H. Ghasemi and F. Rostampour. 2014. The role of oxidative stress in metals toxicity; mitochondrial dysfunction as a key player. Galen Med. J. 3(1): 2–13.

Rodriguez, R.E. and K.S. Kasprzak.1989. Antagonists to metal carcinogens. J. Med. Toxicol. 8(7): 1265–1269.

Roesijadi, G. 1994. Metallothionein induction as a measure of response to metal exposure in aquatic animals. Environ. Health Perspect. 102(12): 91–95.

Roney, N., H.G. Abadin, B. Fowler and H.R. Pohl. 2011. Metal ions affecting the hematological system. Met. Ions Life Sci. 8: 143–155.

Sallsten, G., D. Ellingsen, B. Berlinger, S. Weinbruch and L. Barregard. 2022. Variability of lead in urine and blood in healthy individuals. Environ. Res. 212(Part C): 113412.

Schauder, A., A. Avital and Z. Malik. 2010. Regulation and gene expression of heme synthesis under heavy metal exposure—review. J. Environ. Pathol. Toxicol. Oncol. 29(2): 137–58.

Sekovanić, A., J. Jurasović and M. Piasek. 2020. Metallothionein 2A gene polymorphisms in relation to diseases and trace element levels in humans. Arh. Hig. Rada. Toksikol. 71(1): 27–47.

Shaik, A.P. and K. Jamil. 2008. A study on the ALAD gene polymorphisms associated with lead exposure. Toxicol. Ind. Health. 24(7): 501–506.

Sharaf, A., R. De Michele, A. Sharma, S. Fakhari and M. Oborník. 2019. Transcriptomic analysis reveals the roles of detoxification systems in response to mercury in chromera velia. Biomolecules. 9(11): 647.

Sørensen, M., P.F.R. Schins, O. Hertel and S. Loft. 2005. Transition metals in personal samples of PM2.5 and oxidative stress in human volunteers. Cancer Epidemiol. Biomarkers Prev. 14(5): 1340–1343.

Thompson, J., D.D. Jones and W.H. Beasley. 1977. The effect of metal ions on the activity of delta-aminolevulinic acid dehydratase. Br. J. Ind. Med. 4(1): 32–36.

Wada, T. and J.M. Penninger. 2004. Mitogen-activated protein kinases in apoptosis regulation. Oncogene. 23(16): 2838–2849.

Wang, X., B. Mukherjee and S.K. Park. 2018. Associations of cumulative exposure to heavy metal mixtures with obesity and its comorbidities among U.S. adults in NHANES 2003–2014. Environ. Int. 121(Pt 1): 683–694.

Wirth, J.J. and R.S. Mijal. 2010 Adverse effects of low-level heavy metal exposure on male reproductive function. Syst. Biol. Reprod. 56(2): 147–167.

Woods, J.S. 1995. Porphyrin metabolism as indicator of metal exposure and toxicity. pp. 19–52. *In*: R.A. Goyer and M.G. Cherian [eds.]. Toxicology of Metals. Handbook of Experimental Pharmacology, vol 115. Springer, Berlin, Heidelberg.

Wu, X., S.J. Cobbina, G. Mao, H. Xu, Z. Zhang and L. Yang. 2016. A review of toxicity and mechanisms of individual and mixtures of heavy metals in the environment. Environ. Sci. Pollut. Res. Int. 23(9): 8244–8259. 10.1007/s11356-016-6333-x

Wuana, R.A. and F.E. Okieimen. 2011. Heavy Metals in Contaminated Soils: A Review of Sources, Chemistry, Risks and Best Available Strategies for Remediation. ISRN Ecol. 2011: 402647.

Zamora-Ledezma, C., D. Negrete-Bolagay, F. Figueroa, E. Zamora-Ledezma, M. Ni, F. Alexis and V.H. Guerrero. 2021. Heavy metal water pollution: A fresh look about hazards, novel and conventional remediation methods. Environ. Technol. Innov. 22: 101504.

Zendehdel, R. 2014. Oxidative damage modeling by biomonitoring of exposure to metals for manual metal arc welders. Health Scope. 3(3): e16440.

Zhou, Q., Y. Gu, X. Yue, G. Mao, Y. Wang, H. Su et al. 2016. Combined toxicity and underlying mechanisms of a mixture of eight heavy metals. Mol. Med. Rep. 15(2): 859–866.

Zhou, T., B. Hu, X. Meng, L. Sun, H. Li, P. Xu et al. 2021. The associations between urinary metals and metal mixtures and kidney function in Chinese community-dwelling older adults with diabetes mellitus. Ecotoxicol. Environ. Saf. 226: 112829.

CHAPTER 16

Multibiomarker Approaches

Need and Advantages of Multibiomarker Approaches

The lack of correct assessment of exposure to toxicants, such as metals, is recognized as one of the principal factors contributing to failures in risk assessment (Flora 2014). One of the reasons is the use of traditional models, which cannot estimate the combined effect of exposure to metal mixtures, which is increasingly accepted as closer to the real exposure situation (Ruan et al. 2022). Moreover, such exposures are contributing factor to illness with complex ethiologies, such as cardiovascular and neurodegenerative diseases, diabetes and cancer (Madden 2003, Karri et al. 2016, Wang et al. 2020, Yim et al. 2022). It is not unusual that traditional biomarkers (BM) are not able to identify metal-induced disease susceptibility or contribute to mitigate the development of a disease outcome; in addition, metal exposures can often present non-descript symptoms (Nail et al. 2022). It is also increasingly mentioned that single BMs can hardly capture complex process underlying several illnesses with different disease stages, particularly when the exposure to metal mixtures is a risk or progression factor for the disease (Rachakonda et al., 2004, Quinones and Kaddurah-Daouk, 2009). Considering that in most instances recognizing patterns of changes rather than any single abnormality can dramatically increase diagnostics and monitorization, it is currently accepted that BMs should be combined into panels to increase their predictive power (Mutii 1999, Robin et al. 2013).

Specifically, as risk assessment on metal mixtures is reflected on defending that exposed populations should be studied using a combination of exposure and effect BM (Kakkar and Jaffery 2005). In this view, a more recent approach of using multiple BM in combination has proved to be more potent and accurate than the standalone measurements of individual BMs in the context of several human diseases (Reimann et al. 2019). In the last 20 yr, the possibility of BM investigation has grown immensely due to the advent of new technologies and tools made available to researchers, which allow a myriad of approaches making use of these tools and the possible combinations of them (Araujo et al. 2014).

Multivariate Statistical Methods

BMs can be identified by classical monovariate methods, where each BM is considered as independent (Student's t-test, Mann-Whitney test, etc.), or multivariate methods, which allow to take into consideration the correlation structure of the data, in this case the interactions (Robotti et al. 2013). The last one achieves the best predictive ability, particularly when the challenge is to construct a disease risk from exposure to multiple environmental risk factors (Robotti et al. 2013, Park et al. 2014). For a relatively small number of pollutants, some simpler methods can include the indicator approach, for which one pollutant represents the combined exposure to several pollutants or the source apportionment approach, for which particle constituents are assigned to emission sources using principal component analysis and hierarchical clustering; however, these approaches do not account for a wide range of environmental pollutants (Park et al. 2014). The most relevant applications of statistical techniques were developed specifically for mixtures-based applications and include Environment-Wide Association Studies (EWAS), Environmental Risk Scores (ERS) and Weighted Quantile Sum (WQS) regression (Patel et al. 2010, Merced-Nieves et al. 2021).

   EWAS are used to explore environmental factors associated with health outcomes. This approach is similar to Genome-Wide Association Studies (GWAS), which will be described further , although using the exposome rather than the genome. Despite the challenge that when compared with genetic factors that are usually stable over time, environmental factors have large spatial and temporal heterogeneities, in the future, EWAS and GWAS might be used jointly to guide gene-environment interactions studies (Zheng et al. 2020). An EWAS integrates multiple survey results between chemical and disease using meta-analysis methods, with further validation of the results using other populations (Lee et al. 2020). The process includes two methodological steps. The first one consists of a panel of several unique environmental assays or environmental "loci", measured across cases of diseased subjects and across controls; these yields to environmental factors with significantly high association with the outcome under study, while controlling for multiple hypotheses. In the second step, the found associations are validated using data from other cohorts (Patel et al. 2010). In this way, EWAS provides excellent insights to identify 'top hit' pollutants, constituting a powerful and effective tool to identify potential risk factors for adverse outcomes of concern when multiple candidate chemicals are a starting point (Park et al. 2014, Lee et al. 2020). Using EWAS, hundreds of bio-monitored chemicals were already measured simultaneously in blood or urine for their association, namely with chronic kidney disease. The results obtained from this study helped to identify a list of potential pollutants (which included several metals) with significant association with this disorder. Blood cadmium (Cd) was consistently associated with chronic kidney disease in almost all ranges of albuminuria and glomerular filtration rate (eGFR) manifestations; co-exposure to lead (Pb) and Cd was also suggested to increase the risk of the disease, especially the one defined by albuminuria (Lee et al. 2020). In another study, EWAS was applied analyzing data from 543 environmental factors, which were available from the National Health and

Nutrition Examination Survey (NHNE) and included the amount of pesticides or metals present in urine or blood. Survey-weighted logistic regression was used to associate each one of the environmental attributes with diabetes , while adjusting for age, sex, body mass index, ethnicity and other factors, and also yielded to the identification of several risk environmental factors (Patel et al. 2010).

The Environmental Risk Score (ERS) is a useful tool for characterizing cumulative risk from pollutant mixtures (Park et al. 2017). It is basically a potential summary measure for the effects of mixtures (including metal mixtures) in epidemiologic research and aims to build a predictive risk model. The ERS are estimated as weighted summary measures of the effects of metals, being weights determined by the magnitudes (standardized regression coefficients) of the association between each exposure and the outcome of interest (Park et al. 2017, Ashrap et al. 2021). For common disease pathways, such as oxidative stress, inflammation, epigenetic modification and endocrine disruption, ERS can capture cumulative early biological effects and discriminate individuals who were at increased risk of manifesting various downstream clinical diseases. Specifically, oxidative stress is a well-known common disease pathway linking environmental pollutant exposure to numerous health endpoints. With this view, an interesting example of ERS application is a recent study that linked cumulative risk of oxidative stress due to metal mixtures exposure (20 metal BMs were measured in urine or whole blood) with numerous health endpoints relevant to oxidative stress; these included cancer, type-2 diabetes and others. An ERS of metal mixtures to predict a marker of oxidative stress, Gamma-Glutamyl Transferase (GGT) was also constructed; important metals were identified in relation to GGT, including Cd, dimethylarsonic acid, monomethylarsonic acid, cobalt (Co) and barium (Park et al. 2017).

Weighted Quantile Sum (WQS) regression is another statistical model for multivariate regression in high-dimensional datasets used in studies such as environmental exposures, epi/genomics and metabolomic studies, which will be described later (Renzetti et al. 2021). This tool allows to estimate a body burden index within a set of correlated environmental chemicals, and further estimate the association between the index and an outcome of interest (Czarnota et al. 2015). Therefore, WQS regression has the specific goal of estimating the effect of the mixture as a whole using this index in a generalized linear model to estimate associations with the health outcome (Gennings et al. 2020, Keil et al. 2020). It additionally allows to identify an individual chemical most strongly associated with a health outcome, while adjusting for risk factors, since the contribution of each individual predictor to the overall index effect may be assessed by the relative strength of the weights the model assigns to each variable (Keil et al. 2020, Renzetti et al. 2021, Wheeler et al. 2021). WQS regression have been used in modelling chemical mixtures and cancer risk, identifying "bad actors" in a set of highly correlated environmental chemicals (Carrico et al. 2014, Czarnota et al. 2015). This methodology was also used to assess the effects of metal/metalloid mixtures to Tumour Necrosis Factor (TNF)-α and the kidney function; TNF-α is a cytokine which exerts a direct renal action by regulating hemodynamic and excretory function in the kidney (Mehaffey and Majid 2017). This study showed that Pb, As, Zn, Se and their mixtures may act on TNF-α through interactive mechanisms, offering insights into what primary components of metal mixtures affect inflammation and kidney function during co-exposure to the metals (Luo et al. 2022)

"Omics" Technologies

Exposome

The exposome is a relatively new concept which can be defined as all the exposures of an individual in a lifetime and how these exposures relate to health (NIOSH 2014). In other words, it is a comprehensive evaluation of all exposures and their contribution to disease causation or progression (Chen et al. 2021). Exposomics is the study of the exposome currently being the leading methodology for assessing health impacts of multiple environmental exposures in environmental health studies (NIOSH 2014, Chen et al. 2021). There are three overlapping domains within the exposome, the first one refers to the external exposome: a general external environment, which include factors such as climate, social or stress; a second one which is the specific external environment, which pertains to specific contaminants, diet, physical activity, infections, among other factors; and the third one, the internal environment (internal exposome) which include internal biological factors such as metabolic factors, gut micro flora, inflammation or oxidative stress (Vrijheid 2014). Therefore, exposomics relies on the application of internal and external exposure assessment methods (NIOSH 2014). In the assessment of the external exposome better modelling of exposure can be achieved, for example by using predictive exposure models that combine questionnaire information with BMs (Vrijheid 2014). As regards internal exposome, the use of several "omics" methods has been recommended in exposomics studies to identify the links between exposures and health outcomes, and potentially allows to develop new BMs for exposures and early health effects (Chen et al. 2021).

To provide a definition of "Omics" technologies, these are emerging concepts which refer to collective and high-throughput analyses to detect genes (genomics), mRNA (transcriptomics), proteins (proteomics) and metabolites (metabolomics), among others, in biological samples. The outcomes are integrated through strong systems of biology, bioinformatics and computational tools to study mechanisms, interactions and functions of cell populations', tissues, organs and of the whole organism (Nalbantoglu and Karadag 2019). In this way, "omics" play a role in screening, diagnosis and prognosis of diseases, aid to understand their aetiology and quite importantly, lend themselves to BM discovery (Horgan and Kenny 2011). In the range of exposure to toxics, while determining perturbations in molecular pathways due to chemical stress, this information may ultimately lead to development of novel screens or new BMs, to investigate multiple molecules simultaneously (Kakkar and Jaffery 2005, Horgan and Kenny 2011). In the development of the exposome concept, the contribution of omics techniques is likely to be mainly in their potential to measure profiles (or signatures) of the biological response to complex exposure mixtures or a cumulative exposure experience (Vrijheid 2014). Major challenges in exposomics include the fact that the exposome is dynamic, with exposures varying on an hourly to yearly basis, as well as the fact that a given exposure or dose will not have the same effect during the various age periods (Vrijheid 2014). Pre-natal exposures during pregnancy (maternal exposome) can play a major role and influence the health outcome of a child; past exposures can also cause changes in human health many years later. Hence, both present and past factors must be considered in assessing the exposome, although past measurements are often only available for a limited set of chemicals (Aurich et al. 2021).

Genomics

Genomics is the study of all of a person's genes (the genome), including interactions of these genes with each other and with the person's environment, while toxicogenomics is recognized as a scientific field which looks at how genomes respond to environmental toxicants (NHGRI 2020). Genes whose regulation is consistently disrupted in certain toxic responses are good candidates for genetic BMs (Kakkar and Jaffery 2005). Gene expression profiling can be conducted directly on the blood from preclinical studies or clinical trials (Decristofaro and Daniels 2008). Toxicogenomics may be used for BM discovery because the analysis of large databases of gene expression profiles are included, followed by *in silico* mining for differentially expressed genes. Furthermore, the evaluation and characterization of gene expressions after exposure to toxic insults can lead to the prediction of chemical mixtures toxicity and identify associated mechanisms (Wu et al. 2016). Such an approach is important particularly for environmental pollutants, such as metals, which may contain more than one mechanism of action and may interact with more than one specific site along an Adverse Outcome Pathway (AOP) (Anyanwu et al. 2018). The AOP concept is quite relevant in this context, since is made up of aspects concerning molecular interactions, followed by issues of responses to stress resulting from exposure to the toxicant and, finally, to adverse effects resulting from exposure to the combined mixture (Wu et al. 2016).

Some genome-wide (GWAS) association studies to investigate genetic variants associated with whole blood levels of a range of toxic metals are documented. In one of this works, genetic variants which included single nucleotide polymorphism in genes encoding critical proteins involved in Absorption, Distribution, Metabolism and Excretion (ADME) steps were related to circulating levels of several metals; these metals were aluminium (Al), Cd, Co, copper (Cu), chromium (Cr), mercury (Hg), manganese (Mn), molybdenum, Ni, Pb and zinc (Zn). In other works, novel associations were found between some ion transporters and whole blood Mn concentrations as well as other less well-known genes related to Cd and Hg levels (Ng et al. 2015).

Transcriptomics

While genomics provides an overview of the complete set of genetic instructions provided by the DNA, transcriptomics investigates gene expression patterns (ISAAA 2022). This "omics" comprises the analysis of everything relating to RNAs transcripts, including messenger RNAs (mRNAs), microRNAs (miRNAs) and different types of long noncoding RNAs (lncRNAs), produced by the genotype at a given time; it encompasses their transcription and expression levels, functions, locations, trafficking and degradation. In this way provides a link between the genome, the proteome and the cellular phenotype (Cocolin et al. 2014, Milward et al. 2016). Changes in miRNAs profiles have been widely explored since there is growing evidence that metals might use their toxicity through this path (Wallace et al. 2020). These molecules are small non-coding RNAs, with an average of 22 nucleotides in length, most of them coming from DNA sequences. miRNAs interact with the 3′ UTR of target mRNAs to suppress expression, via decreased translation, deadenylation or degradation of the mRNA (O'Brien et al. 2018, Cardoso et al. 2020); miRNAs seem to be able to interact with other regions, including the 5′ UTR, coding sequence, which is the region of a mRNA that is directly upstream from the initiation codon. Recent studies suggest that miRNAs are shuttled between different subcellular compartments to control the rate of translation and even transcription (O'Brien et al. 2018). As mentioned, environmental factors which include exposure to toxic metals can exert influence on miRNA function leading to aberrant expression. Due to this, the possibility of using miRNAs as sensitive BMs of toxicity or disease prediction has gained much importance in recent years (O'Brien et al. 2018, Wallace et al. 2020). Since multiple miRNAs can be altered following some metals exposure this reduces the use of changes in a single miRNA as a sole BM (Wallace et al. 2020). Therefore, the use of miRNA profiles has proved to be useful, particularly when assessing metal mixtures exposure and/or effects. To cite a few examples, arsenic (As)-treated cells already revealed that 12 miRNAs were up-regulated and 14 were down-regulated; namely, miR-6739-5p, mir-4521, miR-181b-5p, miR-100-5p and miR-3919 were up-regulated and miR-513a-5p was down-regulated. Another study also demonstrated that miR-182-5p suppression was involved in As-induced carcinogenesis (Cardoso et al. 2020, Wallace et al. 2020). The miR-18a, miR-132, and miR-146b levels were additionally found to be key miRNAs involved in responses following Cd exposure (Wallace et al. 2020). It was also shown that early-life exposure to Pb could increase the expression of miRNAs that target proteins associated with Alzheimer's Disease; those included miR-106b (which binds to amyloid-β protein precursor mRNA), miR-29b miR-29b and miR-132 (Masoud et al. 2016).

Epigenomics

The term "epigenetics" refers to the regulation of cell activity and gene expression by mechanisms that do not alter the genetic code (DNA sequence) itself (Forno and Celedón 2019). Epigenetic processes regulate the function of specific cells and tissues over time and in response to the environment, ageing or other factors, even though all cells in the human body contain the same genetic sequence (Wallace et al. 2020). In turn, the epigenome comprises all the chemical compounds and factors added to an individual's genome to regulate the expression of the genes within that genome (Forno and Celedón 2019). Epigenetics is recognized nowadays as a key mediator of environmental response and a target of toxicants and thus, toxicoepigenetics developed to study the relationship between epigenetic modifications and disease status in response to exposure to toxic agents. This rapidly expanding area of research benefits from fast technological advances which enable to acquire and analyze largely increasing volumes of data (Wallace et al. 2020, Le Goff et al. 2022). Again, epigenetic alterations can be used as BMs of effect in response to exposure to environmental toxicants, allowing such BMs the possibility of being used additionally as predictors of disease, when such epigenetic marks are associated with differential gene expression (Wallace et al. 2020). There are several known mechanisms of epigenetic regulation, of which DNA methylation and histone modifications are so far, the best understood (Forno and Celedón 2019, Wallace et al. 2020). Most epigenetic research focused on metals comprised so far, the study of As, Pb, Cd, methylHg, Ni and Cr. These studies had revealed, for example, that chronic As exposures have a dose-response relationship with changes in DNA methylation (including in genes associated with As-mediated diseases) as well as with histone post-translational modifications (Le Goff et al. 2022).

Proteomics

Proteomics is a another 'omics' field, which study the interactions, function, composition and structures of proteins and their cellular activities, comprising three main areas: expression proteomics, functional proteomics and structural proteomics (Al-Amrani et al. 2021). Most proteomic discoveries and efforts to date have been mainly directed towards areas of cancer research, drug and drug target discovery, but also in BMs research; bioinformatics analyses using novel proteomics algorithms allows additionally the management of large and varied data in the process of marker discovery (Husi and Albalat 2014, Al-Amrani et al. 2021). Examining the toxic effects of metals on protein expression can be useful for gaining insight into the biomolecular mechanisms of toxicity and for identifying potential candidate metal-specific protein markers of exposure and response (Kakkar and Jaffery 2005, Luque-Garcia et al. 2011). An efficient proteomics approach for generating scientific data to characterize and predict metal mixture toxicity and to support risk assessment of environmental mixtures, is documented. In this work, the expression of 21 critical toxicity pathway regulators and 445 downstream proteins in human BEAS-2B cells were measured. A high correlation between changes in protein expression and cellular toxic responses to both individual metals and metal mixtures was demonstrated (Ge and Bruno 2017). Another study used a systems toxicology approach to study the impact of the metal mixture, Pb, As and Hg in neurodegenerative diseases. The applied proteomic approach was a unique way to distinguish single and mixtures' effects on hippocampal cells and predict the BMs for mixtures on these cells; additionally, the identified molecules could serve as potential BMs and in the future be included in the biomonitoring of people exposed to metals (Karri et al. 2020). Proteomic profiling of urine and sera of workers occupationally exposed to As and Pb and of a population group residing in an area contaminated by As and Cd allowed discovering new and better BMs of mixed metal exposure; it also permitted the diagnosis of specific physiological alterations such as metal induced renal dysfunction (Luque-Garcia et al. 2011).

Most proteomic analyses have shown that specific groups of proteins, providing the first line of defence against metal-induced reactive oxygen species formation, are differentially regulated in response to metal toxicity; these include several antioxidative enzymes, which are found to be up-regulated, such as superoxide dismutase and proteins involved in glutathione biosynthesis; stress proteins, referred to as heat-shock proteins, are also induced by metals (Bauman et al. 1993, Luque-Garcia et al. 2011).

Notably organisms exposed to metals also exhibit increased expression of several proteins associated with energy production and metabolism, suggesting that higher energy is required to activate the metabolic processes, mainly focused on detoxification (Luque-Garcia et al. 2011).

Metabolomics

Metabolomics is an analytical profiling technique for measuring and comparing large numbers of metabolites present in biological samples, combining analytical chemistry and multivariate data analysis; the repertoire of biochemicals present in cells, tissues and body fluids is known as the metabolome (Quinones and Kaddurah-Daouk 2009, Manchester and Anand 2017). The broad classes of metabolites include amino acids, nucleotides, carbohydrates and lipids (Booth et al. 2011). Metabolomics is an emerging field and can be regarded as the endpoint of the known "omics" cascade and while genomics, transcriptomics or proteomics show the probability that a process may occur, metabolomics provides information about "what is actually happening" (García-Sevillano et al. 2015). In the same way as for the other "omics", metabolomics have the potential to improve current single metabolites-based clinical assessments, by identifying BMs as metabolic signatures that embody global biochemical changes in disease. While mapping in greater detail perturbations in many biochemical pathways and links among these pathways, this information is a promise for BM discovery (Quinones and Kaddurah-Daouk 2009). As metabolomic profiles can reflect external factors, they are becoming increasingly used in studies of metals exposure (environmental toxicometabolomics), as a key to understand and identify cellular or biochemical targets of metals and the underlying physiological responses (Booth et al. 2011, García-Sevillano et al. 2015). Moreover, understanding biochemical networks underlying metabolic homeostasis and their association with exposure to multiple metals may help identify novel BMs, pathways of disease and potential signatures of environmental metal exposure (Sanchez et al. 2021).

Lipidomics

Lipidomics is a sub-discipline of metabolomics and is focused on the systemic analysis of lipids and their interacting partners. Deregulation of lipid profiles have been associated with disease onset and progression, and their characterization has been used in studies searching for novel BMs in many diseases; they include cancer, liver, kidney or cardiovascular disease, diabetes and Alzheimer's disease (Chen et al.

2021). Several epidemiological studies also suggest that abnormal lipid metabolism is associated with environmental chemical exposure, which include exposure to metals (Antonowicz et al. 1998, Kim et al. 2022). While the toxic mechanisms are not yet completely understood, it is acknowledged that they are generally associated with oxidative stress and unsaturated fatty acids are easily affected by metal ions, through lipid oxidation; particularly the production of hydroxyl and peroxynitrite, is well known to cause lipid imbalance (Kim et al. 2022, Zhou et al. 2022). Lipidomics have been applied in several studies focused on metals effects being described as a link between the exposure to Pb, Cd and Hg and dyslipidemia; dyslipidemia is characterized by an imbalance of lipid levels in the blood, and it is caused by excessive entry of lipoproteins into the bloodstream or impaired ability to remove them (Kim et al. 2022). Such a condition is generally defined by Elevated Total Cholesterol (TC), triglyceride (TG), Low-Density Lipoprotein Cholesterol (LDL-C) and a low level of Non-High-Density Lipoprotein Cholesterol (Non-HDL-C), positive associations were found between blood Pb and Hg and urinary Hg concentrations and this serum lipid profile (Kim et al. 2022). Another study assessed lipids of children and adolescents exposed to multiple pollutants from a petrochemical industrial complex, which included As, Cd, Cr, Ni, V, Hg, Pb, Mn, Cu, Sr and thallium. To identify lipid perturbations that could be associated with early health effect BMs, serum acylcarnitine profiles (ACP) were determined (Chen et al. 2021). ACP analysis is performed for the biochemical screening of disorders of Fatty Acid Oxidation (FAO) and organic acid metabolism (Millington et al. 2011). These molecules are included in the carnitine pool, which plays a role in facilitating FAO in mitochondria and peroxisomes; thus, changes in their blood concentrations generally reflect disorders of long-chain FAO (Rinaldo et al. 2008, McCann et al. 2021). Deregulations in serum acylcarnitines can activate inflammatory signalling pathways and have been associated with chronic diseases including cancer, cardiovascular diseases, liver diseases and chronic kidney disease. A study with these exposed children demonstrated changes in acylcarnitines profiles; long-chain acylcarnitines were clustered together and down-regulated in a high exposure group compared to low exposure groups, while a short-chain acylcarnitine (Hexanoylcarnitine, C6) was up-regulated in the high exposure group compared to the low exposure group (Chen et al. 2021). In other studies, novel associations between urinary metal BMs and metabolism networks were also found in fatty acid, energy and amino acid metabolism pathways. Oher results also indicated that individual metabolite associations varied for different metals, specifically effects in amino acid metabolism induced by As, Sb, Se and U and fatty acid and lipid metabolism induced by As, Mo, W, Sb, Pb, Cd and Zn (Sanchez et al. 2021).

Adductomics

Several chemicals can interact with nucleophilic hot spots (susceptible sites to electrophiles) present in DNA, lipids, proteins, RNA and other macromolecules leading to the formation of adducts (Behl et al. 2021). Metal toxicity can induce the formation of adducts; a study already detected adducts with Pb, Cd, Hg, Al, antimony, As, nickel (Ni), strontium, Cu, Mn, Cr and Co (Howard 2009). The formation of covalent adducts, which are irreversible, may play a key role in the onset of several adverse health outcomes (Golime et al. 2019, Behl et al. 2021). Attributed to their significant impact on biological systems, adduct formation can result in deleterious health complications, including diabetes, cancer, birth defects, neurodegenerative, cardio-vascular and autoimmune diseases (Behl et al. 2021). In this context, adductomics is a novel approach and an emerging discipline in toxicological research (Cooke et al. 2018, Golime, et al. 2019). It involves the identification of the nature of the covalent modifications along with the site of adduction within target biomolecules (Behl et al. 2021). Since exposure to electrophilic compounds can directly or indirectly produce reactive oxygen species, adducts of these species can serve as indirect exposure BMs (Golime, et al. 2019). Specifically, as regards to DNA adductomics, this sub-area aims to determine the totality of adducts in the genome being quite important as one of the primary methods to evaluate the genotoxic capability of a chemical compound, because when DNA adducts are not repaired, mutations during cell division may occur (Balbo et al. 2014, Cooke et al. 2018). Examination on exposures to multiple genotoxicants have led researchers to develop a new DNA adductomic approach to screen for many DNA lesions simultaneously as BMs, which provides a very powerful tool for exposome characterization and cancer-based systems toxicology studies prevention studies (Balbo et al. 2014). Currently data from adductomics serve as a guide for regulatory agencies and empowers other stakeholders in taking preventive measures against toxic chemical's exposure (Behl et al. 2021).

Metallomics

Metallomics is defined as a field of science that elucidates all features of the actions, interactions, structures, transports and roles of metals in biological systems (Himeno et al. 2019). In sequence, "metallomes" is a general definition for metalloenzymes, metalloproteins and many other biomolecules containing metal ions (Singh and Verma 2018). However, such definitions are still confusing since other than "metallome" and "metallomics" many related designations have emerged, for instance, ionomics, hetero-atom tagged proteomics or elementomics (García-Sevillano et al. 2015, Singh and Verma 2018). Metallomics also includes metallogenomics, metalloproteomics and metallo-metabolomics; others refer genomics and proteomics as distinct fields that have created tremendous data that can be used in metallomics (Kakkar and Jaffery 2005, Kulkarni et al. 2006, Singh and Verma 2018). Metallomics adds chemical elements/metals to the four building blocks of biomolecules and the fields of their studies: carbohydrates (glycome), lipids (lipidome), proteins (proteome) and nucleotides (genome), and may also include the identification of metallobiomolecules overexpressed or inhibited in living organisms under the action of metals (Maret 2018, Rodríguez-Moro et al. 2021). With the view that metallomics is a field distinct from genomics and proteomics, metallomics can be integrated with genomics, when considering the importance of metals in the structural integrity of DNA and RNA, their role in replication and the fact that several metalloenzymes and metal ions assist the synthesis and the metabolic roles of genes and proteins. Metallomics can also integrated be with proteomics, since metalloproteins comprises nearly 30% of the proteins (Singh and Verma 2018). The term "toxicometallomics" have been used namely in studies focused on the toxicology of exotic metalloids based on speciation studies as well as in the roles of metal transporters in the toxicity induced by Cd, Mn As (Ogra 2009, Himeno et al. 2019). Metallomics demonstrates perturbations in trace and ultra-trace elements of cells, tissues and biofluids, establishing a complete element profile of a sample that can be used as BMs and identify new ones (Araujo et al. 2014, Sanyal et al. 2016). In addition to the role played by metals in structural and catalytic roles in many biochemical reactions, leading to the discovery of new BM candidates, metal-binding proteins may serve as BMs for many diseases (Kulkarni et al. 2006). A metallomic analysis to assess toxic metal burdens in children described that their Pb, Cd and Al burdens were markedly higher than those in their mothers. The levels of the essential metals Zn, magnesium and calcium, which can compete against and antagonize toxic metals, were significantly lower, suggesting serious concern for their neurodevelopment (Yasuda et al. 2021).

Systems biology

To encompass the reality of molecule-atoms interactions in their completeness, combinations of "omics" have been tried, focusing on environment, food and health issues (Rodríguez-Moro et al. 2021). Different "omics" methodologies assess different parts of the complex pathophysiology of complex disease development and progression, the most likely being the analysis of just one "omics" subset providing a skewed, biased and incomplete picture of the underlying biology (Olivier et al. 2019).

The integration of "omics" techniques is called "systems biology" and aims to define the inter-relationships of several or ideally, all the elements in a system, rather than study each element individually. It is necessary to have a deep insight into global metabolism changes caused in bioindicators by metal exposure, such novels approaches have led to improve these observations in a dramatical way. Several investigations have been carried out in environmental monitoring based on "omics" technologies integration, which include proteomics/transcriptomics/metallomics, metallomics/metabolomics and functional genomics/metabolomics, to solve the biological response to contamination caused by metals (García-Sevillano et al. 2015). An example of an integration of genomics, transcriptomics and proteomics and its potential to new BMs discovery is illustrated in Fig. 8.

Final Remarks

This chapter described the use of traditional models based on single biomarkers (BMs) as the main reason explaining why accurate risk assessments are not achieved on combined effect of exposure to metal mixtures or metal contribution as a risk factor or progression factor of multifactorial diseases. Currently, the most relevant applications of novel statistical techniques are developed specifically for mixtures-based applications and include environment-wide association studies, environmental

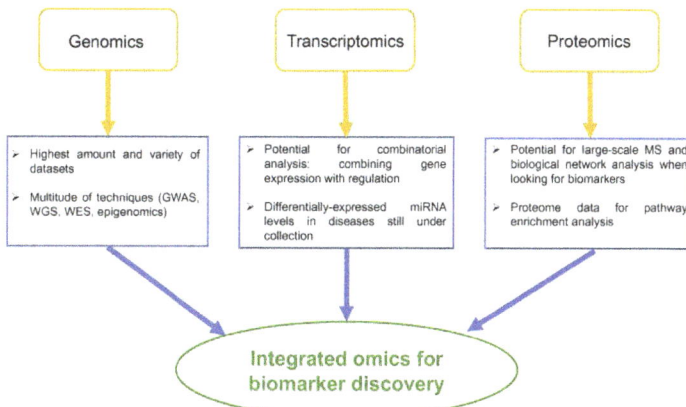

Figure 8. Biomarkers discovery through the integration of genomics, transcriptomics and proteomics (Brooks et al. 2017).

risk scores and weighted quantile sum regression, focused in identifying "bad actors" in a set of environmental chemicals and their contribution to adverse health outcomes. Additionally, exposome is a relatively new concept quite relevant in this context, which can be defined as all the exposures of an individual in a lifetime and how those exposures relate to health; several "omics" methods are highly recommended in exposomics studies. "Omics" technologies are also emerging concepts which refers to collective and high-throughput analyses to detect genes (genomics), mRNA (transcriptomics), proteins (proteomics) and metabolites (metabolomics), among others, in biological samples, the outcomes being integrated through vigorous systems of biology, bioinformatics and computational tools. The advent of "omics" techniques in the last 20 yr brought dramatic improvements in more related risk assessments with real life exposure scenarios, and moreover, provided the possibility for the investigation and discovery of novel BMs. Several investigations have been carried out in environmental monitoring based on various "omics" technologies integration, to decrypt the biological response to contamination caused by metals and find new BMs. Such approaches, with further integration of developed *in silico* methods, is unquestionly the future for new BMs discovery, new uses of these BMs, bringing dramatic improvements in the protection of populations exposed to metals.

References

Al-Amrani, S., Z. Al-Jabri, A. Al-Zaabi, J. Alshekaili and M. Al-Khabori. 2021. Proteomics: Concepts and applications in human medicine. World J. Biol. Chem. 12(5): 57–69

Antonowicz, J., R. Andrzejak and T. Lepetow. 1998. Influence of heavy metals, especially lead, on lipid metabolism, serum alpha-tocopherol level, total antioxidant status, and erythrocyte redox status of copper smelter workers. Fresenius J. Anal. Chem. 361: 365–367.

Anyanwu, O., A.N. Ezejiofor, Z.N. Igweze and O.E. Orisakwe. 2018. Heavy metal mixture exposure and effects in developing nations: An update brilliance. Toxics. 6: 65.

Araujo, T.O., L.T. Costa, J. Fernandes, R.Q. Aucélio and R. Calixto de Campos. 2014. Biomarkers to assess the efficiency of treatment with platinum-based drugs: what can metallomics add? Metallomics. 6(12): 2176–2188.

Ashrap, P., D.J. Watkins, B. Mukherjee, Z. Rosario-Pabón, C.M. Vélez-Vega, A. Alshawabkeh et al. 2021. Performance of urine, blood, and integrated metal biomarkers in relation to birth outcomes in a mixture setting. Environ. Res. 200: 111435.

Aurich, D., O. Miles and E.L. Schymanski. 2021. Historical exposomics and high-resolution mass spectrometry. Exposome. 1(1): osab007.

Balbo, S., R.J. Turesky and P.W. Villalta. 2014. DNA adductomics. Chem. Res. Toxicol. 27: 356–366.

Bauman, J.W., J. Liu and C.D. Klaassen. 1993. Production of Metallothionein and heat-shock proteins in response to metals. Bauman. Fundam. Appl. Toxicol. 21: 15–22.

Behl, T., M. Rachamalla, A. Najda, A. Sehgal, S. Singh, N. Sharma et al. 2021. Applications of adductomics in chemically induced adverse outcomes and major emphasis on DNA adductomics: A pathbreaking tool in biomedical research. Int. J. Mol. Sci. 22(18): 10141.

Booth, S.C., M.L. Workentine, A. Weljie and R.J. Turner. 2011. Metabolomics and its application to studying metal toxicity. Metallomics. 3(11): 1142–1152.

Brooks, J., A. Watson and T. Korcsmaros. 2017. Omics approaches to identify potential biomarkers of inflammatory diseases in the focal adhesion complex. Genom Proteom Bioinf. 15(2): 101–109.

Cardoso, A.P.F., K.T. Udoh and J.C. States. 2020. Arsenic-induced changes in miRNA expression in cancer and other diseases. Toxicol. Appl. Pharmacol. 409: 115306.

Carrico, C.K., C. Gennings, D.C. Wheeler and P. Factor-Litvak. 2014. Characterization of weighted quantile sum regression for highly correlated data in a risk analysis setting. J. Agr. Biol. Environ. Stat. 20(1): 100–120.

Chen, C.-H.S., T.-C. Kuo, H.-C. Kuo, Y.J. Tseng, C.-H. Kuo, T.-H. Yuan et al. 2021. Lipidomics of children and adolescents exposed to multiple industrial pollutants. Environ. Res. 201: 111448.

Cocolin, L. and K. Rantsiou. 2014. Molecular biology in microbiological systems. In: C.A. Batt and M.L. Tortorello [eds.]. Encyclopedia of Food Microbiology. Academic Pfess, Cambridge, USA.

Cooke, M. S., C.-W. Hu, Y.-J. Chang and M.-R. Chao. 2018. Urinary DNA adductomics—A novel approach for exposomics. Environ. Int. 121(2): 1033–1038.

Czarnota, J., C. Gennings and D.C. Wheeler. 2015. Assessment of weighted quantile sum regression for modeling chemical mixtures and cancer risk. Cancer Inform. 14(S2): 159–171.

Decristofaro, M.F. and K.K. Daniels. 2008. Toxicogenomics in biomarker discovery. Methods Mol. Biol. 460: 185–194.

Flora, S.J.S. 2014. Metals. pp. 485–519. In: R.C. Gupta [ed.]. Biomarkers in Toxicology. Academic Press, Cambridge, USA.

Forno, E. and J.C. Celedón. 2019. Epigenomics and transcriptomics in the prediction and diagnosis of childhood asthma: Are we there yet? Front. Pediatr. 2(7): 115.

García-Sevillano, M.A., T. García-Barrera and J.L. Gómez-Ariza. 2015. Environmental metabolomics: Biological markers for metal toxicity. Electrophoresis. 36(18): 2348–2365.

Ge, Y. and M. Bruno. 2017. From Single Metal to Metal Mixtures: Systematic Proteomic Approach to Characterize the Impacts of Chemical Interactions on Protein and Cytotoxicity Responses to Environmental Mixture Exposures. Society of Environmental Toxicology and Chemistry, Denver, USA.

Gennings, C., P. Curtin, G. Bello, R. Wright, M. Arora and C. Austin. 2020. Lagged WQS regression for mixtures with many components. Environ. Res. 186: 109529.

Golime, R., B. Chandra, M. Palit and D.K. Dubey. 2019. Adductomics: A promising tool for the verification of chemical warfare agents' exposures in biological samples. Arch. Toxicol. 93: 1473–1484.

Himeno, S., D. Sumi and H. Fujishiro. 2019. Toxicometallomics of Cadmium, Manganese and Arsenic with special reference to the roles of metal transporters. Toxicol. Res. 35: 311–317.

Horgan, R.P. and L.C. Kenny. 2011. 'Omic' technologies: Genomics, transcriptomics, proteomics and metabolomics. Obstet. Gynaecol. 13: 189–195.

Howard, J.M. 2009. The Detection of DNA Adducts (Risk Factors for DNA Damage). A method for genomic dna, the results and some effects of nutritional intervention. J. Nutr. Environ. Med. 12(1): 19–31.

Husi, H. and A. Albalat. 2014. Proteomics. pp. 147–179. In: S. Padmanabhan [ed.]. Handbook of Pharmacogenomics and Stratified Medicine. Academic Press, Cambridge, USA.

[ISAAA] International Service for the Acquisition of Agri-biotech Applications. 2022. Pocket K No. 15: 'Omics' Sciences: Genomics, Proteomics, and Metabolomics. International Service for the Acquisition of Agri-biotech Applications, Manila, Philippines.

Kakkar, P. and F.N. Jaffery. 2005. Biological markers for metal toxicity. Environ. Toxicol. Pharmacol. 19(2): 335–349.

Karri, V., M. Schuhmacher and V. Kumar. 2016. Heavy metals (Pb, Cd, MeHg, As) as risk factors for cognitive dysfunction: A general review of metal mixture mechanism in brain. Environ. Toxicol. Pharmacol. 48: 203–213.

Karri, V., M. Schuhmacher and V. Kumar. 2020. A systems toxicology approach to compare the heavy metal mixtures (Pb, As, MeHg) impact in neurodegenerative diseases. Food Chem. Toxicol. 139: 111257.

Keil, A.P., J.P. Buckley, K.M. O'Brien, K.K. Ferguson, S. Zhao and A.J. White. 2020. A quantile-based g-computation approach to addressing the effects of exposure mixtures. Environ. Health Perspect. 128(4): 47004.

Kim, D.-W., J. Ock, K.-W. Moon and C.-H. Park. 2022. Association between heavy metal exposure and dyslipidemia among korean adults: From the korean national environmental health survey, 2015–2017. Int. J. Environ. Res. Pub. Health. 19: 3181.

Kulkarni, P.P., Y.M. She, S.D. Smith, E.A. Roberts and B. Sarkar. 2006. Proteomics of metal transport and metal-associated diseases. Chemistry. 12(9): 2410–2422.

Le Goff, A., S. Louvel, H. Boullier and P. Allard. 2022. Toxicoepigenetics for risk assessment: Bridging the gap between basic and regulatory science. Epigenet. Insights. 15: 25168657221113149.

Lee, J., S. Oh, H. Kang, S. Kim, G. Lee, L. Li et al. 2020. Environment-wide association study of CKD. Clin. J. Am. Soc. Nephrol. 15(6): 766–775.

Luo, K.H., H.P. Tu, C.H. Yang, C.C. Yang, T.H. Chen and H.Y. Chuang. 2022. Use of generalized weighted quantile sum regressions of tumor necrosis factor alpha and kidney function to explore joint effects of multiple metals in blood. Int. J. Environ. Res. Pub. Health. 19(12): 7399.

Luque-Garcia, J.L., P. Cabezas-Sanchez and C. Camara. 2011. Proteomics as a tool for examining the toxicity of heavy metals. Trends Analyt. Chem. 30(5): 703–716.

Madden, E.F. 2003. The role of combined metal interactions in metal carcinogenesis: A review. Rev. Environ. Health. 18(2): 91–109

Manchester, M. and A. Anand. 2017. Metabolomics: Strategies to define the role of metabolism in virus infection and pathogenesis. Adv. Virus Res. 98: 57–81.

Maret, W. 2018. Metallomics: The science of biometals and biometalloids. Adv. Exp. Med. Biol. 1055: 1–20.

Masoud, A.M., S.W. Bihaqi, J.T. Machan, N.H. Zawia and W.E. 2016. Early-Life Exposure to Lead (Pb) alters the expression of microRNA that target proteins associated with alzheimer's disease. J. Alzheimers Dis. 51(4): 1257–1264.

McCann, M.R., M.V.G. De la Rosa, G.R. Rosania and K.A. Stringer. 2021. L-Carnitine and Acylcarnitines: Mitochondrial biomarkers for precision medicine. Metabolites. 14; 11(1): 51.

Mehaffey, E. and D.S.A. Majid. 2017. Tumor necrosis factor-α, kidney function, and hypertension. Am. J. Physiol. Renal Physiol. 313(4): F1005–F1008.

Merced-Nieves, F.M., M. Arora, R.O. Wright and P. Curtin. 2021. Metal mixtures and neurodevelopment: recent findings and emerging principles. Curr. Opin. Toxicol. 26: 28–32.

Millington, D.S. and R.D. Stevens. 2011. Acylcarnitines: Analysis in plasma and whole blood using tandem mass spectrometry. Methods Mol. Biol. 708: 55–72.

Milward, E.A., A. Shahandeh and H. Hondermarck. 2016. Transcriptomics. pp. 160–165. *In*: R.A. Bradshaw and P.D. Stahl [eds.]. Encyclopedia of Cell Biology. Academic Press, Cambridge, USA.

Mutti, A. 1999. Biological monitoring in occupational and environmental toxicology. Toxicol. Lett. 108: 77–89.

Nail, A.N., A.P.F. Cardoso, M. Banerjee and J.C. States. 2022. Circulating miRNAs as biomarkers of toxic heavy metal exposure. pp. 63–88. *In*: S.C. Sahu [ed.]. Genomic and Epigenomic Biomarkers of Toxicology and Disease: Clinical and Therapeutic Actions. John Wiley & Sons, Hoboken, USA.

Nalbantoglu, S. and A. Karadag. 2019. Insight into the OMICS technologies and molecular medicine. *In*: S. Nalbantoglu and H. Amri [eds.]. Molecular Medicine. IntechOpen, London, UK.

Ng, E., P.M. Lind, C. Lindgren, E. Ingelsson, A. Mahajan and A. Morris et al. 2015. Genome-wide association study of toxic metals and trace elements reveals novel associations. Hum. Mol. Genet. 24(16): 4739–4745.

[NHGRI] National Human Genome Research Institute. 2020. A Brief Guide to Genomics. National Institute of Health, National Human Genome Research Institute, Bethesda, USA.

[NIOSH] National Institute for Occupational Safety and Health. 2014. Exposome and Exposomics. National Institute for Occupational Safety and Health, Washington, USA.

O'Brien, J., H. Hayder, Y. Zayed and C. Peng. 2018. Overview of MicroRNA biogenesis, mechanisms of actions, and circulation. Front. Endocrinol. (Lausanne). 9: 402.

Ogra, Y. 2009. Toxicometallomics for research on the toxicology of exotic metalloids based on speciation studies. Anal. Sci. 25(10): 1189–1195.

Olivier, M., R. Asmis, G.A. Hawkins, T.D. Howard and L.A. Cox. 2019. The need for multi-omics biomarker signatures in precision medicine. Int. J. Mol. Sci. 20: 4781.

Park, S.K., Y. Tao, J.D. Meeker, S.D. Harlow and B. Mukherjee. 2014. Environmental risk score as a new tool to examine multi-pollutants in epidemiologic research: an example from the NHANES study using serum lipid levels. PlosOne. 9(6): e98632.

Park, S.K., Z. Zhao and B. Mukherjee. 2017. Construction of environmental risk score beyond standard linear models using machine learning methods: application to metal mixtures, oxidative stress and cardiovascular disease in NHANES. Environ. Health. 16(1): 102.

Patel, C.J., Bhattacharya and A.J. Butte. 2010. An Environment-Wide Association Study (EWAS) on type 2 diabetes mellitus. PLoS One. 5(5): e10746.

Quinones, M.P. and R. Kaddurah-Daouk. 2009. Metabolomics tools for identifying biomarkers for neuropsychiatric diseases. Neurobiol. Dis. 35(2): 165–176.

Rachakonda, V., T.H. Pan and W.D. Le. 2004. Biomarkers of neurodegenerative disorders: how good are they? Cell Res. 14(5): 347–358.

Reimann, E., F. Lättekivi, M. Keermann, K. Abram, S. Kõks, K. Kingo et al. 2019. Multicomponent biomarker approach improves the accuracy of diagnostic biomarkers for psoriasis vulgaris. Acta Derm. Venereol. 99(13): 1258–1265.

Renzetti, S., P. Curtin, A.C. Just, G. Bello and C. Gennings. 2021. CHEAR Data Center (Dept. of Environmental Medicine and Public Health, Icahn School of Medicine at Mount Sinai). https://cran.r-project.org/web/packages/gWQS/vignettes/gwqs-vignette.html.

Rinaldo, P., T. Cowan and D. Matern. 2008. Acylcarnitine profile analysis. Genet. Med. 10: 151–156.

Robin, X., N. Turck, A. Hainard, N. Tiberti, F. Lisacek, J.-C. Sanchez et al. 2013. PanelomiX: A threshold-based algorithm to create panels of biomarkers. Transl. Proteom. 1(1): 57–64.

Robotti, E., M. Manfredi and E. Marengo 2013. Biomarkers discovery through multivariate statistical methods: A review of recently developed methods and applications in proteomics. J. Proteomics Bioinf. S3: 003.

Rodríguez-Moro, G., S. Ramírez-Acosta, B. Callejón-Leblic, A. Arias-Borrego, T. García-Barrera and J.-L. Gómez-Ariz. 2021. Environmental metal toxicity assessment by the combined application of metallomics and metabolomics. Environ. Sci. Pollut. Res. 28: 25014–25034.

Ruan, F., J. Zhang, J. Liu, X. Sun, Y. Li, S. Xu et al. 2022. Association between prenatal exposure to metal mixtures and early childhood allergic diseases. Environ. Res. 206: 112615.

Sanchez, T.R., X. Hu, J. Zhao, V. Tran, N. Loiacono, Y.-M. Go et al. 2021. An atlas of metallome and metabolome interactions and associations with incident diabetes in the Strong Heart Family Study. Environ. Int. 157: 106810.

Sanyal, J., S.S.S.J. Ahmed, H.K.T. Ng, T. Naiya, E. Ghosh, T. Banerjee et al. 2016. Metallomic Biomarkers in Cerebrospinal fluid and Serum in patients with Parkinson's disease in Indian population. Sci. Rep. 6: 35097.

Singh, V. and K. Verma. 2018. Metals from cell to environment: Connecting Metallomics with other omics. Open J. Plant Sci. 3(1): 1–14.

Vrijheid, M. 2014. The exposome: a new paradigm to study the impact of environment on health. Thorax. 69: 876–878.

Wallace, D.R., Y.M. Taalab, S. Heinze, B.T. Lovaković, A. Pizent, E. Renieri et al. 2020. Toxic-metal-induced alteration in mirna expression profile as a proposed mechanism for disease development. Cells. 9(4): 901.

Wang, X., B. Mukherjee, C.A. Karvonen-Gutierrez, W.H. Herman, S. Batterman, S.D. Harlow et al. 2020. Urinary metal mixtures and longitudinal changes in glucose homeostasis: The Study of Women's Health Across the Nation (SWAN). Environ. Int. 145: 106109.

Wheeler, D.C., S. Rustom, M. Carli, T.P. Whitehead, M.H. Ward and C. Metayer. 2021. Assessment of grouped weighted quantile sum regression for modeling chemical mixtures and cancer risk. Int. J. Environ. Res. Public Health. 18(2): 504.

Wu, X., S.J. Cobbina, G. Mao, H. Xu, Z. Zhang and L. Yang. 2016. A review of toxicity and mechanisms of individual and mixtures of heavy metals in the environment. Environ. Sci. Pollut. Res. Int. 23(9): 8244–8259. 10.1007/s11356-016-6333-x.

Yasuda, H., T. Tsutsui and K. Suzuki. 2021. Metallomics analysis for assessment of toxic metal burdens in infants/children and their mothers: early assessment and intervention are essential and their mothers: Early assessment and intervention are essential. Biomolecules. 11(1): 6.

Yim, G., Y. Wang, C.G. Howe and M.E. Romano. 2022. Exposure to metal mixtures in association with cardiovascular risk factors and outcomes: A scoping review. Toxics. 10(3): 116.

Zheng, Y., Z. Chen, T. Pearson, J. Zhao, H. Hu and M. Prosperi. 2020. Design and methodology challenges of environment-wide association studies: A systematic review. Environ. Res. 183: 109275.

Zhou, Z., Y.-Y. Zhang, R. Xin, X.-H. Huang, Y.-L. Li, X. Dong et al. 2022. Metal ion-mediated pro-oxidative reactions of different lipid molecules: revealed by nontargeted lipidomic approaches. J. Agric. Food Chem. 2022 in press.

Index

For Product Safety Concerns and Information please contact our EU
representative GPSR@taylorandfrancis.com
Taylor & Francis Verlag GmbH, Kaufingerstraße 24, 80331 München, Germany

www.ingramcontent.com/pod-product-compliance
Lightning Source LLC
Chambersburg PA
CBHW060338220326
41598CB00023B/2750

9 781032 039404